# Springer Series on Atomic, Optical, and Plasma Physics

Volume 115

The Springer Series on Atomic, Optical, and Plasma Physics covers in a comprehensive manner theory and experiment in the entire field of atoms and molecules and their interaction with electromagnetic radiation. Books in the series provide a rich source of new ideas and techniques with wide applications in fields such as chemistry, materials science, astrophysics, surface science, plasma technology, advanced optics, aeronomy, and engineering. Laser physics is a particular connecting theme that has provided much of the continuing impetus for new developments in the field, such as quantum computation and Bose-Einstein condensation. The purpose of the series is to cover the gap between standard undergraduate textbooks and the research literature with emphasis on the fundamental ideas, methods, techniques, and results in the field.

More information about this series at http://www.springer.com/series/411

Michael Keidar
Editor

# Plasma Cancer Therapy

Springer

*Editor*
Michael Keidar
A. James Clark Professor of Engineering
Mechanical and Aerospace Engineering
School of Engineering and Applied Science
The George Washington University
Washington, DC, USA

ISSN 1615-5653 ISSN 2197-6791 (electronic)
Springer Series on Atomic, Optical, and Plasma Physics
ISBN 978-3-030-49965-5 ISBN 978-3-030-49966-2 (eBook)
https://doi.org/10.1007/978-3-030-49966-2



This Springer imprint is published by the registered company Springer Nature Switzerland AG
The registered company address is: Gewerbestrasse 11, 6330 Cham, Switzerland

# Foreword

As an immunologist and a practicing hematologist/oncologist, my knowledge of cold atmospheric plasma (CAP), until recently, was none existent at all. Two events changed that perspective: first, the co-localization in the same building (Science and Engineering Hall) of my cancer immunology research laboratory with several engineering labs, which in turn led to the second and most important event, the opportunity of meeting Dr. Michael Keidar, an aerospace engineer at the forefront of many of the therapeutic applications of cold atmospheric plasma (CAP), who is also the Editor of this timely and comprehensive book, *Plasma Cancer Therapy*. I still remember my initial interactions with Michael in 2016 during one of the first meetings of the recently established Cancer Engineering and Technology (CET) Program at our Cancer Center. In addition to learning from him that beyond the three states of matter, that I remember from my high school years, there was a fourth one—plasma—I became fascinated with his didactic lecture explaining the great potential of CAP in biomedicine and in particular as a novel anticancer agent. Given my background in cancer immunotherapy, I was also intrigued about the potential effects—if any—of CAP upon immune cells. I approached Michael at the end of his presentation and learned from him that little was known at that time about the effects of CAP upon the immune system. Since then, and by taking advantage of our different scientific backgrounds, we have established a complementary collaboration that not only led us to unveil previously unknown stimulatory effects of CAP upon immune cells but also helped me to increase my knowledge of CAP from "none" to "some." This limited knowledge improved even further after attending a three-day meeting, invited by Dr. Keidar, entirely dedicated to the fundamentals and therapeutic applications of CAP. In the 2019 Workshop on Plasma for Cancer Therapy in Antwerp, Belgium, I was extremely impressed by the depth and breadth of CAP research and the lectures by experts in this field, several of them contributors in this book.

During the past several years, there has been a significant increase in the CAP's knowledge that has been highlighted in a variety of settings, from scientific forums to basic science and translational/clinical publications. In particular, the cancer field has witnessed a rapid growth in CAP-related research. In this book, Dr.

Keidar and collaborators have brought together the most updated and relevant information regarding plasma cancer therapy. The fundamentals of plasma, atmospheric discharges, plasma jet, quasi-neutrality plasma, and other basic concepts are provided in a very didactic and easy to read in Chap. 1. This is followed by three chapters focused on CAP sources for cancer, diagnostics, and an overview of cancer applications, respectively. Next, an in-depth review of the biochemistry and microbiology of plasma in cancer therapy, plasma-activated media, and plasma–cell interactions simulations is provided in Chaps. 5–7. Although I am biased given my own interest in cancer immunotherapy, Chap. 8, *plasma based immunotherapy*, is timely and of high translational impact. Having said that, the topic of adaptive plasma and machine learning (Chap. 9) is perhaps the most fascinating and innovative. Chapter 10 expanded the applications of CAP beyond cancer cells into plasmid DNA and viruses, the later very timely given the current COVID-19 viral pandemic and the need for innovative antiviral treatments. Finally, Chaps. 11 and 12 will "walk the readers through" important mechanisms of cancer therapy, recent advances in plasma cancer therapy, and the clinical studies/applications emanating from basic/translational CAP research.

As you would gather from the different chapters of this book, these are exciting times for the field of plasma therapy. Novel applications of CAP are increasing at a rapid pace our therapeutic armamentarium against cancer and other diseases. The significant advances made in the past several years represent just the beginning of a promising road to the future that ultimately will lead us to a better understanding of CAP and its role in cancer treatment. For me, this book has also increased my knowledge of CAP from "none existent to fair."

Eduardo M. Sotomayor
The Dr. Cyrus Katzen Family Director
of the George Washington University Cancer Center,
Washington, DC, USA
Professor, Department of Medicine, George Washington
University School of Medicine and Health Sciences,
Washington, DC, USA

# Contents

# Contributors

**Neil D. Almeida** The George Washington University, Washington, DC, USA

**Georg Bauer** Institute of Virology, Medical Center - University of Freiburg, Freiburg, Germany

**Isak I. Beilis** Tel Aviv University, Tel Aviv, Israel

**Sander Bekeschus** INP Greifswald, Leibniz Institute for Plasma Science and Technology, Greifswald, Germany
Department of ZIK Plasmatis, Leibniz Institute for Plasma Science and Technology (INP), Greifswald, Germany

**Annemie Bogaerts** Research group PLASMANT, Department of Chemistry, University of Antwerp, Antwerp, Belgium

**Eun Ha Choi** Plasma Bioscience Research Center (PBRC), Kwangwoon University, Seoul, South Korea

**Carles Corbella** The George Washington University, Washington, DC, USA

**Alexander Fridman** Drexel University, Philadelphia, PA, USA

**Eda Gjika** The George Washington University, Washington, DC, USA

**Pepijn Heirman** Research group PLASMANT, Department of Chemistry, University of Antwerp, Antwerp, Belgium

**Masaru Hori** Center for Low-Temperature Plasma Sciences, Nagoya University, Nagoya, Japan

**Michael Keidar** Mechanical and Aerospace Engineering, School of Engineering and Applied Science, The George Washington University, Washington, DC, USA

**Mounir Laroussi** Plasma Engineering and Medicine Institute, Old Dominion University, Norfolk, VA, USA

**Taeyoung Lee** Mechanical and Aerospace Engineering, The George Washington University, Washington, DC, USA

**Li Lin** The George Washington University, Washington, DC, USA

**XinPei Lu** School of Electrical and Electronic Engineering, Huazhong University of Science and Technology, Wuhan, People's Republic of China

**Alisa Malyavko** The George Washington University, Washington, DC, USA

**Vandana Miller** Department of Microbiology and Immunology, Drexel University College of Medicine, Philadelphia, PA, USA

**Lan Lan Nie** School of Electrical and Electronic Engineering, HuaZhong University of Science and Technology, Wuhan, People's Republic of China

**Jamoliddin Razzokov** Research group PLASMANT, Department of Chemistry, University of Antwerp, Antwerp, Belgium

**Kenneth Sack** The George Washington University, Washington, DC, USA

**Jonathan H. Sherman** The George Washington University, Department of Neurological Surgery, Washington, DC, USA
West Virginia University, Martinsburg, WV

**Hiromasa Tanaka** Center for Low-Temperature Plasma Sciences, Nagoya University, Nagoya, Japan

**Jonas Van der Paal** Research group PLASMANT, Department of Chemistry, University of Antwerp, Antwerp, Belgium

**Dayun Yan** The George Washington University, School of Engineering and Applied Science, Washington, DC, USA

**Xu Yan** Beijing Neurosurgical Institute, Beijing, People's Republic of China

**Maksudbek Yusupov** Research group PLASMANT, Department of Chemistry, University of Antwerp, Antwerp, Belgium

# Chapter 1
# Introduction: Plasma for Cancer Therapy

**Michael Keidar**

## Contents



**Abstract** Plasma is an ionized gas that is typically formed under high-temperature laboratory conditions. Recent progress in atmospheric plasmas has led to non-thermal or cold atmospheric plasma (CAP) devices with ion temperatures close to room temperature. CAP demonstrated significant potential in cancer therapy. In contrast to many existing anti-cancer approaches, CAP is a selective anti-cancer modality. This book written by key researchers in the field will provide a comprehensive analysis and overview of the state-of-the-art plasma-based cancer therapy. We will also present some recent advances in the field, the primary challenges, and future directions. This book is meant for a broad audience, from students to engineers and scientists, who are interested in the emerging field of plasma medical application.

Plasma is the fourth state of matter that can be defined as an ionized gas. Ionization is the process of stripping electrons from atoms and molecules leading to formation of negatively and positively charged particles [1]. Depending on relative density of charged particles one can distinguish between the weakly and strongly ionized gases. A weakly ionized gas is characterized by a relatively small fraction of charged particles and its behavior can be largely described by neutral gas laws while one requires electrodynamics to properly describe the strongly ionized media.

M. Keidar (✉)
Mechanical and Aerospace Engineering, School of Engineering and Applied Science, The George Washington University, Washington, DC, USA
e-mail: keidar@gwu.edu



M. Keidar (ed.), *Plasma Cancer Therapy*, Springer Series on Atomic, Optical, and Plasma Physics 115, https://doi.org/10.1007/978-3-030-49966-2_1

Let us call a physical state of an ionized gas in which the densities of positively and negatively charged particles are approximately equal as a quasi-neutrality state. To that end, plasma can be defined as an ionized gas, which satisfies the quasi-neutrality condition. Mott-Smith described the history of introducing the term "plasma" in his letter to Nature. "*..Langmuir.. pointed out that the 'equilibrium' part of the discharge acted as a sort of sub-stratum carrying particles of special kinds, like high-velocity electrons from thermionic filaments, molecules and ions of gas impurities. This reminds him of the way blood plasma carries around red and white corpuscles and germs. So he proposed to call our 'uniform discharge' a 'plasma'.*" [2]

As it is mentioned above plasma is characterized by quasi-neutrality, i.e., number densities of positively and negatively charged particles are approximately equal. As such, any perturbation in the plasma region such as a shift of electrons with respect to the ions causes separation of charges. The charge separation produces an electric field that works to restore the unperturbed plasma state. To this end, let us consider the quasi-neutrality condition in the plasma of a high-density discharge. As an example, a fully ionized plasma is formed with an electron density of about $10^{22}$ m$^{-3}$ and in the volume with a characteristics size of about 1 mm.

As the electrons move away, only a net positive charge of ions will remain. It is plausible to assume that at the electron timescale, ions will not be moving since their mass is much larger. Due to charge separation, the induced electric field E will act to reduce this charge separation and to accelerate electrons back. Let us estimate an electric field utilizing the Gauss theorem that is applied to rectangular region:

$$\oint E \cdot \mathrm{d}s = \frac{\oint q \, \mathrm{d} V}{\varepsilon_0}$$

where $q$ is the charge in volume $V$, $s$ is the surface area, and $\varepsilon_0$ is the permittivity:

$$\oint q \mathrm{d}V = \mathrm{e}N_\mathrm{e} x \Delta y \Delta z \quad \text{and} \quad \oint E \mathrm{d}s = -E_x \Delta y$$

where $N_e$ is the electron density. As a result, one can estimate the electric field as follows:

$$E_x = -\frac{eN_e x}{\varepsilon_0}$$

where $x$ is the separation region.

Based on this argument one can conclude that if the plasma's quasi-neutrality is violated due to charge separation at the characteristic distance of about 1 mm, an electric field of about $10^{11}$ V/m would appear. This means that the voltage difference (potential drop) of about $10^8$ V will be set at a distance of about 1 mm. Such a large electric field will work to restore the charge neutrality. However, if a relatively small plasma volume is considered, such an electric field and potential drop might not be strong enough to affect the particle motion and restore the quasi-neutrality. Thus, the quasi-neutrality condition can be violated at the small scale. The characteristics scale at which the charge separation can exist is called the Debye length.

Accordingly, the electron motion affects the microscale electric field and, as a result, electrons oscillate around ions. Such oscillations are signature of plasma temporal behavior. As the electrons move, only a net positive charge of ions will remain. As such one can assume that at the electron timescale, ions will not be moving since their mass is much larger. Due to charge separation, the induced electric field E will act to reduce this charge separation and to accelerate electrons back. The electrons will gain kinetic energy and their inertia will cause them to pass ("overshoot") their original position. As a result, the plasma will become charge separated again and an electric field will be formed again in the opposite direction. If there is no damping mechanism (for instance, collisions), these oscillations will continue forever with the frequency [3]

$$\omega_{\mathrm{p}} = \sqrt{\frac{\mathrm{e}^2 N_{\mathrm{e}}}{\varepsilon_0 m_{\mathrm{e}}}}$$

where $e$ is the elementary charge, $N_e$ is the electron density, $\varepsilon_0$ is the permittivity of vacuum, and $m_e$ is the electron mass.

In the absence of collisions and thermal effects, this frequency uniquely determines the oscillations in the plasma. These oscillations are localized and do not propagate. Thermal effects lead to propagation of the oscillations, i.e., the generation of a plasma wave or Langmuir wave. In turn, collisions can lead to the damping of these oscillations and Langmuir wave propagation.

In this book we will discuss the application of atmospheric plasmas, i.e., plasmas created in air conditions. When the discharge voltage is applied free electrons naturally existing in the gas are attracted towards the anode. Traveling in the electric field, these electrons gain energy and also collide with atoms and molecules. Such collision might lead to ionization when electron energy is larger than the ionization potential of atom and molecules. This process multiplies electrons and is referred to as avalanche. During the avalanche, not all electrons will ionize, but more electrons have lower energy and can cause excitation of other species. This makes the avalanche region luminous, which is also referred to as the "plasma bullet" for its bullet-like appearance. On the other hand, positive ions remain almost stationary during the timescale associated with electron motion since they are much heavier than electrons. Therefore, a positive ion cloud appears and turns into a new anode by attracting the free electrons at the front to start another avalanche. The process is also known as the "ionization wave" because it is a peak of ionization rate (plasma bullet) moving forward. Due to its importance for plasma application in cancer therapy, this process will be described in detail in Chap. 2.

## 1.1 Plasma Types

Over the last 70 years the development of plasma physics and engineering was associated with various applications, such as lighting sources, current interrupters, thermonuclear fusion, and plasma accelerators; nowadays, plasma applications range from plasma processing, space propulsion, nanotechnology, and most recently plasma medicine [1].

Under the ordinary conditions on Earth, plasma is rather a very rare phenomenon, but in the universe, cold solid bodies are an exception. Most of the matter in the universe is ionized and thus in the plasma state (about 99.9%). Plasma in the universe is produced by various mechanisms. In the stars, the neutral atoms are ionized due to high temperatures. Interstellar gases are ionized due to the ultraviolet radiation from the stars. Plasmas produced in nature and laboratory plasmas are characterized by a wide range of temperatures and pressures. Using the Debye length and the plasma frequency, which depend on the electron number density and electron temperature, it is possible to classify plasmas as rarified, dense, classical, and quantum. It spans from high-temperature plasmas in thermonuclear fusion reactors to room-temperature cold plasmas in some biomedical applications. Figure 1.1 shows the diagram of various natural and laboratory plasmas with their typical plasma density and electron temperatures. This book focuses exclusively on non-thermal atmospheric plasmas (at the bottom of temperature range as shown in Fig. 1.1).

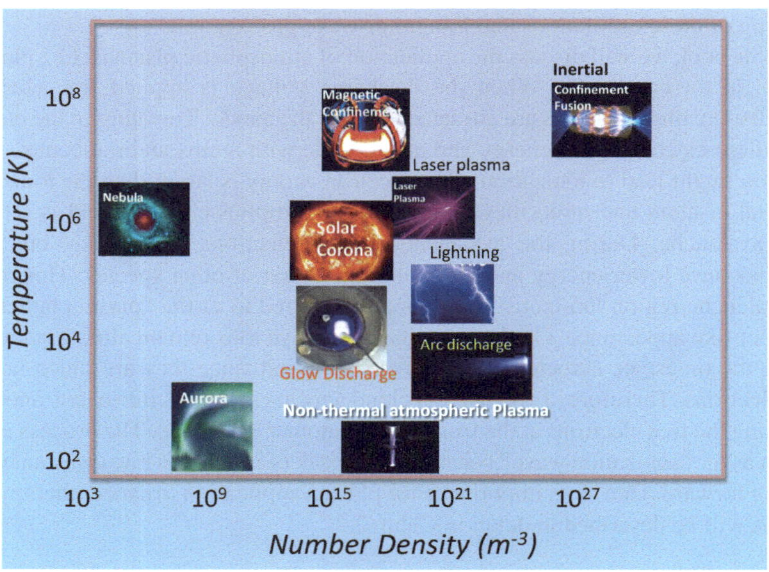


**Fig. 1.1** Plasma density–temperature diagram showing various natural and laboratory plasmas

It can be seen from Fig. 1.1 that plasmas are typically generated in high-temperature conditions. Recent progress in atmospheric plasmas has led to the creation of cold plasmas with ion temperature close to room temperature. The unique chemical and physical properties of cold atmospheric plasmas have enabled its recent applications in biomedicine including sterilization, polymer material preparation, wound healing, tissue or cellular removal, and dental drills. This led to establishment of a new scientific field called plasma medicine. *Plasma medicine* is a new and rapidly growing field combining plasma physics, engineering, medicine, and bioengineering. To that end, non-thermal or cold atmospheric plasmas (CAP) can be developed. Due to non-thermal nature of CAP, it can offer a minimally invasive surgery that allows specific cell targeting without influencing the whole tissue. CAP interaction with tissue may allow specific cell removal without significant necrosis. Overall, CAP triggered effects include cell detachment without affecting cell viability, controllable cell death, modified cell migration, etc. [4–6]. The presence of the plasma can promote chemical reactions that would have the desired effect. Chemical reactions can be promoted by tuning the pressure, gas composition, and applied power.

Nowadays CAP emerges as a possible new modality for cancer treatment [4, 5]. High level understanding of CAP interaction with cells and tissue depends on the notion that chemical elements of the CAP are potentially toxic, such as reactive oxygen species (ROS), which might promote a "plasma killing effect," while others such as reactive nitrogen species (RNS) could produce a "plasma healing" effect. Forming various combinations of these species might provide a great potential for activation of specific signaling pathways in cells. CAP treatment possesses powerful lethal capabilities against tumor cells both in vitro and in vivo and just as importantly, the normal counterpart cells have been shown to be less sensitive to the same CAP treatment [6]. All these aforementioned effects are believed to be related to the plasma chemistry. While different species are produced as a result of plasma treatment, the role of other plasma effects such as charged particles and electric field still remains elusive. In addition, what makes plasma unique is its ability to self-organize and to form coherent structures. These coherent structures could instantaneously modulate electric field, ROS/RNS, and charged particles. Thus, as a result of self-organization, plasma adaptivity to specific cells through tailoring its composition in situ can be possible. This unique feature makes plasma interaction with cells intrinsically selective. Some recent progress in developing adaptive plasmas is described in this chapter.

## 1.2 Brief History of Plasma Medicine

CAP application for bacteria inactivation started from mid-1990s [Ref. 7]. This research consequently led to many applications of CAP for surface sterilization and bacteria deactivation. Stoffels et al. [8] studied the plasma needle device and demonstrated the promising potential of the cold plasma in biomedical applications.

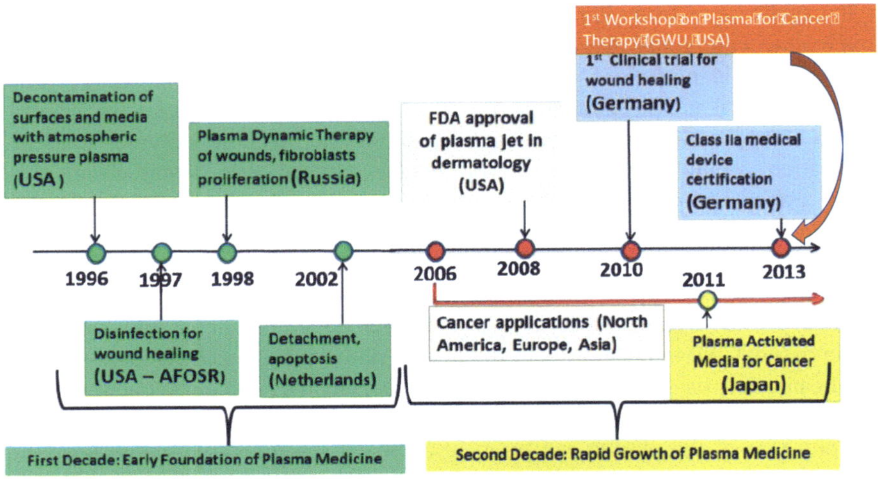


**Fig. 1.2** Timeline illustrating major milestones of plasma medicine including plasma application in cancer therapy. From M. Laroussi, Plasma, 2018, 1, 47–60; doi:10.3390/plasma1010005

It became clear that plasma can interact with organic materials without causing thermal/electric damage to the surface [7]. In recent years, several new devices have been developed enabling generation of a non-thermal or cold plasma [4]. The understanding about the increasing role of CAP in biomedical applications motivates development of a variety of reliable and user-friendly plasma sources. To this end, plasma devices have to meet many requirements such as low temperature, stable operation at atmospheric pressure, and no risk of arcing. In terms of CAP devices, Laroussi and Lu [9] described the operation of a cold plasma plume using helium as carrier gas. To this end, it has been demonstrated that the plasma plume can be touched by bare hands and can come in contact with skin and dental gums without causing any heating or painful sensation. The device later received the name "plasma pencil." Fridman et al. [10] demonstrated that cold plasmas can promote blood coagulation and tissue sterilization. It was shown previously that thermal plasma treatment is very beneficial in terms of blood coagulation and sterilization, but it induces significant damage. Many devices have been developed in Europe and Asia. Below we illustrate the timeline of important milestones in plasma medicine (Fig. 1.2). An important milestone is the First International Workshop on Plasmas for Cancer Treatment.

Many of the topics were discussed at this inaugural which took place at the George Washington University Convention Center in Washington DC, USA, March 25–26, 2014. Since then annual workshop series has been continued and timeline and locations are illustrated in Fig. 1.3.

Nowadays it becomes clear that plasma interaction with cells and tissue can elicit multiple pathways. The variety of different effects of plasma can be explained by their complex chemical composition and variations in the way that CAP is

**Fig. 1.3** Timeline and locations of International Workshop of Plasmas for Cancer Treatment

IWPCT

| | |
|---|---|
| IWPCT 2014 : Washington D.C., USA | 25-26 March 2014 |
| IWPCT 2015 : Nagoya, Japan | 16-17 March 2015 |
| IWPCT 2016 : Washington D.C., USA | 11-12 April 2016 |
| IWPCT 2017 : Paris, France | 27-28 March 2017 |
| IWPCT 2018 : Greifswald, Germany | 20-21 March 2018 |
| IWPCT 2019 : Antwerp, Belgium | 1-3 April 2019 |
| IWPCT 2020 : Raleigh NC, USA | 23-25 March 2020 |

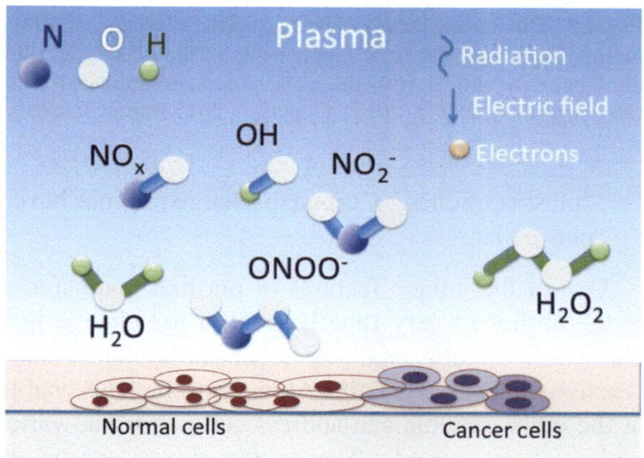


**Fig. 1.4** Plasma interfaces the living cells. Components of cold atmospheric plasma

generated. CAP is a cocktail containing a variety of reactive oxygen species (ROS), reactive nitrogen species (RNS), charged particles, electric field, UV, etc. (as shown in Fig. 1.4).

This variety leads to a variety of effects mentioned above. In general, the CAP sources can be classified into two major groups according to the principal mechanism of generation and application [6, 11]. Schematically two approaches are shown in Fig. 1.5.

(a) *Direct plasmas* employ living tissue or organs as one of the electrodes, and thus, living tissue directly participates in the active discharge plasma processes. Some current may flow through living tissue in the form of small conduction current, displacement current, or both. Conduction current has to be limited to avoid any thermal effects or electrical stimulation of the muscles. The dielectric barrier discharge (DBD) is a typical example of direct plasma.

(b) *Indirect plasmas* are produced between two electrodes and are then transported to the area of application entrained in a gas flow. There is a great variety of different configurations of indirect plasma sources exist in the size, type of gas, and power. They range from very narrow "plasma needles," plasma jets to larger

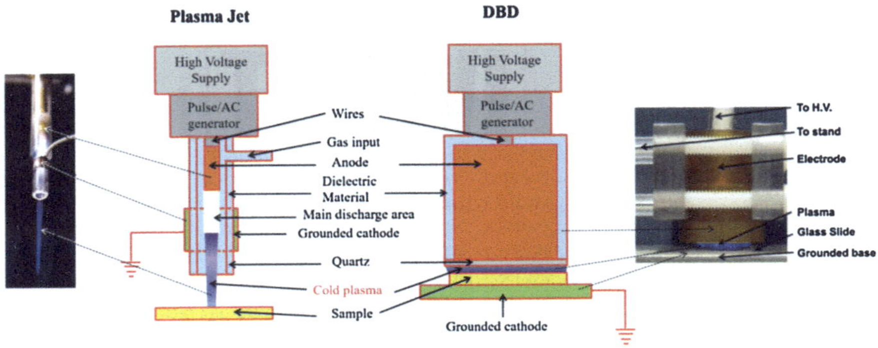


**Fig. 1.5** The plasma jet and dielectric barrier discharge (DBD) are two main CAP devices used in plasma medicine. The same components in the plasma jet (direct discharge) and DBD (indirect discharge) are drawn with the same colors. Reproduced with permission from D Yan, et al., Oncotarget, 8.9, 15977 (2017). Copyright 2017 Impact Journals

"plasma torches." Corona discharge plasmas have similar properties to that of a plasma jet.

One of the unique features of plasmas compared to other sources of reactivity is the ability to very rapidly (within ns) change the reactive species production pathways, thereby enabling feedback systems that customize *in real time* the reactivity delivered to cells. At one extreme, this enables personalized medicine and at the other extreme can address cell and tissue variability while treating a single patient as an example. That is, the plasma can be engineered to affect particular cellular or tissue functions. Plasmas can also *self-organize and form coherent structures*, which modulate the electric field, along with production and delivery of ROS/RNS and charged particles. This self-organization adds another level of control and engineering capability. To this end, we will briefly describe adaptive plasmas approach.

Possible ideas related to the adaptive CAP (ACAP) are shown schematically in Fig. 1.6. ACAP device relays on real time monitoring of plasma–living tissue interaction and adaptation of plasma parameters based on a feedback mechanism [12, 13]. Related idea of plasma device feedback control and managed delivery of plasma dose was described elsewhere [14]. As a result of differential action of ACAP on cancer and normal cells it might be possible to manage selective plasma treatment of cancer cells. In the case of a direct feedback or self-adaptation, plasma interaction with cells could lead to self-organization via transition between various discharge modes (see below). As an example of feedback-based ACAP, recent work [15] utilized a RealTime™ assay and provided evidence that cellular response to plasma treatment can be monitored in real time that is a pre-cursor for adaptation. For example, some implementation of an adaptive concept that is able to adjust its composition to obtain optimal outcomes through its interaction with healthy and non-healthy cells was recently described [16].

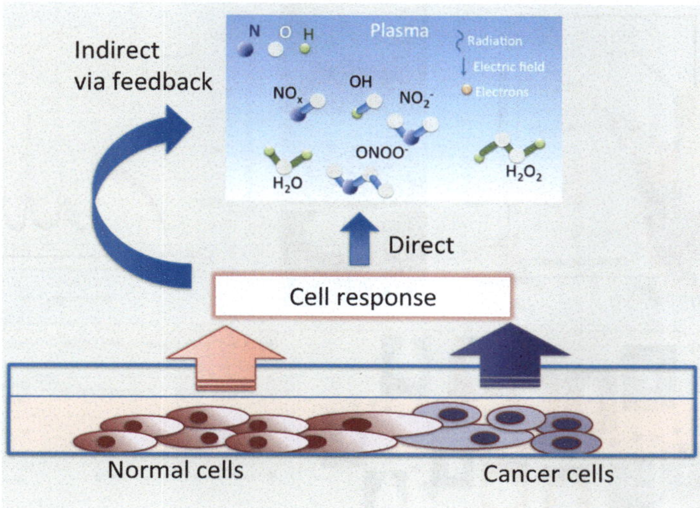


**Fig. 1.6** Adaptive cold atmospheric plasma (ACAP) concept based on differential treatment of cancer and normal cells. Feedback system modulates the chemical composition of plasma leading to adaptation. Direct feedback or plasma self-adaptation is based on self-organization and pattern formation

## 1.3 Self-Organization and Relevance to Plasma-Based Cancer Therapy

Self-organization at various discharge modes has been a focus of recent study [17]. It was demonstrated that complex structures can be formed at the plasma–liquid medium interface dependent on discharge current as shown in Fig. 1.7. One can notice that in this particular setup the discharge current changes drive the pattern formation above the dioinized water (DI) as shown in Fig. 1.7d. Recall that four specific types of discharge modes can be noticed. In particular, stage I is the low current discharge (glow discharge) having a single filament. Discharge current increase (stage II) leads to temperature rise of the tungsten cathode and high heat radiation (see Fig. 1.7c). Stage III is an unstable intermediate state in which discharge oscillates between two (II and IV) stages displaying either high heat radiation (as typical in stage II) or the multi-filament pattern (as typical in stage IV). It was observed that the multi-filament discharge (stage IV) is generally stable.

It should be pointed out that the effect of different discharge modes on production of RONS and resulting treatment of cancer cells was analyzed [18]. It was observed that selective cancer treatment is possible at some discharge conditions. Utilizing the transition between discharge modes one can envision a possibility to control

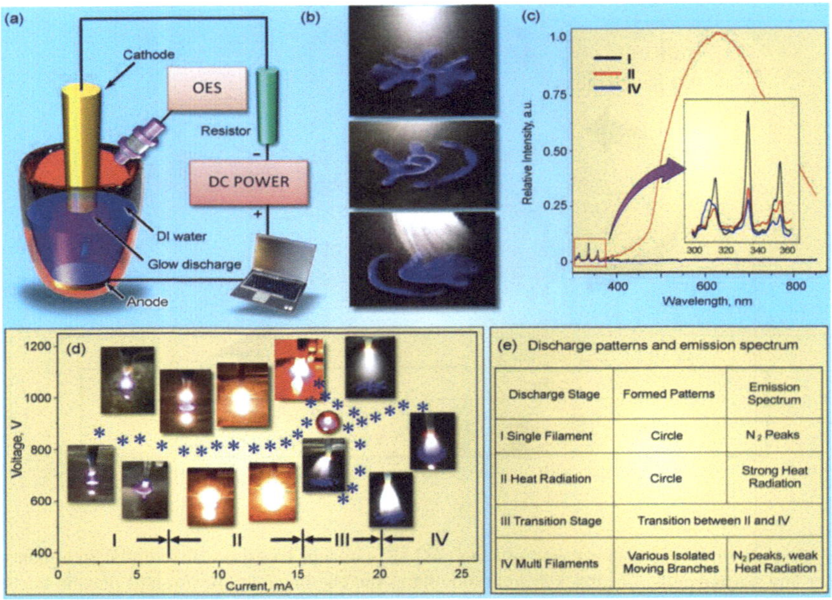


**Fig. 1.7** Discharge self-organization. (**a**) Schematics of the atmospheric glow micro-discharge setup. (**b**) Various discharge patterns above the liquid media. (**c**) Typical optical emission spectra for discharge modes I, II, and IV. (**d**) Current–voltage characteristics. Insets show the self-organized patterns. (**e**) Summary of the relations between discharge modes and self-organized patterns

treatment using plasma and adaptation of plasma treatment to a particular area or specific cell type.

In this book various plasma devices will be considered (see Chaps. 2 and 3 for detailed description). In this chapter we will present a very brief outline of jet and DBD plasmas.

Many results shown in this book are related to the helium plasma jet device. The typical atmospheric plasma jet device is shown schematically in Fig. 1.8. The gun is equipped with a pair of high-voltage (HV) electrodes—a central electrode (which is isolated from direct contact with plasma by ceramics) and an outer ring electrode, as shown schematically in Fig. 1.8a. The electrodes are connected to the HV resonant transformer with voltage 2–10 kV and frequency of about 10–30 kHz. A typical photograph of the plasma jet is shown in Fig. 1.8a. The visible well-collimated plasma jet has a typical length of about 4–5 cm dependent on flow rate and discharge voltage. The length of the plasma jet varied with gas flow. In particular, the increase of the helium feeding results in jet elongation, while further increase in the flow rate caused jet shortening and appearance of a turbulent tail at the jet's end. In general, the increase in the HV amplitude applied to the electrodes results in increase in plasma jet intensity but does not affect the plasma jet diameter.

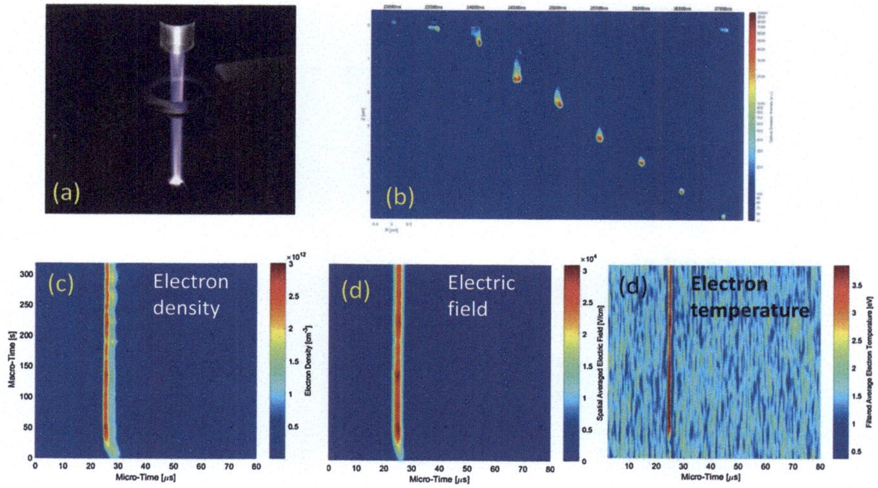


**Fig. 1.8** (**a**) Photograph of the RF discharge (6 kV pick to pick voltage) and non-equilibrium atmospheric plasma jet (**b**) Series of fast photographs of the plasma jet indicating streamer propagation. (**c**–**e**) Temporal evolution of electron density, electric field, and electron temperature

A typical microwave scattering signal from the jet, ICCD images of the jet (taken at times indicated by the rectangular bars; the plasma bullet is moving down) are all shown in Fig. 1.8b. Figure 1.8c–e presents temporal evolution of the electron density, electric field, and electron temperature. The discharge consists of a series of elementary breakdown events inside the discharge tube followed by the development of a streamer propagating outside the tube in open air. The streamer propagates axially about 4–5 cm dependent on helium flow rate and voltage with speed of about $2 \times 10^6$ cm/s along the He flow ejected to ambient air until it decays at $t \approx 5$ μs as shown in a series of instant photographs in Fig. 1.8b.

Measurements of plasma density in the streamer channel conducted using Rayleigh Microwave Scattering (RMS) [19] facility yield averaged value of $n_e$ along the streamer channel of about $3 \times 10^{12}$ cm$^{-3}$. Note the plasma ionization degree in the jet is very low ~$5 \times 10^{-7}$ (gas density at 1 atmosphere and 300 K is around $2 \times 10^{19}$ cm$^{-3}$).

Streamers observed by high-speed cameras in atmospheric jets (see Fig. 1.8b) were characterized by a new method of stopping ("scattering") by means of external DC potential [20–22]. Interaction of the streamer with the DC potential created by the ring located at $z = 3$ cm and dependence of streamer length L as a function of the ring potential $U_r$. This technique allows measurement of the electric field in the plasma jet as shown in Fig. 1.8d. As the electric field around the streamer head is governed by difference $U_h - U_r$, where $U_h$ is potential of streamer head, one can consider the condition $U_r = U_h$ as sufficient to stop propagation of the streamer. Electron temperature was estimated using the recently developed procedure [23] and shown in Fig. 1.8e.

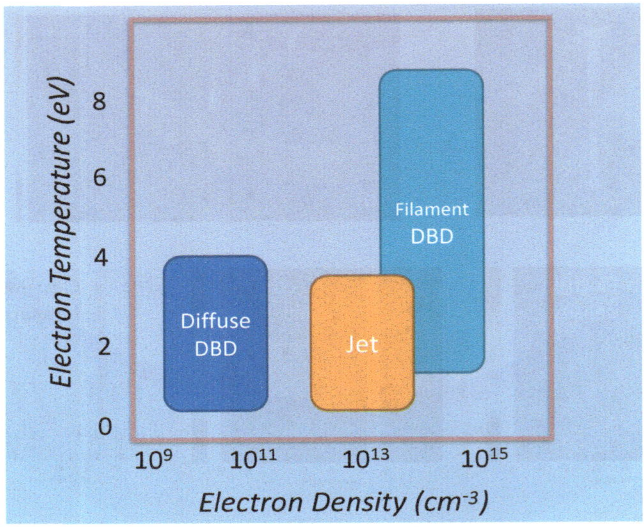


**Fig. 1.9** Electron density and temperature ranges in DBD and plasma jets

Dielectric barrier discharge (DBD) devices typically exhibit two electrodes where a dielectric material covers at least one of these. By applying an AC high voltage of varying kV range a plasma discharge is formed between the two electrodes [24, 25]. Dependent on the discharge parameters both filamentary and diffuse modes are possible in DBD [23]. Typical plasma parameters for both modes and plasma jet are shown in Fig. 1.9. One can see that electron density and electron temperature in the diffuse mode are close to that of the plasma jets.

## References

1. M. Keidar, I.I. Beilis, *Plasma Engineering*, 2nd edn. (Elsevier-Academic Press, Oxford, 2018)
2. H.M. Mott-Smith, History of "plasmas". Nature **233**(5316), 219 (1971)
3. F. Chen, *Plasma Physics and Controlled Fusion* (Springer, New York, 2006)
4. M. Laroussi, X. Lu, M. Keidar, Perspective: the physics, diagnostics, and applications of low temperature plasma sources used in plasma medicine. J. Appl. Phys. **122**, 020901 (2017)
5. M. Keidar, A. Shashurin, O. Volotskova, M.A. Stepp, P. Srinivasan, A. Sandler, B. Trink, Cold atmospheric plasma in cancer therapy. Phys Plasmas **20**, 057101 (2013)
6. M. Keidar, Plasma for cancer treatment. Plasma Sources Sci. Technol. **24**(3), 033001 (2015)
7. M. Laroussi, IEEE Trans. Plasma Sci. **24**, 1188 (1996)
8. E. Stoffels, A.J. Flikweert, W.W. Stoffels, G.M.W. Kroesen, Plasma needle: a non-destructive atmospheric plasma source for fine surface treatment of (bio)materials. Plasma Sources Sci. Technol. **11**(4), 383 (2002)
9. M. Laroussi, X. Lu, Room-temperature atmospheric pressure plasma plume for biomedical applications. Appl. Phys. Lett. **87**, 113902 (2005)

10. G. Fridman, G. Friedman, A. Gutsol, A.B. Shekhter, V.N. Vasilets, A. Fridman, Applied plasma medicine. Plasma Process. Polym. **5**, 503–533 (2008)
11. K.-D. Weltmann, E. Kindel, T. Von Woedtke, M. Hähnel, M. Stieber, R. Brandenburg, Pure Appl. Chem. **82**, 1223–1237 (2010)
12. M. Keidar, Therapeutic approaches based on plasmas and nanoparticles. J. Nanomed. Res. **3**(2), 00052 (2016)
13. M. Keidar, D. Yan, I.I. Beilis, B. Trink, J.H. Sherman, Plasmas for treating cancer: opportunities for adaptive and self-adaptive approaches. Trends Biotechnol. **36**(6), 586–593 (2018)
14. D. Gidon, D.B. Graves, A. Mesbah, Plasma Sources Sci. Technol. **26**, 085005 (2017)
15. E. Gjika, S. Pal-Ghosh, A. Tang, M. Kirschner, G. Tadvalkar, J. Canady, M.A. Stepp, M. Keidar, Adaptation of operational parameters of cold atmospheric plasma for in vitro treatment of cancer cells. ACS Appl. Mater. Interfaces **10**(11), 9269–9279 (2018)
16. Y. Lyu, L. Lin, E. Gjika, T. Lee, M. Keidar, Mathematical modeling and control for cancer treatment with cold atmospheric plasma jet. J. Phys. D: Applied Phys. **52**, 185202 (2019)
17. Z. Chen, S. Zhang, I. Levchenko, I.I. Beilis, M. Keidar, In vitro demonstration of cancer inhibiting properties from stratified self-organized micro-discharge plasma-liquid interface. Sci. Rep. **7**, 1–17 (2017)
18. Z. Chen, L. Lin, E. Gjika, X. Cheng, J. Canady, M. Keidar, Selective treatment of pancreatic cancer cells by plasma-activated saline solutions. IEEE Trans. Radiat. Plasma Med. Sci. **2**(2), 116–120 (2018)
19. A. Shashurin, M.N. Shneider, A. Dogariu, R.B. Miles, M. Keidar, Temporary-resolved measurement of electron density in small atmospheric plasmas. Appl. Phys. Lett. **96**(17), 171502 (2010)
20. A. Shashurin, M. Keidar, Experimental approaches for studying non-equilibrium atmospheric plasma jets. Phys. Plasmas **22**, 122002 (2015)
21. A. Shashurin, M.N. Shneider, M. Keidar, Measurements of streamer head potential and conductivity of streamer column in the cold nonequilibrium atmospheric plasmas. Plasma Sources Sci. Technol. **21**, 034006 (2012)
22. L. Lin, M. Keidar, Cold atmospheric plasma jet in an axial DC electric field. Phys. Plasmas **23**, 083529 (2016)
23. L. Lin, Y. Lyu, M. Shneider, M. Keidar, Average electron temperature estimation of streamer discharge in ambient air. Rev. Sci. Instrum. **89**, 113502 (2018)
24. N. Gherardi, G. Gouda, E. Gat, A. Ricard, F. Massines, Transition from glow silent discharge to micro-discharges in nitrogen gas. Plasma Sources Sci. Technol. **9**, 340 (2000)
25. R. Brandenburg, Dielectric barrier discharges: progress on plasma sources and on the understanding of regimes and single filaments. Plasma Sources Sci. Technol. **26**, 053001 (2017)

# Chapter 2
# Cold Atmospheric Pressure Plasma Sources for Cancer Applications

**Mounir Laroussi, Lan Lan Nie, and XinPei Lu**

**Contents**



**Abstract** In this Chapter, the physics and engineering of low-temperature plasma sources used in plasma medicine are covered. First, an overall review of the major plasma devices is presented. This is followed by a detailed description of one of the major plasma sources used in plasma oncology, the Nonequilibrium Atmospheric Pressure Plasma Jet (N-APPJ).

## 2.1 Review of Cold Atmospheric Pressure Plasma Sources for Cancer Applications

Cold plasmas exhibit nonequilibrium energetic conditions where the electrons have high energy while the heavy particles (ions and neutral) maintain low energy. This way, the plasma, which is weakly ionized, can have relatively low temperature, as

---

M. Laroussi (✉)
Plasma Engineering and Medicine Institute, Old Dominion University, Norfolk, VA, USA
e-mail: mlarouss@odu.edu

L. L. Nie · X. Lu
School of Electrical and Electronic Engineering, Huazhong University of Science and Technology, Wuhan, People's Republic of China



M. Keidar (ed.), *Plasma Cancer Therapy*, Springer Series on Atomic, Optical, and Plasma Physics 115, https://doi.org/10.1007/978-3-030-49966-2_2

low as room temperature. The main sources discussed here are the dielectric barrier discharge (DBD), the resistive barrier discharge (RBD), and the nonequilibrium atmospheric pressure plasma jets (N-APPJs). All these sources allow for the production of reactive species, such as reactive oxygen species (ROS) and reactive nitrogen species (RNS), which are believed to play key roles in the interaction of cold plasma with cells and tissues. In addition, charged species and relatively high electric fields generated by these sources can be implicated in some of the biological effects.

## *2.1.1 Dielectric Barrier Discharge (DBD)*

### History

Theodose du Moncel was the first to discover in 1855 that discharge can be generated in the gap separating two conducting plates covered by two glass plates [1]. To drive the discharge, he used a Ruhmkorff coil, which is an induction coil that allowed for the generation of high AC voltages from a low voltage DC source. This was followed, 2 years later, by the work of Werner von Siemens who reported on the application a dielectric barrier discharge (DBD) to generate ozone [2]. Siemens' design had a cylindrical geometry with tin foils as electrodes and glass as a dielectric. In the 1930s, Von Engel tried, not so successfully, to generate an atmospheric pressure nonequilibrium plasma without using a dielectric barrier, by controlling the temperature of the cathode [3]. Finally, in the late 1980s and early 1990s, reports on the generation of nonequilibrium, diffuse atmospheric pressure plasma using the DBD were published [4–6]. Planar electrode geometry and sinusoidal voltages in the kV at frequencies in the kHz range were used. The performance of the DBD was later improved when repetitive fast rise time voltage pulses with pulse widths in the nanoseconds-microseconds range were employed [7–9]. With the application of repetitive short pulses, the applied energy is preferentially coupled to the electron population and, therefore, better control of the electron energy distribution function (EEDF) is established [8–10]. DBDs have been widely used in surface-processing applications and more recently, starting around the mid-1990s, in biomedical applications, i.e., plasma medicine [11–14].

### Operation of the Dielectric Barrier Discharge

The Dielectric Barrier Discharge was the first device used to generate large volume nonequilibrium atmospheric pressure diffuse plasma. Many decades of use allowed a good understanding and improvement of its operation [4–6, 15–23]. DBDs use a dielectric material, such as glass or alumina, to cover at least one of two electrodes. The electrodes are powered by high voltages (several kV) at frequencies in the kHz range. Plasma generated by DBDs can be used for surface processing, as

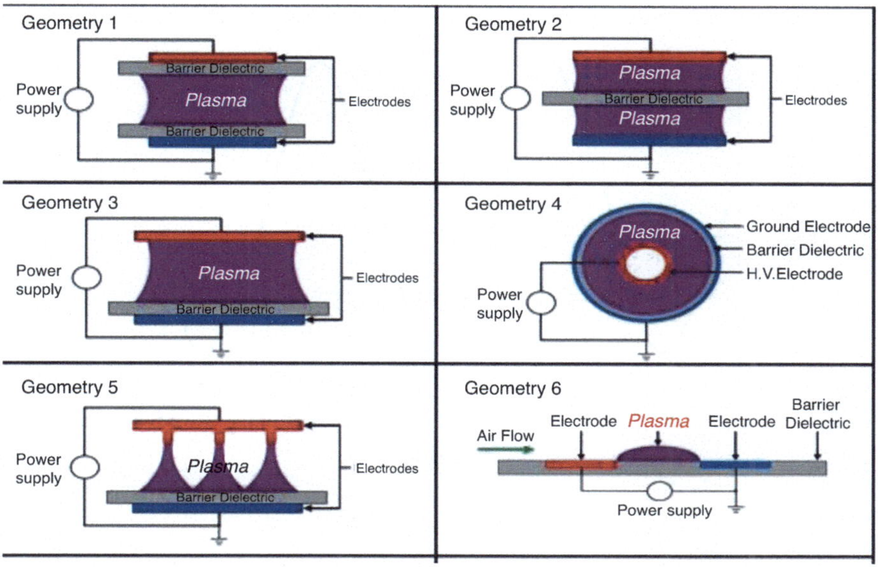


**Fig. 2.1** Various DBD configurations

flow control actuators, and for the generation of ozone [16, 24, 25]. Since the mid-1990s when it was reported that cold plasma generated by DBDs inactivates bacteria efficiently, DBDs have also been used extensively for biomedical applications [11–14].

DBDs can be designed in various geometrical configurations. Figure 2.1 shows a few of them.

DBDs have been mostly energized by high sinusoidal voltages with amplitudes in the kV range and frequencies of a few kHz, or by repetitive high-voltage short pulses (ns - μs). The electrode arrangement is generally contained within a chamber to allow for the introduction of the working gas mixture. Surface charges accumulate on the dielectric surface as soon as a discharge is ignited, which creates an electrical potential that counteracts the externally applied voltage. This results in a pulsed current waveform and a self-limitation of the discharge current. Figure 2.2 shows a typical current–voltage characteristic of the DBD. Generally, DBDs produce filamentary plasmas, but under some conditions, they can produce homogeneous plasmas. Under the homogeneous operating regime, the current waveform exhibits a single pulse per half cycle, as shown in Fig. 2.2.

The electron energy distribution function (EEDF) defines/controls the chemistry in the plasma. Therefore, in order to achieve an increase of ionization and excitation, short high-voltage pulses have been used. Pulses with widths less than the characteristic time of the onset of the glow-to-arc transition maintain a stable nonequilibrium low-temperature plasma. Laroussi, Lu et al. [9] reported that two discharges occur for every applied voltage pulse. The first discharge, termed as

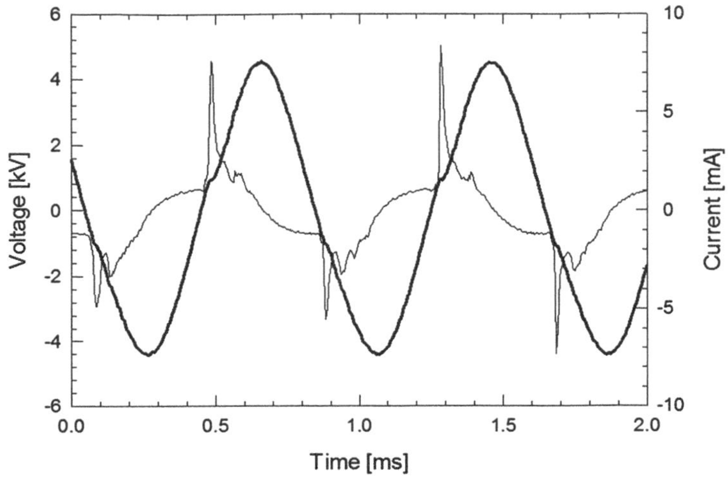


**Fig. 2.2** Current–voltage characteristics of a homogeneous DBD

**Fig. 2.3** Current–voltage characteristic of a pulsed DBD [9]

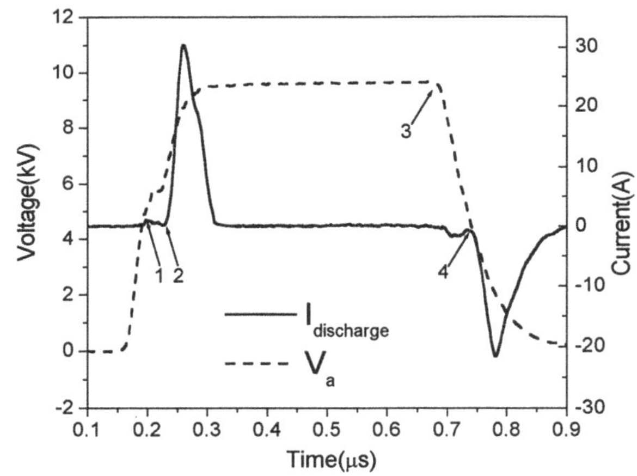


primary discharge, was ignited at the pulse rising edge while the second discharge, termed as secondary discharge, was ignited during the pulse falling edge. Figure 2.3 shows a typical current–voltage characteristic of a pulsed DBD.

### *2.1.2 Resistive Barrier Discharge (RBD)*

DBDs require operation at frequencies in the kHz range and above. This necessitates the use of high-voltage amplifiers that can operate in this range of frequencies. To extend the operating frequency range below kHz, few methods were proposed. For

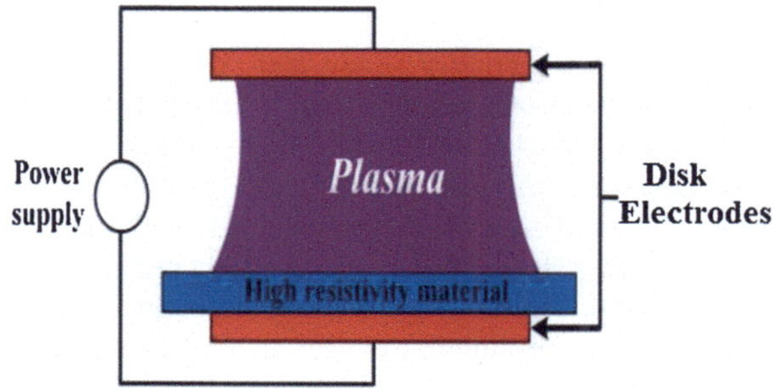


**Fig. 2.4** Schematic of the resistive barrier discharge (RBD)

example, Okazaki and coworkers used a dielectric wire mesh electrode to generate a discharge at a frequency of 50 Hz [19]. Laroussi and coworkers used a high resistivity layer/film to cover one of the electrodes in a device they referred to as the Resistive Barrier Discharge (RBD) [26]. The RBD can be operated with low frequencies extending all the way to DC. The film barrier usually has a resistivity of few MΩ.cm. The high resistivity film plays the role of a distributed resistive ballast which inhibits the discharge from localizing and the current from reaching high values. This way the glow-to-arc transition is avoided. If helium is used as the operating gas a diffuse plasma can be maintained at electrodes separation gaps up to 5 cm. If a helium/air mixture is used, the plasma becomes filamentary even for small gap distances [26]. Figure 2.4 shows a schematic of the RBD.

Under both AC and DC excitation, the discharge current is a series of pulses, making the RBD a self-pulsed discharge, similar to the DBD. The pulsing of the discharge current can be explained by the combined resistive and capacitive nature of the device. When the discharge is ignited, a current starts flowing and the equivalent capacitance of the electrodes becomes charged to the point where most of the applied voltage appears across the resistive layer covering the electrodes. The voltage across the gap becomes too small to maintain a plasma. When the discharge turns off, the equivalent capacitor discharges through the resistive layer. This leads to a rapid decrease of the voltage across the resistive layer and an increase in the voltage across the gas, initiating the discharge again [27].

The RBD has been used for various industrial and biological applications including the inactivation of bacteria for sterilization/decontamination purposes [28, 29].

### *2.1.3 Nonequilibrium Atmospheric Pressure Plasma Jets (N-APPJs)*

Nonequilibrium atmospheric pressure plasma jets (N-APPJs) are devices that can emit low-temperature plasma plumes in the surrounding air. They proved to be very useful for various plasma processing applications including biomedical applications

[14, 30, 31]. One of the attractive features of these plasma sources is the fact that they can generate a stable and controllable thin column of plasma outside the confinement of electrodes and into the surrounding environment. Because the plasma propagates away from the high-voltage region and into a region where there is no externally applied high voltage, the plasma is electrically safe and does not cause electrical shock/damage to the treated sample or exposed cells or tissues. However, the plasma plume does exhibit a very high instantaneous and local electric field at its tip. This field plays a role in the propagation of the plasma plume in the surrounding environment and can also affect the sample under treatment. Fig. 2.5 shows various designs of N-APPJs and Fig. 2.6 shows photographs of two devices, the plasma pencil and the kINPen.

The plasma plumes generated by N-APPJs are usually launched in ambient air, and as such, they provide reactive chemistry that can be exploited in biomedical applications. Reactive oxygen species (ROS), such as O, OH, $O_2^-$, and reactive nitrogen species (RNS), such as NO, $NO_2$, which can affect biological cells are produced by these devices. N-APPJs have recently been extensively used for the inactivation of pathogenic bacteria, for plasma-aided wound healing, dental applications, and the destruction of cancer cells and tumors.

The plasma plumes generated by N-APPJs are not continuous volumes of plasma but are discrete plasma packets/bullets propagating at high velocities, up to $10^5$ m/s [32, 33]. Figure 2.7 shows the plasma bullets velocity as a function of time along with ICCD images of the bullets. Extensive experimental and modeling work allowed the elucidation of the mechanisms governing the generation and propagation of these plasma bullets [34–45]. Lu and Laroussi first proposed a photoionization model to explain the dynamics of the plasma bullet [33]. However, it was also shown that the high electrical field at the head of the plume plays a crucial role in the propagation process. The strength of the electric field was experimentally measured by various investigators and was found to have an average value in the 10-30 kV/cm range [46–48].

The cold plasma sources described above produce a rich "cocktail" of chemically reactive species including reactive oxygen species (ROS) and reactive nitrogen species (RNS), which are known to have important biological implication [50]. As an illustration, Figs. 2.8 and 2.9 show two important species generated by plasma jets. Figure 2.8 shows the radial distribution of hydroxyl, OH, generated by a plasma pencil for different gas flow rates. When in contact with cells the hydroxyl radical (OH) causes peroxidation of unsaturated fatty acids which are a major component of the lipids constituting the cell membrane. Fig. 2.9 shows the axial distribution of nitric oxide, NO, generated by an RF jet at different applied powers. NO has various biological effects including participation in vascular homeostasis, neurotransmission, antimicrobial defense, and immune system regulation.

Other agents generated by cold plasma sources are suspected to play active roles in biological applications. These include charges species, UV radiation, and electric fields. In fact, N-APPJs exhibit a high electric field at the tip of the plasma

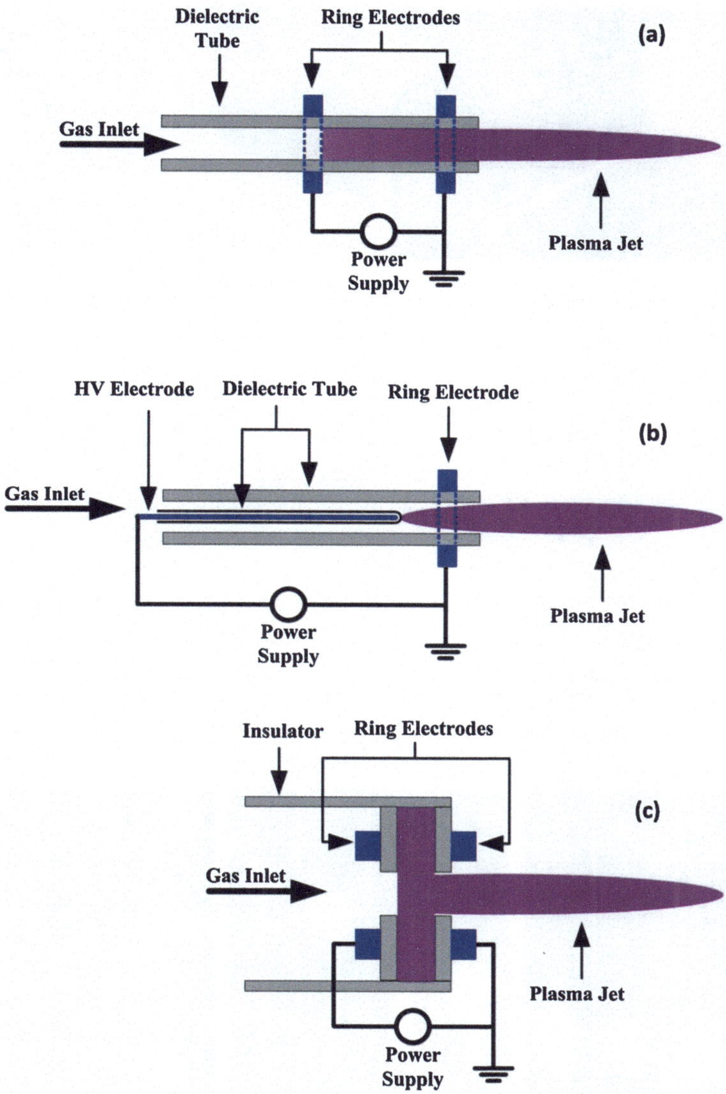


**Fig. 2.5** Three different designs of N-APPJs. (**a**) Two outer ring electrodes, (**b**) An inner insulated electrode and an outer ring, (**c**) Two flat rings on perforated dielectric disks

plume. This electric field can cause electroporation of cell membranes, allowing large molecules (including ROS and RNS) to enter the cells. Figure 2.10 shows an example of the electric field generated at the head of the plume of a pulsed plasma jet as a function of axial position and pulse time.

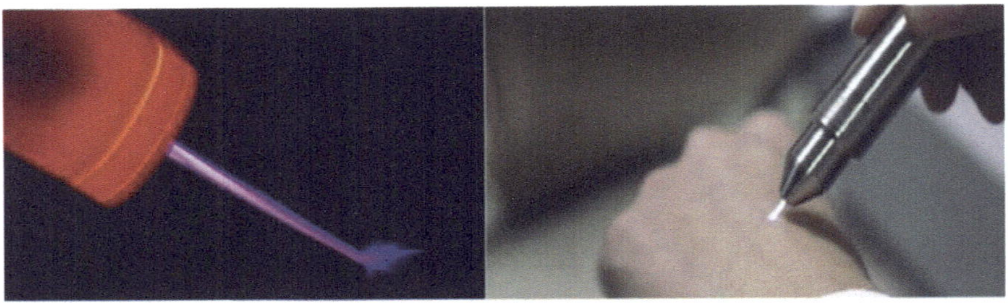

**Fig. 2.6** Photos of two N-APPJs in operation: Plasma pencil (left) and the kINPen

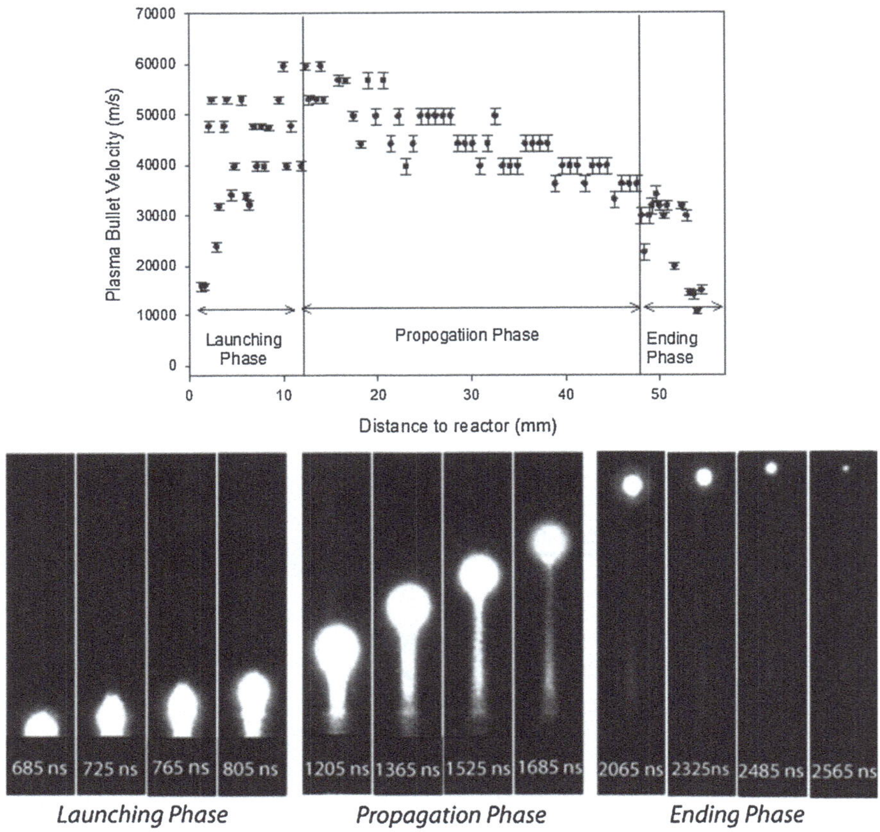


**Fig. 2.7** Plasma bullet velocity as a function of distance and ICCD images [49]

## Summary

Nonequilibrium atmospheric pressure plasma sources have introduced a paradigm-shifting technology in biomedicine. Early work from the mid-1990s to the early

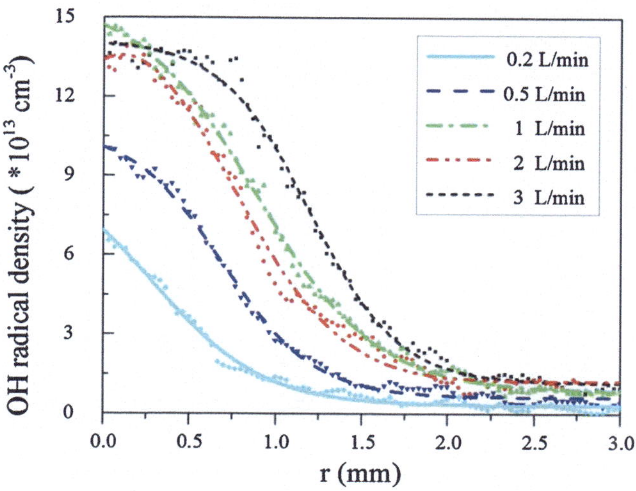


**Fig. 2.8** Radial distribution of the OH density for various gas flow rates. The plasma jet (gas: helium with 115 ppm of $H_2O$) has an insulated single pin electrode inside a quartz tube and the plasma plume is applied on top of a water film [51]

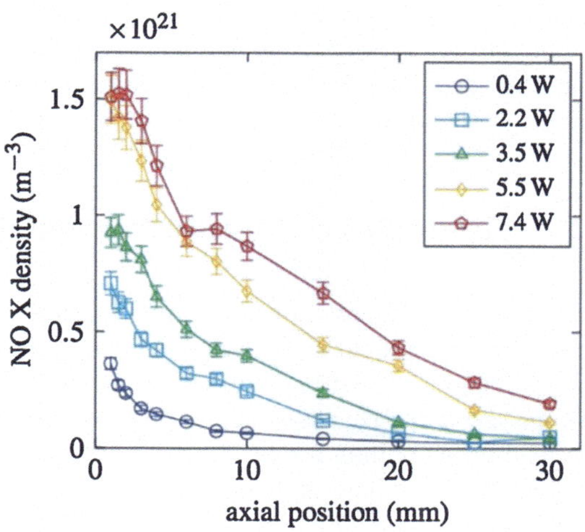


**Fig. 2.9** Axial profiles of NO concentration for different discharge power. Plasma jet sustained with a RF power of 3.5 W. The plasma is operated at a frequency of 13.6 MHz, pulsed at 20 kHz with 20% duty cycle. Working gas is Ar at 1 L/min with 2% air admixture [52]

2000s showed that these cold plasma devices can be used to affect biological cells in a controlled fashion, with effects ranging from inactivation of bacteria to proliferation of fibroblasts to induction of apoptosis. These effects are mediated by the active agents generated by cold plasma, which include ROS, RNS, UV radiation, charged species, and electric fields. These agents act alone and/or synergistically. The ability to modulate biological outcomes by controlling the type and concentrations of the reactive species, the magnitude of the electric field, etc. makes cold plasma a very attractive technology upon which novel medical therapies

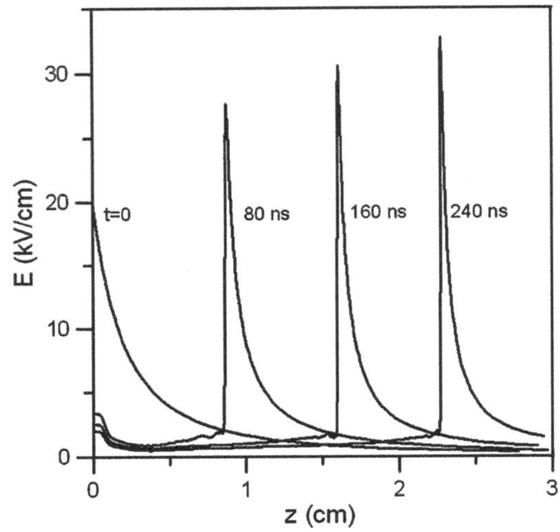


**Fig. 2.10** Electric field profile of a pulsed plasma jet at various axial positions and at different times of the applied pulse [40, 53]

can be developed. Of particular interest to this book are the cancer applications. Here, investigators have shown that cold plasma can be used successfully to kill cancer cells and shrink tumors [54, 55]. Experimental in vitro and in vivo tests on numerous types of adherent and nonadherent cancer cells showed that cold plasma has the potential to be used in the treatment of some cancers either on its own or as adjuvant therapy.

## 2.2 Atmospheric Pressure Nonequilibrium Plasma Jets

Atmospheric pressure nonequilibrium plasma jets (APNP-Js) have received vigorous developments in recent years, which generates low-temperature plasma plume in the open air [56–61]. Without the limitation of the treatment objects, this type of gas discharge becomes the hot point for its advanced plasma chemistry, flexible structure, and pollution-free. Based on the above advantages, it has great application potential in various fields, like biomedicine [12, 62–68], material science [69–76], environment [77, 78], and so on. Normally, the plasma devices that can propagate in the open air are called APNP-Js. Another key feature of APNP-Js is its nonequilibrium, where the electron temperature is much higher than the temperature of the heavy particles. In a nonequilibrium plasma, since the electron mass is much smaller than the ion mass (less than 1/1000), the electrical energy coupled into the plasma is mainly absorbed by the electrons, so the electron temperature is significantly higher than the ion temperature [53]. The nonequilibrium of APNP-Js greatly expands its application range. For example, when the plasma jet is extremely unbalanced, the electron temperature is about $10^4$ K and the gas temperature is room temperature. This nonequilibrium make the plasma is rich in reactive particles such as the OH, O, $O_3$, and NO which is suitable for biomedical applications [79–81].

However, there is no uniform classification standard until now, and many kinds of classification method and various names are used. For example, it can be classified by the type of working gas such as the air (or $N_2$, $O_2$) plasma jets and the noble gas plasma jets; or it can be classified by the number of the electrodes or by if there is the dielectric cover the electrode [53].

In this Section, the plasma jets are classified by the difference of the driven power and then make a more detailed division according to each category. First of all, they are divided into four parts, (1) Radio frequency driven jets; (2) Microwave power driven plasma jets; (3) AC and Pulsed DC driven plasma jets with several kilohertz; and (4) DC driven plasma jets. For the third category, the APNP-Js would been further divided into dielectric barrier discharge (DBD) jets, DBD-like jets, and floating ground electrode jets according to the presence or absence of the dielectric and the ground electrode. For the DC driven plasma jets, the APNP-Js have been classified into single electrode jets and double electrodes jets according to the number of the electrode. What's more, some work on the multi-power supply driven jets has been down, such as the AC with RF driven APNP-Js; double pulsed DC driven APNP-Js; pulsed DC with AC driven APNP-Js.

### *2.2.1 Radio Frequency Driven Jets*

One of the early APNP-Js, developed by Hicks' group, is shown in Fig. 2.11 [82, 83]. It consists of two concentric electrodes made of stainless steel. Radio frequency power working at 13.56 MHz is applied to the inner electrode and the outer electrode is grounded. It is also named dielectric-free electrode (DFE) jets is due to its no need of a dielectric material between two electrodes [53]. Helium with oxygen or other reactive gases the mixed together and fed into the annular space between the two electrodes. Under the standard operating conditions, the RF power ranged approximately from 40 W to 500 W, and higher power leads to arcing. In order to avoid arcing, the cooling water is needed which can keep the plasma temperature at the range of 100–150 °C. Besides the cooling jacket, the stable operation condition uses greater gas flow rates such as He flow rates 25 L/min, $O_2$ concentrations of up to 3.0% by volume, $CF_4$ concentrations of up to 4.0% by volume.

This kind of DFE jet has several notable characteristics, first, the stable operation conditions must be met otherwise arcing is unavoidable. Second, the power delivered to the plasma is higher and the gas temperature is higher than the room temperature or the cell culture temperature. Third, for this DFE jet, the peak voltage is a few hundred volts and the electric field within the discharge gap is relatively low and its direction is perpendicular to the gas flow direction which means that the electric field in the plasma plume region is even lower. And the plasma plume out of the nozzle is driven by the gas flow rather than by the electrical field. On the other hand, because a relatively high power can be delivered to the plasma and the gas temperature is relatively high, the plasma is very reactive. This kind of plasma jet is suitable for applications such as material treatment as long as the material to be treated is not very sensitive to high temperatures.

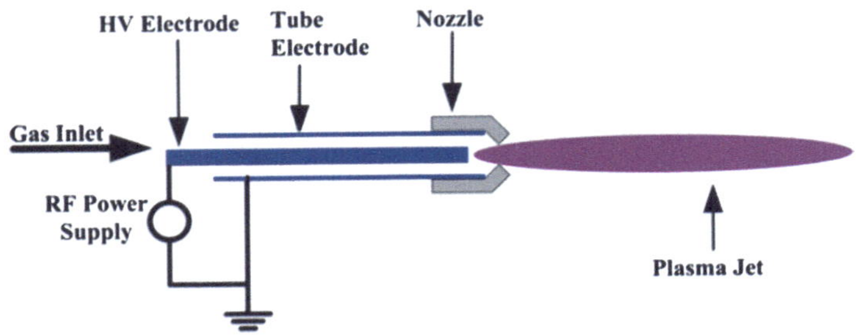


**Fig. 2.11** Cross-sectional view of the RF driven APNP-Js [82]

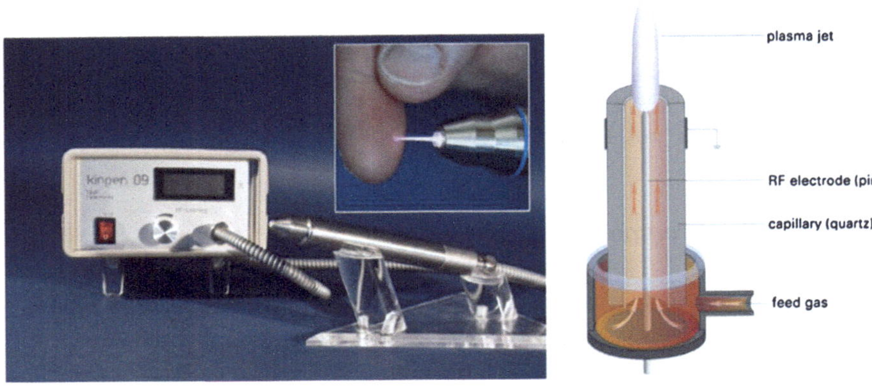


**Fig. 2.12** Photo of argon APNP-J driven by RF power supply

For the biomedical application, the plasma temperature must be lower than 40 °C and it must be safety for human touching. Many researchers have developed the RF driven plasma jets which were suitable for biomedical application [84–88]. K. D. Weltmann et al. added the dielectric between the high-voltage electrode and the ground electrode which make the discharge more stable and decrease the RF power to get lower gas temperature [89]. The commercial Argon RF device was named KINpen with different operating frequencies from 1 MHz to 40.68 MHz as shown in Fig. 2.12 The gas temperature is usually less than 40 °C. The plasma-dissipated power is approximately 1 W, with a device input power of approximately 20 W. Electron densities were expected to be up to the order of $10^{14}$ $cm^{-3}$ in the streamer head at electron temperatures of up to 4 eV.

Liu et al. have obtained a 4 cm long helium cold atmospheric pressure plasma jet with pulsed radio frequency excitation as shown in Fig. 2.13 [90]. The plasma generator consisted of an 8-cm length glass tube and the same length copper electrode along the tube center. The diameter of copper electrode is 2 mm. The inner and outer diameter of the quartz tube is 5.3 mm and 8 mm. Helium fed into the tube with 5 L/min and the plasma flushed out of the generator. The pulsed RF

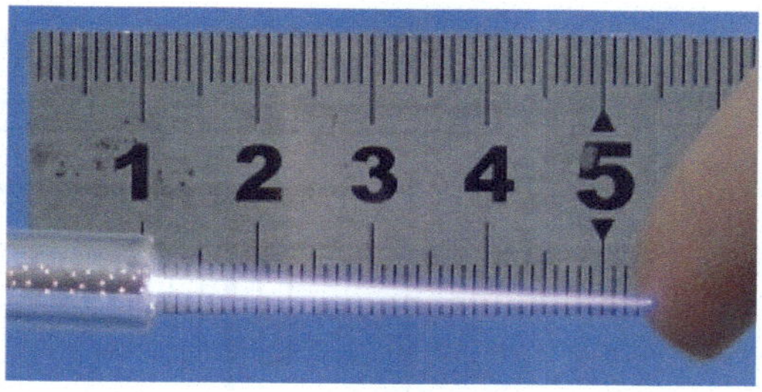


**Fig. 2.13** Photo of APNP-J driven by pulsed RF power supply

voltage applied to the inner electrode with the radio frequency at 12.8 MHz and the pulsed frequency at 4 kHz. The gas temperature of the plasma plume is about the room temperature. This pulsed RF plasma jets operated at more stable state and it has overcome the easier transition to arcing.

### *2.2.2 Microwave Power-Driven Plasma Jets*

Due to the high incident power (kW level) required to excite microwave discharge at atmospheric pressure and super-atmospheric pressure, a high-power pulse-modulated microwave source is needed for energy saving. However, it is difficult to use and control. Low-power microwave excitation and modulation of atmospheric gas discharge for a specific application is a technical problem that needs to be overcome. There have been many designs for discharge sources for microwave excitation [91–97], but the microwave power for this part of the work is mostly on the order of kW. Zhang et al. [93, 94] reported the use of a symmetric slotted antenna and the microwave discharge device realized 10 kW atmospheric pressure microwave discharge. German and Japanese researchers [95, 96] reported their experimental works of atmospheric pressure argon microwave jets which is excited by several kW power, and the periodic filament discharge characteristics of the jet have been discussed. Normally, for plasma biomedical and some temperature sensitive processing applications, the plasma temperature must be controlled, which is closely related to the microwave power.

Now, there are four typical kinds of microwave driven plasma jet devices with lower than 100 W microwave power, such as (1) coaxial transmission line resonators (CTLRs), [98, 99] (2) microstrip line resonators, [100, 101] (3) surface wave resonators [102, 103], and (4) hairpin shaped launchers [104]. There are some research studies on discharge characteristics of microwave jets. The earlier studies on CTLR operated at microwave frequencies (900 MHz and 2 GHz) were reported by Iza [105], and the devices showed great advantages such as simplicity, robustness, small ignition power, and high-power efficiency. In 2009, Choi et al. investigated the design, fabrication, and characterization of two microwave-excited microplasma sources based on coaxial transmission line resonators (CTLRs) operating at 900 MHz and 2.45 GHz [106]. In 2011, Seo et al. found that

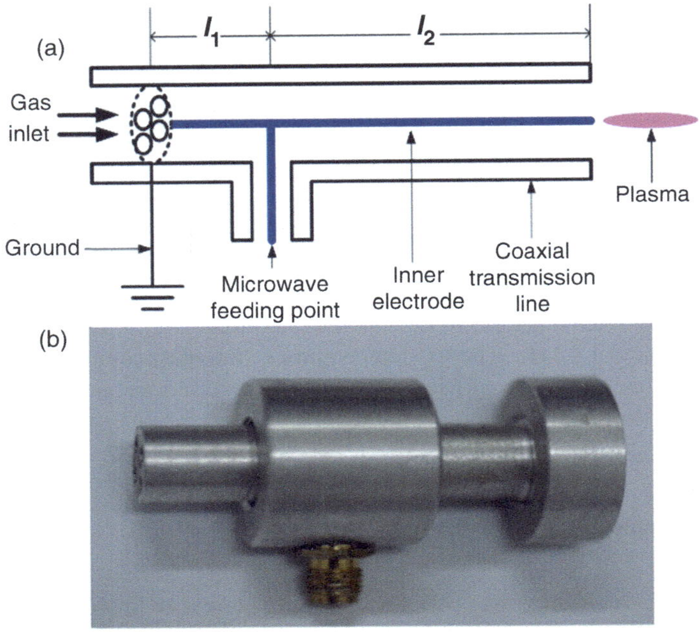


**Fig. 2.14** The structure of the microwave plasma jet

microwave plasma had better electrical properties such as no electric shock for safety, high discharge current for effective chemical reactions, clean waveforms for homogenous plasma, and high power coupling efficiency by comparing two different driving frequency discharges at 13 kHz (low frequency) and 877 MHz (microwave frequency) [107]. In 2017, Chen et al. studied the argon microwave plasma jet by experiments with simulation and found that there were bullet-shaped ionization fronts ahead of this kind of plasma jet which is resonantly excited by the local enhanced electric field [108].

In 2019, Nie et al. designed a microwave plasma device with resonance frequency of 2.45 GHz [109]. The output power of the microwave source were 5 W and 10 W. The device has a short end at one side and an open end at the opposite side of the coaxial transmission line, as shown in Fig. 2.14. The diameter of the inner electrode was 1 mm, and the inner radius of the outer electrode was 4 mm. The total length of the device was approximately one-quarter of the wavelength, which was about 30 mm. The five gaseous reactive oxygen and nitrogen species (RONS) density including OH, O, $O_3$, NO, and $NO_2$ density are measured synchronously while the plasma jet is treating a biological tissue under different discharge parameters (gas composition, gas flow, skin humidity, and the output power). It has been found that the gas composition and the output power are two key factors to regulate the dose of RONS concentrations. For OH concentration, water vapor in the gas flow plays an important role. By changing the microwave power and the gas composition, the

OH concentration changes from $0.4 \times 10^{14}$ cm$^{-3}$ to $5.54 \times 10^{14}$ cm$^{-3}$. O is mostly affected by the $O_2$ percentage in Ar and 1% percentage is the optimal value. $O_3$ is proportional to $O_2$ percentage in Ar and 2% $O_2$ can get maximum $O_3$ density of $6.9 \times 10^{16}$ cm$^{-3}$. NO is in the range of $2 \times 10^{13}$ cm$^{-3}$ to $5.4 \times 10^{14}$ cm$^{-3}$ and $NO_2$ is in the range of $5.5 \times 10^{14}$ cm$^{-3}$ to $5 \times 10^{15}$ cm$^{-3}$. The humidity of the skin increasing has a positive effect on the concentration of OH, $O_3$, and NO while a negative effect on the O and $NO_2$ concentrationa. For the normal skin with 40% humidity, the concentrations of the long lifetime species like $O_3$ and $NO_2$ is almost $10^{1\sim2}$ times of the other reactive species.

### 2.2.3 AC and Pulsed DC Driven Plasma Jets

#### DBD Jets

For DBD jets, as shown in Fig. 2.15a–e, there are many different configurations and all devices have a commonality, and there is dielectric layer(s) between the high-voltage electrode and the ground electrode, as shown in Fig. 2.15a, which was reported by Teschke et al. [32], the jet consists of a dielectric tube with two

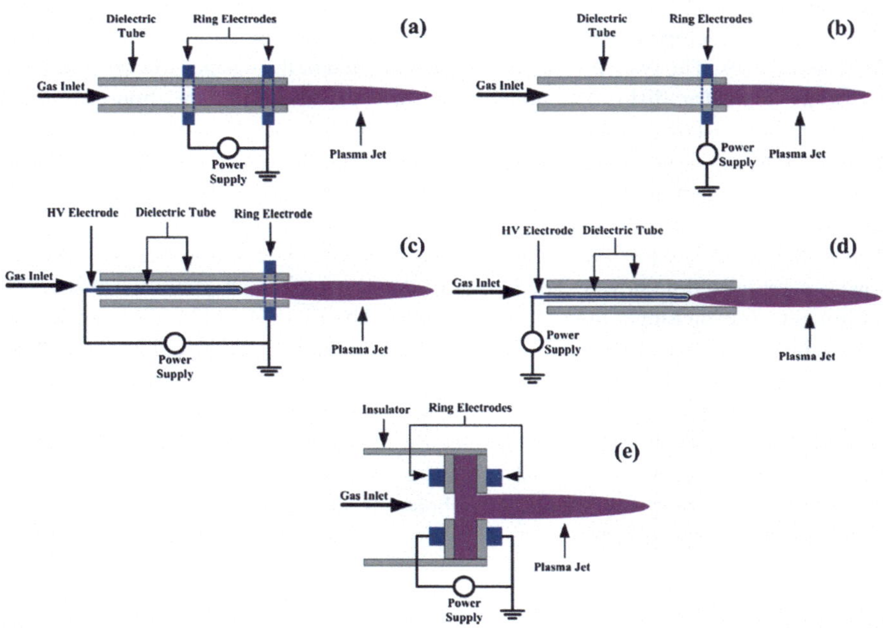


**Fig. 2.15** Schematics of various DBD plasma jets. (**a**) Two outer ring electrodes, (**b**) one single outer ring electrode, (**c**) an inner insulated electrode and an outer ring, (**d**) one inner insulated electrode, (**e**) two flat rings on perforated dielectric disks

metal ring electrodes on the outer side of the tube. When a working gas (He, Ar) flows through the dielectric tube and kHz high-voltage (HV) power supply is turned on, a cold plasma jet is generated in the surrounding air. This plasma jet operation condition is easy to control with gas flow velocity less than 20 m/s. The power consumed by the plasma was only several watts, so the gas temperature is close to the room temperature. The discharge current was several mA to several tens mA which is safe for human body and cells. The plasma plume that appears homogeneous by the naked eye is actually "bullet"-like propagating along the gas flow at more than $10^4$ m/s speed.

Figure 2.15b eliminates one ring electrode [110], when the jet is used acting on other objects, the objects play as the (virtual) ground electrode. Compared with two electrodes as shown in Fig. 2.12a, the discharge inside the dielectric tube is weakened. Figure 2.15c choose the pin electrode as the HV electrode and the electrode is covered by dielectric layer with one end closed [111]. With this configuration, the electric field along the plasma plume is enhanced. Walsh and Kong's studies show that a high electric field along the plasma plume is favorable for generating long plasma plumes and more active plasma chemistry [36]. Figure 2.15d further removes the ground ring electrode of Fig. 2.15c [112], so the discharge inside the tube is also weakened. On the other hand, a stronger discharge inside the discharge tube (as in the case of Fig. 2.15a and c) helps the generation of more reactive species. With the gas flow, the reactive species with relatively long lifetimes may also play an important role in various applications. The configuration of Fig. 2.15e, developed by Laroussi and Lu [31], is different from the previous four DBD jet devices. The two ring electrodes are attached to the surface of two centrally perforated dielectric disks. The holes in the center of the disks are about 3 mm in diameter. The distance between the two dielectric disks is about 5 mm. With this device, a plasma plume of up to several centimeters in length can be obtained. This plasma jet is known as the Plasma Pencil and was the earliest bio-tolerant plasma jet developed for biomedical applications.

All the DBD jet devices discussed above can be operated either by kHz ac power or by pulsed dc power. The length of the plasma jet can easily reach several centimeters or even longer than 10 cm, as reported by Lu et al. [111]. This capability makes the operation of these plasma jets easy and practical. There are several other advantages of the DBD jets. First, due to the low power density delivered to the plasma, the gas temperature of the plasma remains close to room temperature. Second, because of the use of the dielectric, there is no risk of arcing whether the object to be treated is placed far away or close to the nozzle. These two characteristics are very important for applications such as plasma medicine, where safety is a strict requirement.

### DBD-Like Jets

All the plasma jet devices shown in Fig. 2.13 are named DBD-like jets. There is a dielectric layer between the HV electrode and the ground electrode, while

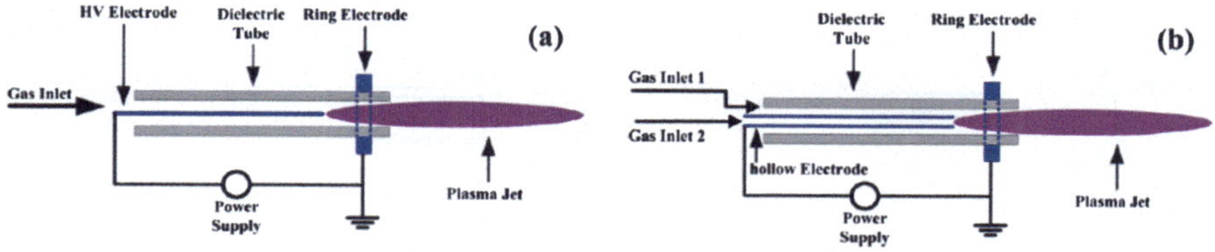


**Fig. 2.16** Schematics of DBD-like plasma jets. (**a**) One inner needle electrode and one outer ring electrode, (**b**) one inner hollow tube electrode and one outer ring electrode

the plasma plume is in contact with an electrically conducting (a nondielectric material) object, especially a ground conductor, the discharge is actually running between the HV electrode and the object to be treated (ground conductor). For such a circumstance, it no longer operates as a DBD. The devices shown in Fig. 2.16 can be driven by kHz ac power, or by pulsed dc power.

Figure 2.16a the HV electrode is a bare metal needle electrode and in Fig. 2.16b, the HV electrode has been replaced by a hollow electrode [113, 114]. The main difference of two devices is that the device in Fig. 2.16b has two gas flow channel. Normally, gas inlet 1 is for a noble gas and gas inlet 2 is used for a reactive gas such as $O_2$ flow. It was found that the plasma plume is much longer with this kind of gas control than that using a pre-mix gas mixture with the same percentage [113]. The role (and advantage) of the ring electrode in Fig. 2.16a and b is the same as in the case of DBD jets.

When the DBD-like plasma jets are used for plasma medicine applications, the object to be treated could be cells or whole tissue. In this case, these types of jet devices should be used carefully because of the risk of arcing. On the other hand, if it is used for treatment of conductive materials, since there is no dielectric, more power can easily be delivered to the plasma. Therefore, as long as arcing is carefully avoided, the DBD-like jets have their own advantages.

### Floating Electrode Jets

The floating electrode (SE) jets are shown in Fig. 2.17a–c. Figure 2.17a and b are similar to the DBD-like jets except there is no ring electrode on the outside of the dielectric tube. The dielectric tube only plays the role of guiding the gas flow.

The plasma jets shown in Fig. 2.17a and b can be driven by kHz AC and pulsed DC power. However, when they are used to work on the conductive objects, the discharge is easy to translate to arc. Considering the safety issues especially for biomedical application, the device configurations in Fig. 2.17a and b are not suitable for direct action on living organisms [115]. In order to overcome this problem, Lu et al. developed a similar SE jet, as shown in Fig. 2.17c [116]. The capacitance C and resistance R are about 50 pF and 60 kΩ, respectively. The resistor and capacitor are used for controlling the discharge current and voltage on the hollow electrode

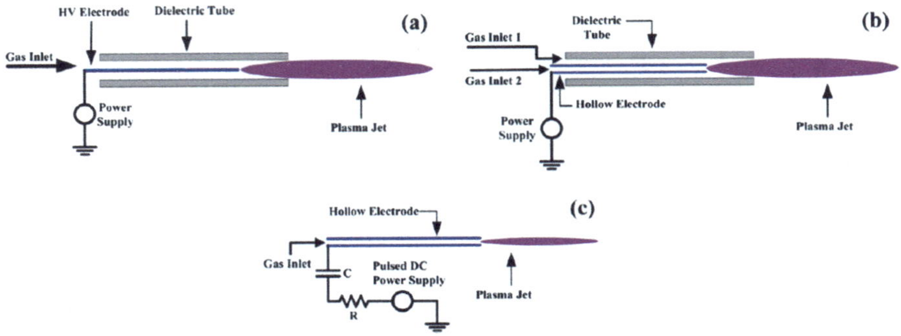


**Fig. 2.17** Schematics of SE plasma jets. (**a**) One inner needle electrode, (**b**) One inner hollow tube electrode, (**c**) One hollow tube electrode without outer dielectric tubing

(needle). This jet is driven by a pulsed dc power supply with a pulse width of 500 ns, repetition frequency of 10 kHz and amplitude of 8 kV. The advantage of this jet is that the plasma plume or even the hollow electrode can be touched without any risk of injury, making it suitable for plasma medicine applications.

Based on the configuration of Fig. 2.17c, one application is the device made by Lu et al. which is used for root canal treatment [116]. Due to the narrow channel geometry of a root canal, which typically has a length of few centimeters and a diameter of 1 mm or less, the plasma generated by a plasma jet is not efficient to deliver reactive agents into the root canal for disinfection. Therefore, to have a better killing efficacy, a plasma needs to be generated inside the root canal, whereupon reactive agents, including the short-lifetime species, such as charged particles, could play some role in the killing of bacteria. As shown in Fig. 2.18, the plasma can be generated inside the root canal.

## *2.2.4 DC Driven Plasma Jets*

### Single Electrode (SE) Plasma Jets

The single-electrode plasma jets are characterized by only one high-voltage electrode, and the object to be treated acts as another electrode. The plasma is produced between the high-voltage electrode and the item being processed. In 2010, Bussiahn et al. developed a single-electrode argon APNP-J device with a negative DC drive [117]. The schematic of the device is shown in Fig. 2.19 [117]. The high-voltage electrode is a hollow needle electrode that is directly connected to a negative DC high-voltage power supply. When the input voltage is −14 kV and the volumetric flow rate is 0.5 L/min, filamentous argon APNP-J is produced in the narrow root canal. The gas temperature of argon APNP-J is 300 K. Although the input voltage is DC high voltage, the discharge current is actually in the form of a pulse with a

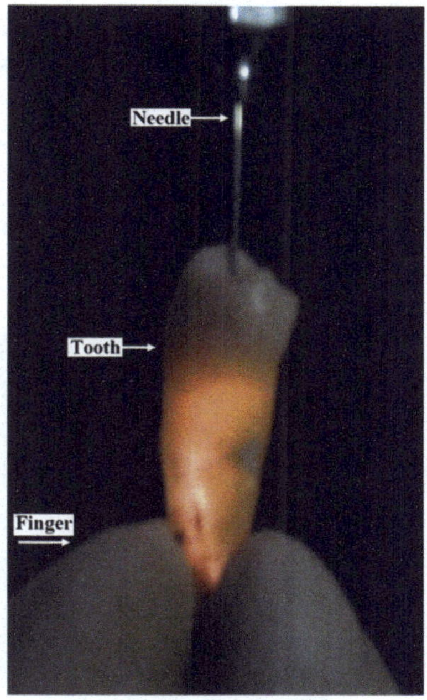


**Fig. 2.18** Photograph of a plasma generated in the root canal of a tooth using the jet of Fig. 2.17c [116]

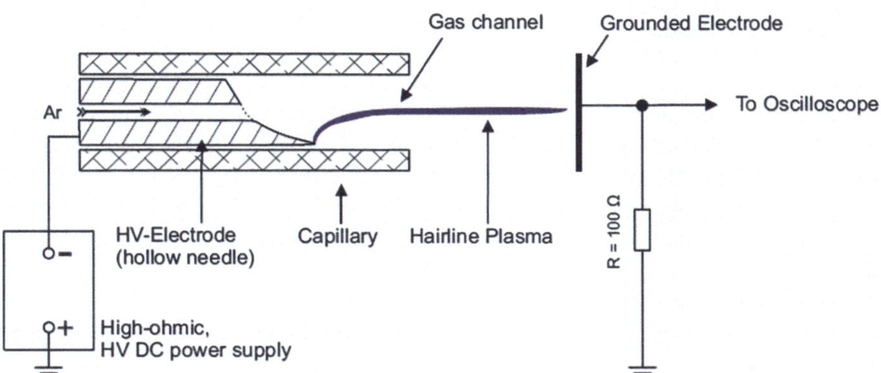


**Fig. 2.19** Schematic of the hairline plasma device

pulse repetition rate of 1–3 kHz, a current peak of 1.4 A, and a pulse width of 9.8 ns. The average power consumption of argon APNP-J is 0.1 ~ 0.5 W. Bio-inactivation experiments showed that the typical pathogen Escherichia coli NTCC 10538 in the mouth was effectively inactivated by argon plasma jets.

In the same year, in order to use cheaper air as a working gas and completely remove the complicated gas supply system, Wu and Lu developed a single-electrode air APNP-J device driven by a positive DC high-voltage power supply [118]. The schematic of the device is shown in Fig. 2.20a [118]. The high-voltage electrode is

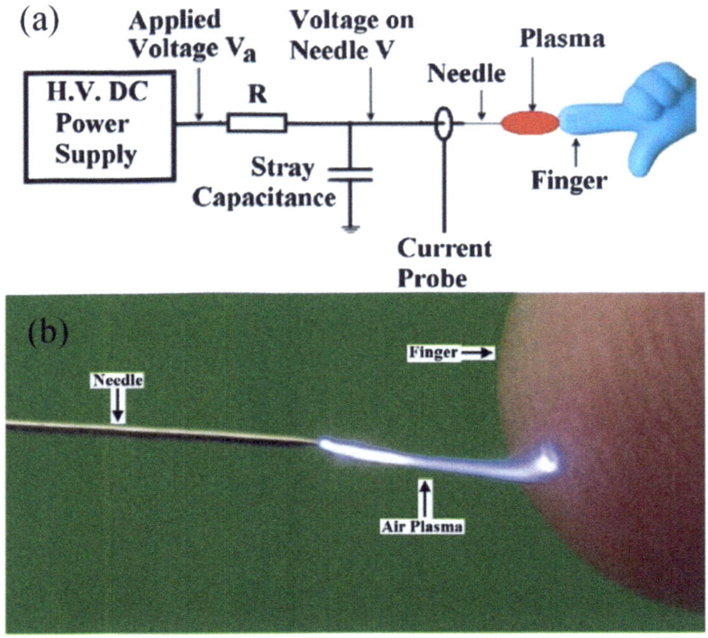


**Fig. 2.20** (**a**) Schematic of air APNP-J device and (**b**) the photo of air plasma touched by a finger

a bare solid needle electrode that is connected to a high-voltage DC power supply through a large resistor with 130 MΩ resistance value. The stainless steel tip has a radius of 100 μm. The mass of the whole device is less than 1 kg. When the object to be treated is a finger, normal temperature air APNP-J can be generated between the needle tip and the finger, as shown in Fig. 2.20b [118]. The length of the air plasma plume can be up to 2 cm. The air plasma can be in safe contact with the human body without electric shock and burning sensation, and even the fingers can be safely placed on the high-voltage needle electrode. Figure 2.21 is the waveform diagram of the tip voltage and discharge current on the needle electrode (Fig. 2.21b is a partial enlarged view of Fig. 2.21a) [118]. The results show that the discharge current is actually in the form of periodic pulse. The current pulse repetition rate is approximately 25 kHz with a pulse width of 100 ns and a current peak of 17 mA. This short-pulse, low-current self-sustaining discharge mode ensures that the human body can be safely contacted. Inactivation experiments showed that the air APPJ was able to effectively inactivate Enterococcus faecalis bacteria that caused oral treatment failure.

In order to enlarge the cross section of the air APPJ, the high-voltage needle electrode was placed in a quartz tube. Figure 2.22a is the schematic diagram of the APPJ device [119]. The filament-like air APPJ section increases rapidly as the volumetric air flow rate in the tube is >1 L/min. Subsequently, an air APPJ of 0.5 cm (length) × 1 cm (width) was further successfully produced in the form of an array,

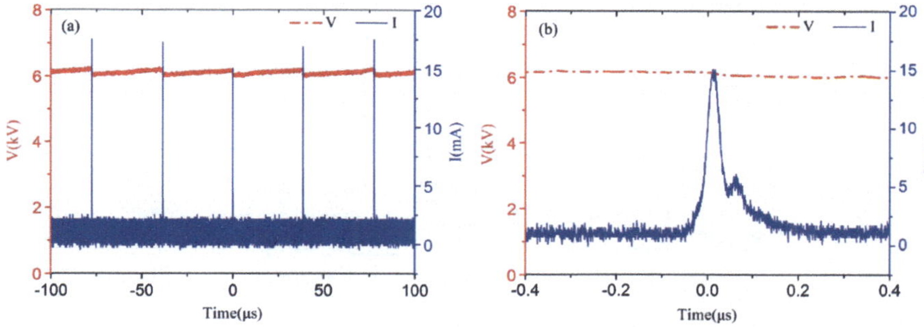


**Fig. 2.21** (**a**) Current and voltage waveforms on the needle (**b**) is a partial enlarged view

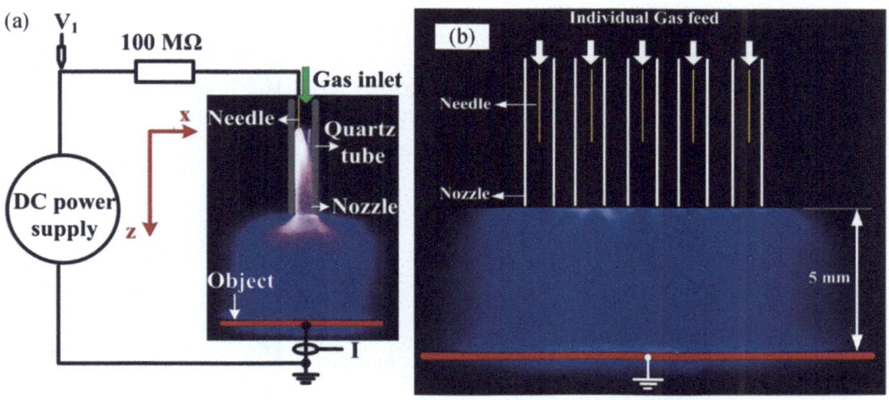


**Fig. 2.22** (**a**) Schematic of air discharge device and (**b**) the photo of air plasma jet array

as shown in Fig. 2.22b [119]. It is worth noting that the air APPJ in the array overlap each other, and the spectral measurements show that the plasma radiation intensity is uniform, which is significantly different from the inert gas APPJ array [120, 121]. In this way, the quartz tube and the gas flow play an important role in increasing the cross section of the air APNP-J. Depending on the application, air APNP-Js arrays of different sizes can be arranged as desired. This is meaningful for large-area material handling and biomedical applications.

It is worth mentioning that recently, Pei and Lu developed a battery-powered, self-contained, ambient air APPJ device [122] (called plasma flashlight). Figure 2.23a is a schematic diagram of the APNP-J device [122]. The air APNP-J is driven by a 12 V battery and can be boosted to 10 kV after DC-DC conversion. The high-voltage pin electrode is connected to a high-voltage DC power supply through a current limiting resistor with a resistance of 50 MΩ. Figure 2.23b is a photo of a plasma flashlight [122]. Normal temperature air APNP-J is produced between the object to be treated and the high-voltage electrode, which is safe for touching. The device does not depend on the external power supply and the air supply circuit,

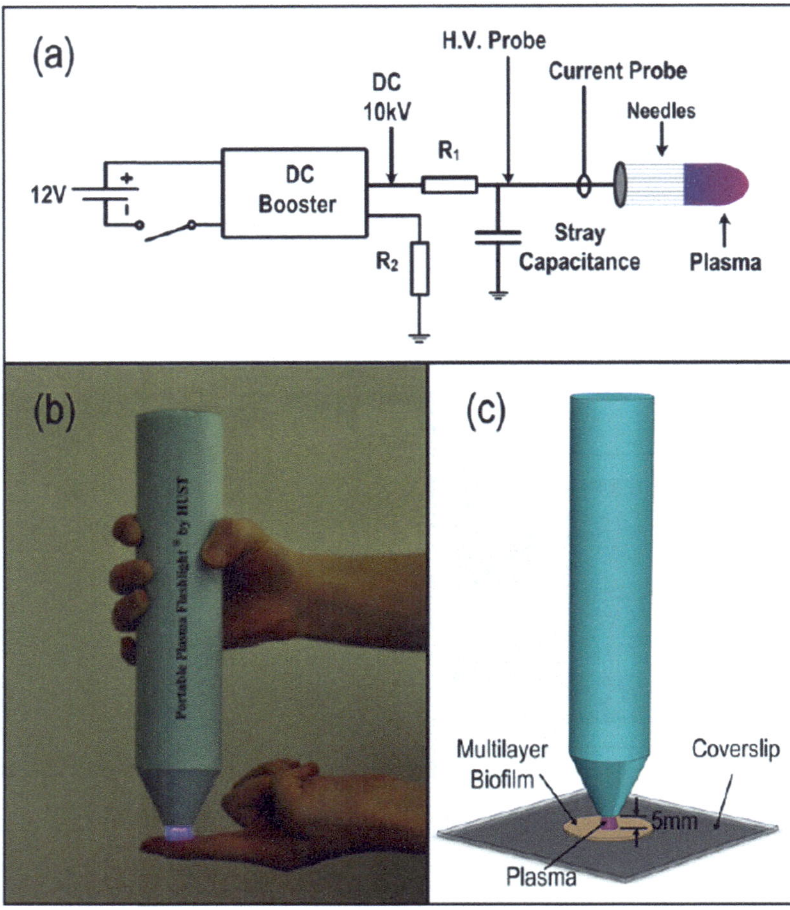


**Fig. 2.23** (**a**) Schematic of the plasma jet setup, (**b**) photograph of the portable handheld plasma flashlight device, and (**c**) schematic of the biofilm treatment

making the entire experimental device small, lightweight, and highly adaptable. The experiment found that the discharge current is actually a self-pulse mode, which is similar to Fig. 2.21. The current pulse peak is about 6 mA, the pulse width is about 100 ns, the pulse repetition rate is about 20 kHz, and the power consumed on the plasma is about 60 mW. As shown in Fig. 2.23c [122], a plasma flashlight was applied to the biofilm inactivation experiment. The ambient air APNP-J was found to effectively penetrate the Enterococcus faecalis biofilm with a thickness of 25.5 μm. This portable plasma flashlight has great application prospects in the absence of AC power, such as natural disasters and wound sterilization in remote areas.

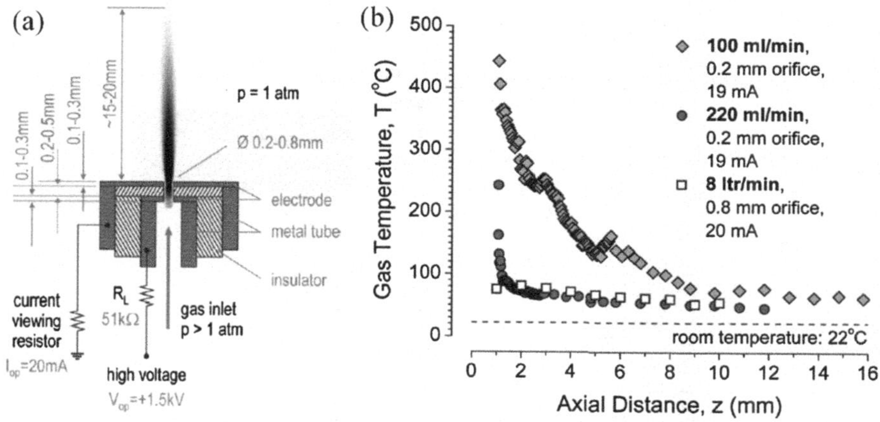


**Fig. 2.24** (**a**) Schematic of air APNP-J device and (**b**) gas temperature along the plasma plume for different gas flow rate

Single-electrode air or argon APNP-Js is similar to positive corona discharge, but it is different from positive corona discharge such as the discharge current is a periodic ns pulse, the discharge mode is stable, and it is not easy to convert to glow or arc discharge.

### Double Electrodes Plasma Jets

The two-electrode APNP-Js driven by a DC power source usually discharges between the high-voltage electrode and the ground electrode, and then the airflow ejects the plasma outside air. In 2008, Kolb et al. developed a two-electrode air APNP-J device [123]. The schematic of the device is shown in Fig. 2.24a [123]. The two electrodes are separated by an insulating medium with a thickness of 0.2–0.5 mm. There is a circular hole with a diameter of 0.2–0.8 mm at the center of the medium. The high-voltage electrode is connected to the high-voltage DC power supply through a 51 kΩ resistor, and the center hole on the ground electrode is the same size as the dielectric medium. When the voltage on the high-voltage electrode is 1.5–2.5 kV, gas breakdown occurs between the two electrodes to form a glow discharge, and the plasma plume injected into the air can reach a length of 1.5 cm. The temperature of the gas is determined by the velocity of the gas flow and the distance from the hole, as shown in Fig. 2.24b [123]. When the gas flow rate is 100 mL/min, the gas temperature drops from 500 °C at the nozzle to 100 °C at 1.5 cm; when the gas volume flow rate is >220 mL/min, the gas temperature outside the nozzle is close to room temperature. When the current is 20 mA, the power consumed on the plasma is about 10 W.

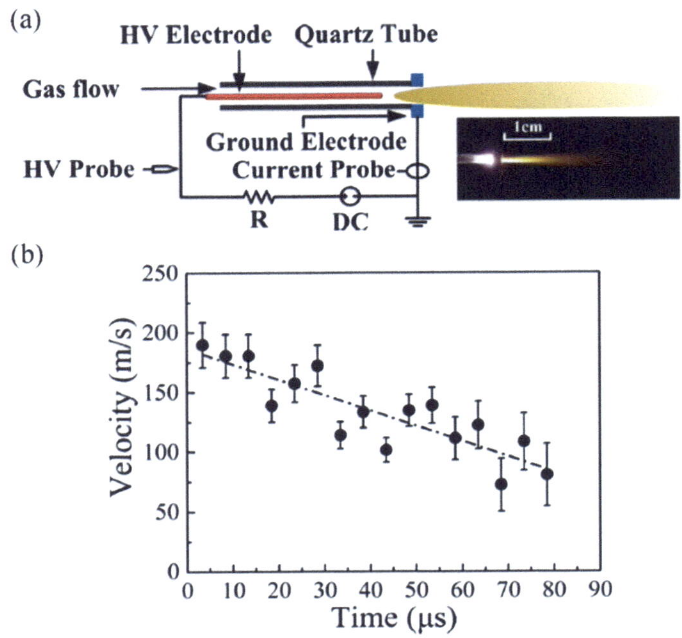


**Fig. 2.25** (**a**) Schematic of discharge device and (**b**) the propagation velocity of APNP-J vs time

In addition to the glow discharge mode, Xian and Lu found that the DC-driven two-electrode nitrogen and air APNP-J also have a self-sustained pulse mode, and further studied its propulsion mechanism [124]. The APNP-J device used in the experiment is shown in Fig. 2.25a [124]. The device is similar to that developed by Kolb et al. except that the high-voltage electrode is a solid needle with a distance of 2.5 mm from ground and a current limiting resistor of 20 MΩ. Figure 2.25b shows the speed of nitrogen APNP-J with the time [124]. From the discharge dynamic process taken by the high-speed camera, it can be found that the propagation process of the two-electrode nitrogen APNP-J is similar to the "plasma bullet" propagation, but the propagation speed is not $10^4$–$10^5$ m/s which is close to the airflow about 100 m/s. This finding demonstrates that nitrogen APNP-J is gas-driven rather than electric field driven in this two-electrode self-pulse discharge mode.

In 2013, Deng and Nikiforov et al. studied DC-driven nitrogen and air APNP-J devices [125]. The experimental setup is similar to Fig. 2.25a. Two discharge modes were also found as self-sustained pulse discharge mode and glow discharge mode. In the two discharge modes, the gas temperature of the plasma between the two electrodes is about 1000 °C, and the gas temperature of the APNP-Js outside the nozzle is close to room temperature. The air APNP-Js can generate high-density NO particles up to $1.1 \times 10^{16}$ cm$^{-3}$.

The DC-driven single-electrode APNP-Js is similar to the DC corona discharge. The discharge has a self-pulse mode. This dissipated power is <1 W. There are two discharge modes for the DC-driven two-electrode APNP-Js: glow mode and self-pulse mode. Xian et al. have demonstrated that in self-pulsing mode, nitrogen and air APNP-Js are not driven by electric field, but are driven by airflow [124].

The DC-driven two-electrode APNP-Js consumes less than 10 W power. In general, the structural characteristics of the DC-driven APNP-Js contain two parts, one is that the high-voltage electrode is generally connected to the high-voltage DC power supply through a current limiting resistor which, is used to limit the current growth, effectively suppress the mode transition, and obtain a stable discharge mode; the other is that the higher gas flow velocity can effectively lower the gas temperature and suppress thermal instability. However, the presence of the current-limiting resistor reduces the efficiency of power supply. In single electrode air APNP-Js, the power consumed in the resistor is 0.3 W, and the power consumed on the plasma is 0.1 W [118]; in the two-electrode APNP-Js, the power consumed in the resistor is 20 W, and is consumed in The power on the plasma is <10 W [123]. Therefore, most of the power is actually consumed by the resistor, which reduces the coupling efficiency between the power source and the plasma. It is worth noting that the two-electrode APNP-Js consumes more power (which is about 10 times) than the single-electrode APNP-Js. However, due to the low cost of DC high-voltage power supply (such as battery drive), the overall power required for discharge is relatively small, the discharge structure is simple and effective, which make the handheld and portable APNP-Js be realized, so the DC-driven APNP-Js is applied. These kinds of devices are still appealing.

## 2.2.5 *Multipower-Driven APNP-Js*

### AC with RF-Driven APNP-Js

In order to achieve higher activity of APNP-Js, Cao et al. used dual power supply [126]. A schematic diagram of the discharge device is shown in Fig. 2.26a [126]. The jet device consists of three electrodes, and the upstream capillary tube is inserted into the quartz tube and directly connected to the RF power source (frequency is 5.5 MHz). One ring ground electrode is wrapped at the quartz nozzle, and one plate electrode is placed downstream, and the plate electrode is connected to the AC power source with a driving frequency of 30 kHz. Figure 2.26b shows the change in APNP-J length under dual-supply interactions [126]. It is clear that the dual power supply APNP-J is longer than the single RF power drive. In addition, the experiment found that the spectral radiation intensity is stronger when the dual power supply is driven, but the gas temperature is not significantly increased. The authors believe that the plasma generated by the RF drive provides electrons for the downstream AC discharge, resulting in a higher plasma number density and

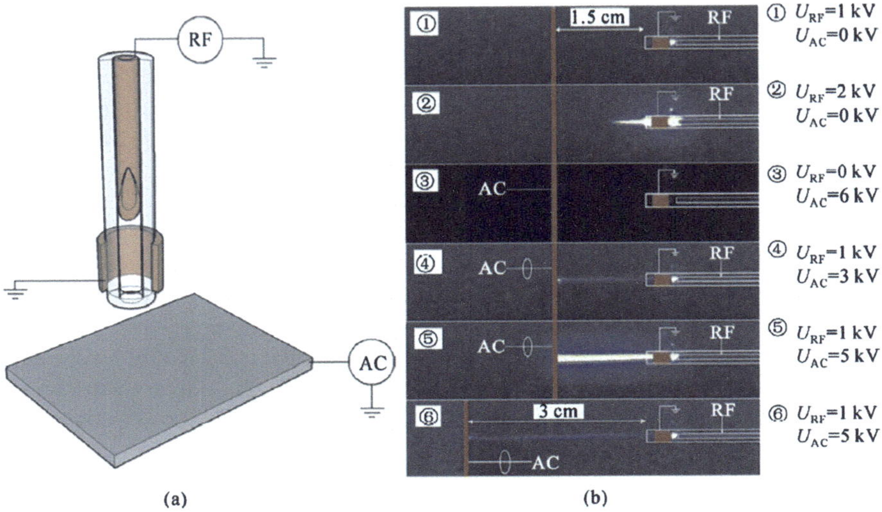


**Fig. 2.26** (**a**) Schematic of discharge device driven by RF and ac power supply, and (**b**) the photo of APNP-Js (with a helium gas flow rate of 4 L/min, the distances between the tube and the grounded electrode are 1.5 cm and 3 cm, respectively)

longer plasma jet for AC discharge, while ensuring that the gas temperature is not significantly improved.

In order to further study the interaction between kHz-level AC dielectric barrier discharge and RF-driven APNP-J, Li et al. designed a special argon APNP-J device [127]. As shown in Fig. 2.27a and b [127], the APNP-J unit consists of a glass tube and 3 electrodes. The two coaxial 1 mm thick glass tubes are 2.5 mm apart which have been sealed at both ends. The two copper foils are 20 mm and 21 mm wide and are wrapped around the outer glass tube and connected to AC high voltage (frequency 50 kHz) and RF high voltage (frequency 13.56 MHz). A grounded tungsten electrode (3 mm diameter) is inserted into the inner glass tube. When the argon volume flow rate is 5 L/min and the voltages are applied, argon APNP-J can be generated in the tube. Figure 2.27c shows the voltage and current characteristics of APNP-J with or without kHz AC power supply [127]. It has been found that the kHz AC dielectric barrier discharge can effectively reduce the voltage required to produce a stable and uniform RF-driven APNP-J. By exchanging the locations of the AC and RF, a same result was obtained. This indicates that the seed electrons generated by the AC discharge are not blown by the airflow to the RF region. By adjusting the distance between the AC electrode and the RF electrode, it is found that the RF-driven uniform APNP-J can be reduced only when the distance between the two electrodes is <18 mm and the AC-driven filament discharge can reach the RF region.

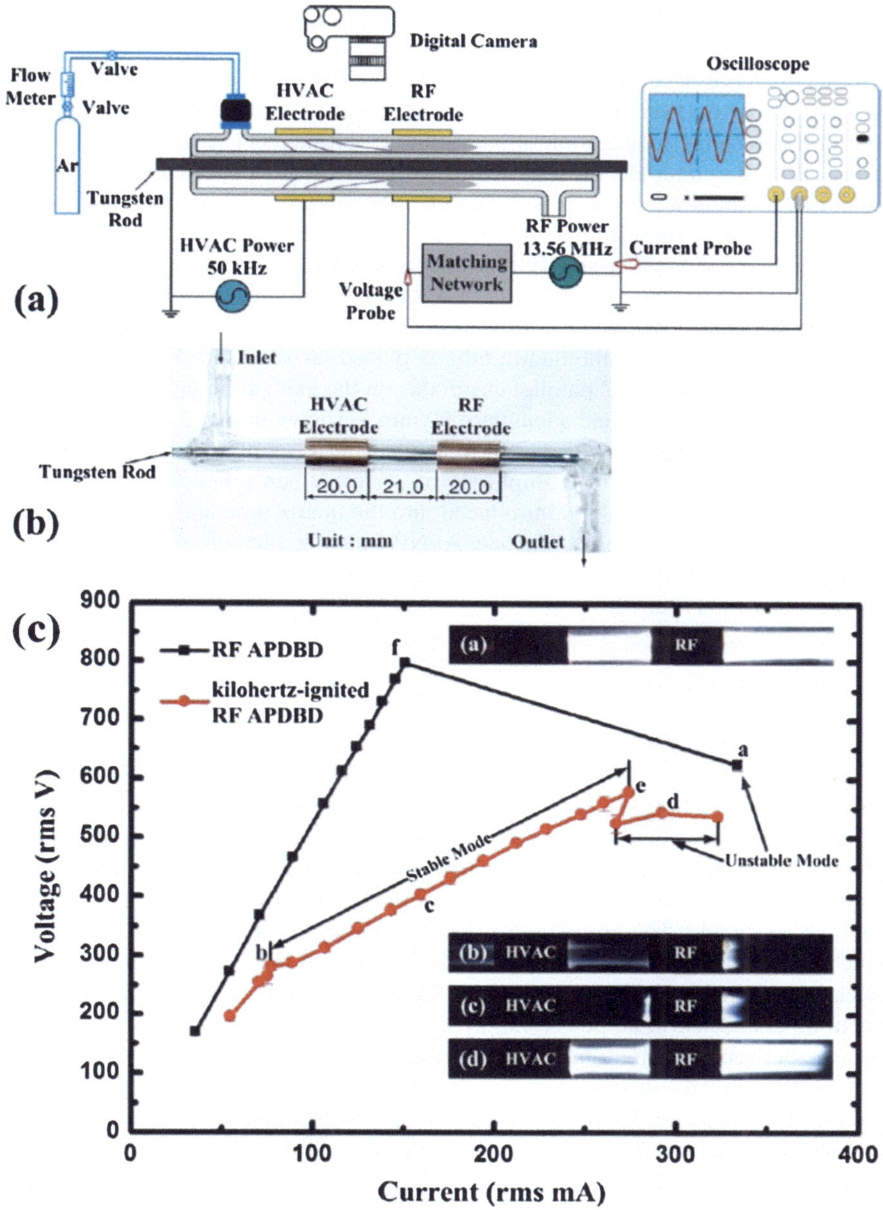


**Fig. 2.27** (**a**) Schematic of discharge device, (**b**) the photo of experimental device, (**c**) and the I-V character curve driven by RF power with or without kH ac power applied (the inset images corresponds to the photos of discharge at a, b, c, and d, respectively)

### Double-Pulsed DC-Driven APNP-Js

In order to more accurately control the relative change in phase and amplitude between the two power supplies, Walsh et al. used two pulsed power supplies to control the dynamic propagation of atmospheric plasma jets [128, 129]. As shown in Fig. 2.28, the main electrode structure is as follows, a copper electrode with a width of 10 mm is wrapped around a quartz tube with an inner diameter of 2 mm and a distance of 10 mm from the nozzle. The electrode is directly connected to the pulsed high-voltage power supply (the fixed pulse width is 1950 ns, with a pulse frequency of 5 kHz and an amplitude of 3.9 kV). There are two kinds of external electrode structures: one is 14 mm from the nozzle, and a ring electrode (10 mm in diameter) coaxial with the quartz tube is placed, as shown in Fig. 2.28a [128, 129]; the other is to place 2 parallel electrodes on the axis of the quartz tube which have a distance of 10 mm and a length of 60 mm as shown in Fig. 2.28b [128, 129]. The external electrodes are connected to another pulse supply (with a pulse width of 300 ns and a pulse voltage amplitude of 3 kV). When a helium gas having a volume flow rate of 3 L/min is introduced into the quartz tube and a pulse voltage is applied to the high-voltage electrode, APNP-J having a length of several cm can be produced outside the nozzle. Figure 2.28c) shows two pulse voltage and current waveform curves [128, 129].

### AC and Pulsed DC Driven APNP-Js

Li et al. have studied the APNP-Js driven by AC and pulsed DC power and they have changed the phase shift between the two powers [130]. The APNP-Js structure is shown as Fig. 2.29a. The main configuration is composed of three electrodes and a clear description is given in the following. The HV electrode made of copper wire, which is connected to the ac power supply, is inserted into 4.5-cm-long quartz tube 1 with one end closed. The inner and outer diameters of quartz tube 1 are 1 and 2 mm, respectively. The ground electrode and pulsed dc ring electrode, which are attached to the outer surface of quartz tube 2, are also made of copper wire. The inner and outer diameters of quartz tube 2 are 5 and 6 mm, respectively. The HV electrode connected to ac power supply and tube 2 are aligned along tube 1. The distance between the tip of the pulsed dc ring electrode and the ground electrode is about 1 cm, while the distance between the tip of the pulsed dc ring electrode and ac HV electrode is about 0.5 cm. Figure 2.29b shows the applied voltage waveforms for both ac and pulsed dc power supplies at the phase difference of 0.

It is found that when the pulsed dc voltage is added on the negative ac half period at the phase shift of about 90°, the gas temperature reached their maximum as shown in Fig. 2.30a. On the other hand, the ground state O atom concentration has two peaks; one of the peaks also appears when the phase shift is about 90° as shown in Fig. 2.30b. When the pulsed dc voltage is applied on the positive half period at the phase shift of 270°, the O concentration reaches another peak and the peak value is higher than that when the ac or the pulsed dc power supply works alone.

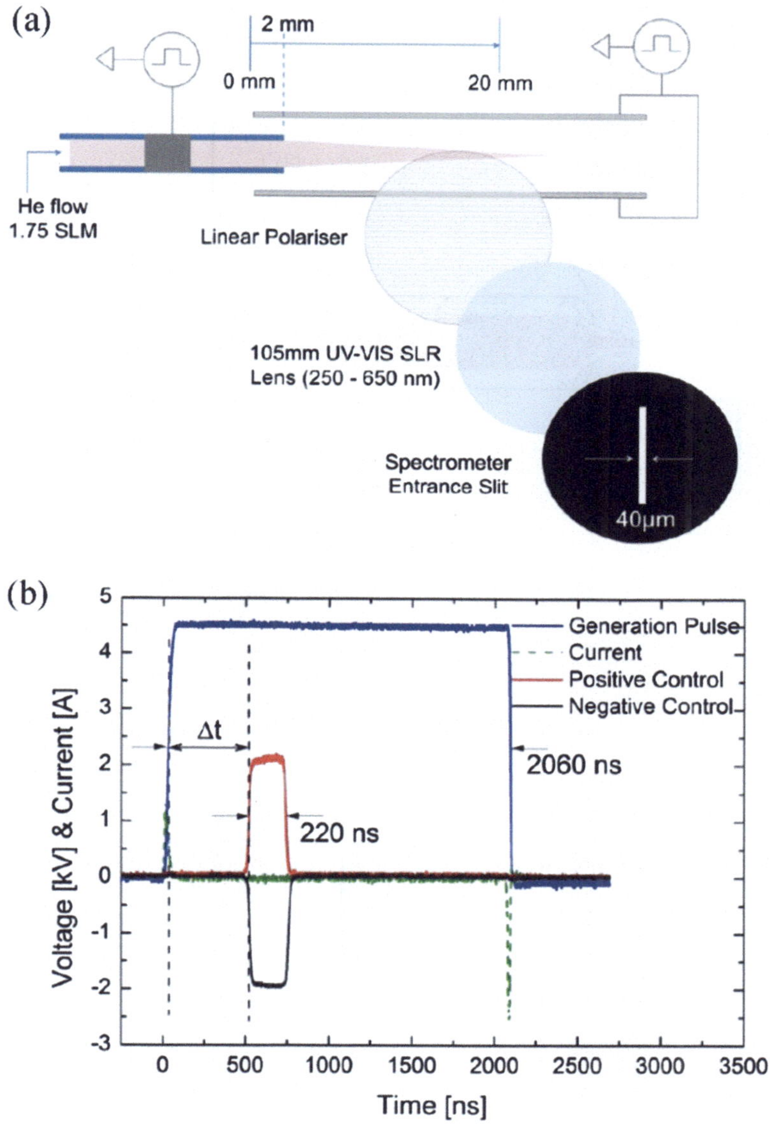


**Fig. 2.28** (**a**) Schematics of discharge device with two different electrode configuration, and (**b**) current–voltage waveforms

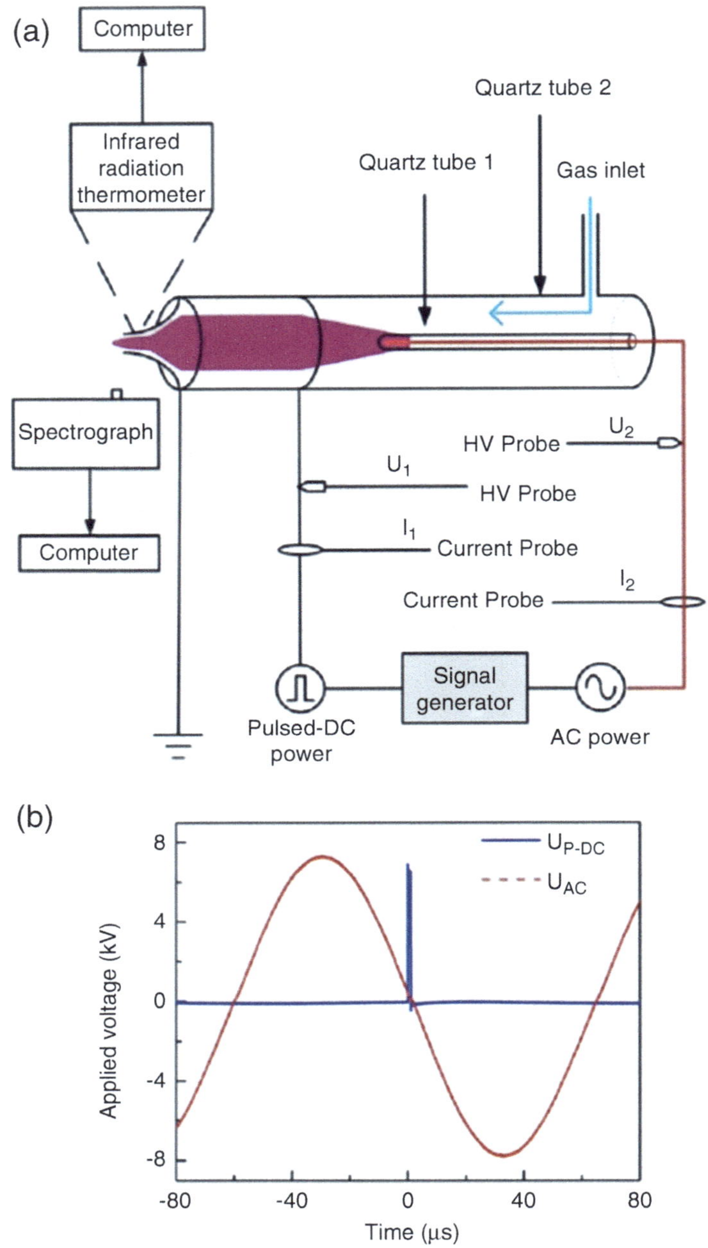


**Fig. 2.29** (**a**) The schematic of the AC and pulsed DC driven APNP-Js; the phase shift between the AC voltage and pulsed DC voltage with 0° (**b**)

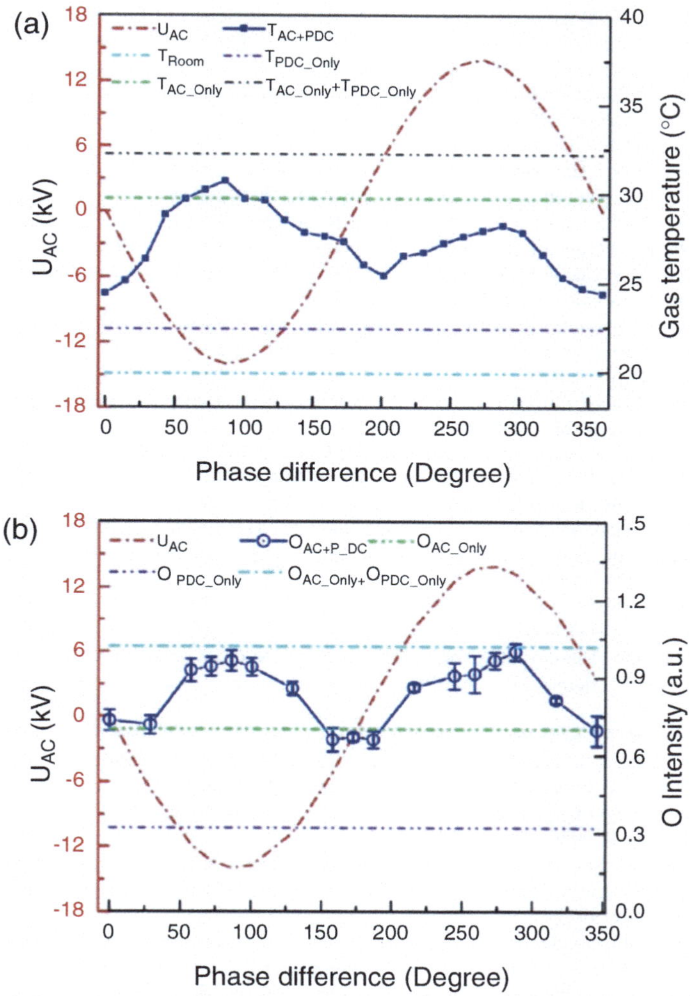


**Fig. 2.30** (**a**) The gas temperature and (**b**) the O emission intensity changes with the different phase shift

# References


1. Th. Du Moncel, Notice sur l'appareil d'induction electrique de Ruhmkorff et sur les experiences que l'on peut faire avec cet instrument, Hachette et Cie publishers (1855)
2. W. von Siemens, Poggendorfs. Ann. Phys. Chem. **12**, 66 (1857)
3. A. Von Engel, R. Seelinger, M. Steenbeck, Über die Glimmentladung bei hohen Drucken. Z. Phys. **85**, 144 (1933)
4. S. Kanazawa, M. Kogoma, T. Moriwaki, S. Okazaki, Stable glow at atmospheric pressure. J. Phys. D. Appl. Phys. **21**, 838 (1988)

5. F. Massines, C. Mayoux, R. Messaoudi, A. Rabehi, P. Ségur, in Experimental study of an atmospheric pressure glow discharge application to polymers surface treatment. Proc. GD-92, Swansea, UK, vol. 2 (1992), pp. 730–733
6. J.R. Roth, M. Laroussi, C. Liu, in Experimental generation of a steady-state glow discharge at atmospheric pressure. Proceedings of the IEEE International Conference on Plasma Science (1992), pp. 170–171
7. R.P. Mildren, R.J. Carman, Enhanced performance of a dielectric barrier discharge lamp using short-pulsed excitation. J. Phys. D. Appl. Phys. **34**, 3378 (2001)
8. X. Duten, D. Packan, L. Yu, C.O. Laux, C.H. Kruger, DC and pulsed glow discharges in atmospheric pressure air and nitrogen. IEEE Trans. Plasma Sci. **30**, 178 (2002)
9. M. Laroussi, X. Lu, V. Kolobov, R. Arslanbekov, Power consideration in the pulsed DBD at atmospheric pressure. J. Appl. Phys. **6**, 3028 (2004)
10. X. Lu, M. Laroussi, Temporal and spatial emission behavior of homogeneous dielectric barrier discharge driven by unipolar sub-Microsecond square pulses. J. Phys. D. Appl. Phys. **39**, 1127 (2006)
11. M. Laroussi, Sterilization of contaminated matter with an atmospheric pressure plasma. IEEE Trans. Plasma Sci. **24**, 1188 (1996)
12. M. Laroussi, Low temperature plasmas for medicine? IEEE Trans. Plasma Sci. **37**, 714 (2009)
13. G. Fridman, G. Friedman, A. Gutsol, A.B. Shekhter, V.N. Vasilets, A. Fridman, Applied plasma medicine. Plasma Process. Polym. **5**, 503 (2008)
14. K.-D. Weltmann, E. Kindel, T. von Woedtke, M. Hähnel, M. Stieber, R. Brandenburg, Atmospheric-pressure plasma sources: prospective tools for plasma medicine. Pure Appl. Chem. **82**, 1223 (2010)
15. R. Bartnikas, Note on discharges in helium under AC conditions. Br. J. Appl. Phys. (J. Phys. D.) Ser. **2**(1), 659 (1968)
16. K.G. Donohoe, The development and characterization of an atmospheric pressure nonequilibrium plasma chemical Reactor, Ph.D. thesis, California Institute of Technology, Pasadena, CA (1976)
17. T. Yokoyama, M. Kogoma, T. Moriwaki, S. Okazaki, The mechanism of the stabilized glow plasma at atmospheric pressure. J. Phys. D. Appl. Phys. **23**, 1128 (1990)
18. F. Massines, A. Rabehi, P. Decomps, R.B. Gadri, P. Ségur, C. Mayoux, Experimental and theoretical study of a glow discharge at atmospheric pressure controled by a dielectric barrier. J. Appl. Phys. **8**, 2950 (1998)
19. S. Okazaki, M. Kogoma, M. Uehara, Y. Kimura, Appearance of a stable glow discharge in air, argon, oxygen and nitrogen at atmospheric pressure using a 50 Hz source. J. Phys. D. Appl. Phys. **26**, 889 (1993)
20. N. Gherardi, G. Gouda, E. Gat, A. Ricard, F. Massines, Transition from glow silent discharge to micro-discharges in nitrogen gas. Plasma Sources Sci. Technol. **9**, 340 (2000)
21. N. Gheradi, F. Massines, Mechanisms controlling the transition from glow silent discharge to streamer discharge in nitrogen. IEEE Trans. Plasma Sci **29**, 536 (2001)
22. J.J. Shi, X.T. Deng, R. Hall, J.D. Punnett, M. Kong, Three modes in a radio frequency atmospheric pressure glow discharge. J. Appl. Phys. **94**, 6303 (2003)
23. F. Massines, N. Gherardi, N. Naude, P. Segur, Glow and Townsend dielectric barrier discharge in various atmosphere. Plasma Phys. Controlled Fusion **47**, B557 (2005)
24. U. Kogelschatz, Silent discharges for the generation of ultraviolet and vacuum ultraviolet Excimer radiation. Pure Appl. Chem. **62**, 1667 (1990)
25. U. Kogelschatz, B. Eliasson, W. Egli, Dielectric-barrier discharges: principle and applications. J. Phys. IV **7**(C4), 47 (1997)
26. M. Laroussi, I. Alexeff, J.P. Richardson, F.F. Dyer, The resistive barrier discharge. IEEE Trans. Plasma Sci **30**, 158 (2002)
27. X. Wang, C. Li, M. Lu, Y. Pu, Study on atmospheric pressure glow discharge. Plasma Sources Sci. Technol. **12**, 358 (2003)
28. M. Laroussi, J.P. Richardson, F.C. Dobbs, Effects of non-equilibrium atmospheric pressure plasmas on the heterotrophic pathways of bacteria and on their cell morphology. Appl. Phys. Lett. **81**, 772 (2002)

29. M. Laroussi, D.A. Mendis, M. Rosenberg, Plasma interaction with microbes. New J. Phys. **5**(1), 41 (2003)
30. M. Laroussi, T. Akan, Arc-free atmospheric pressure cold plasma jets: ~a review. Plasma Process. Polym. **4**, 777 (2007)
31. M. Laroussi, X. Lu, Room temperature atmospheric pressure plasma plume for biomedical applications. Appl. Phys. Lett. **87**, 113902 (2005)
32. M. Teschke, J. Kedzierski, E.G. Finantu-Dinu, D. Korzec, J. Engemann, High-speed photographs of a dielectric barrier atmospheric pressure plasma jet. IEEE Trans. Plasma Sci. **33**, 310 (2005)
33. X. Lu, M. Laroussi, Dynamics of an atmospheric pressure plasma plume generated by submicrosecond voltage pulses. J. Appl. Phys. **100**, 063302 (2006)
34. N. Mericam-Bourdet, M. Laroussi, A. Begum, E. Karakas, Experimental investigations of plasma bullets. J. Phys. D. Appl. Phys. **42**, 055207 (2009)
35. B.L. Sands, B.N. Ganguly, K. Tachibana, A streamer-like atmospheric pressure plasma jet. Appl. Phys. Lett. **92**, 151503 (2008)
36. J.L. Walsh, M.G. Kong, Contrasting characteristics of linear-field and cross-field atmospheric plasma jets. Appl. Phys. Lett. **93**, 111501 (2008)
37. Q. Xiong, X. Lu, J. Liu, Y. Xian, Z. Xiong, F. Zou, C. Zou, W. Gong, J. Hu, K. Chen, X. Pei, Z. Jiang, Y. Pan, Temporal and spatial resolved optical emission behaviors of a cold atmospheric pressure plasma jet. J. Appl. Phys. **106**, 083302 (2009)
38. E. Karakas, M. Koklu, M. Laroussi, Correlation between helium mole fraction and plasma bullet propagation in low temperature plasma jets. J. Phys. D. Appl. Phys. **43**, 155202 (2010)
39. G.V. Naidis, Modeling of plasma bullet propagation along a helium jet in ambient air. J. Phys. D. Appl. Phys. **44**, 215203 (2011)
40. G.V. Naidis, Modeling of streamer propagation in atmospheric pressure helium plasma jets. J. Phys. D. Appl. Phys. **43**, 402001 (2010)
41. M. Yousfi, O. Eichwald, N. Merbahi, N. Jomma, Analysis of ionization wave dynamics in low-temperature plasma jets from fluid modeling supported by experimental investigations. Plasma Sources Sci. Technol. **21**, 045003 (2012)
42. J.-P. Boeuf, L. Yang, L. Pitchford, Dynamics of guided streamer (plasma bullet) in a helium jet in air at atmospheric pressure. J. Phys. D. Appl. Phys. **46**, 015201 (2013)
43. E. Karakas, M. Laroussi, Experimental studies on the plasma bullet propagation and its inhibiton. J. Appl. Phys. **108**, 063305 (2010)
44. J. Jarrige, M. Laroussi, E. Karakas, Formation and dynamics of the plasma bullets in a non-thermal plasma jet: influence of the high voltage parameters on the plume characteristics. Plasma Sources Sci. Technol. **19**, 065005 (2010)
45. Y. Sakiyama, D.B. Graves, J. Jarrige, M. Laroussi, Finite elements analysis of ring-shaped emission profile in plasma bullets. Appl. Phys. Lett. **96**, 041501 (2010)
46. A. Begum, M. Laroussi, M.R. Pervez, Atmospheric pressure helium/air plasma jet: breakdown processes and propagation phenomenon. AIP Adv. **3**, 062117 (2013)
47. G.B. Stretenovic, I.B. Krstic, V.V. Kovacevic, A.M. Obradovic, M.M. Kuraica, Spatio-temporally resolved electric field measurements in helium plasma jet. J. Phys. D. Appl. Phys. **47**, 102001 (2014)
48. A. Sobota, O. Guaitella, E. Garcia-Caurel, Experimentally obtained values of electric field of an atmospheric pressure plasma jet impinging on a dielectric surface. J. Phys. D. Appl. Phys. **46**, 372001 (2013)
49. M. Laroussi, X. Lu, M. Keidar, Perspective: The physics, diagnostics, and applications of atmospheric pressure low temperature plasma sources used in plasma medicine. J. Appl. Phys. **122**, 020901 (2017)
50. X. Lu, G.V. Naidis, M. Laroussi, S. Reuter, D.B. Graves, K. Ostrikov, Reactive species in non-equilibrium atmospheric pressure plasma: generation, transport, and biological effects. Phys. Rep. **630**, 1–84 (2016)
51. Y.F. Yue, S. Mohades, M. Laroussi, X. Lu, Measurements of plasma-generated hydroxyl and hydrogen peroxide concentrations for plasma medicine applications. IEEE Trans. Plasma Sci. **44**, 2754 (2016)

52. A.F.H. van Gessel, K. Alards, P. Bruggeman, NO production in an RF plasma jet at atmospheric pressure. J. Phys. D. Appl. Phys. **46**, 265202 (2013)
53. X. Lu, M. Laroussi, V. Puech, On atmospheric pressure non-equilibrium plasma jets and plasma bullets. Plasma Sources Sci. Technol. **21**, 034005 (2012)
54. M. Laroussi, From killing bacteria to destroying cancer cells: twenty years of plasma medicine. Plasma Process. Polym. **11**, 1138 (2014)
55. M. Keidar, A. Shashurin, O. Volotskova, M.A. Stepp, P. Srinivasan, A. Sandler, B. Trink, Cold atmospheric plasma in cancer therapy. Phys. Plasmas **20**, 057101 (2013)
56. S.E. Babayan, J.Y. Jeong, A. Schutze, V.J. Tu, M. Moravej, G.S. Selwyn, R.F. Hicks, Deposition of silicon dioxide films with a nonequilibrium atmospheric-pressure plasma jet. Plasma Sources Sci. Technol. **10**, 573–578 (2001)
57. K.D. Weltmann, E. Kindel, T. von Woedtke, M. Haehnel, M. Stieber, R. Brandenburg, Atmospheric-pressure plasma sources: prospective tools for plasma medicine. Pure Appl. Chem. **82**, 1223–1237 (2010)
58. H. Ayan, E.D. Yildirim, D.D. Pappas, W. Sun, Development of a cold atmospheric pressure microplasma jet for freeform cell printing. Appl. Phys. Lett. **99**(11), 111502 (2011)
59. J.J. Lowke, Plasma predictions: past, present and future. Plasma Sources Sci. Technol. **22**(2), 023002 (2013)
60. S. Samukawa, M. Hori, S. Rauf, K. Tachibana, P. Bruggeman, G. Kroesen, J.C. Whitehead, A.B. Murphy, A.F. Gutsol, S. Starikovskaia, U. Kortshagen, J.-P. Boeuf, T.J. Sommerer, M.J. Kushner, U. Czarnetzki, N. Mason, The 2012 plasma roadmap. J. Phys. D. Appl. Phys. **45**(25), 253001 (2012)
61. M. Laroussi, Low temperature plasma-based sterilization: overview and state-of-the-art. Plasma Process. Polym. **2**, 391–400 (2005)
62. G. Fridman, A. Gutsol, A.B. Shekhter, V.N. Vasilets, A. Fridman, Applied plasma medicine. Plasma Process. Polym. **5**, 503–533 (2008)
63. R. Diwan, F.M. Debta, A. Deoghare, S. Ghom, A. Khandelwal, S.D. Sikdar, A.G. Ghom, Plasma therapy: an overview. J. Indian Acad. Oral Med. Radiol. **23**, 120–123 (2011)
64. T. von Woedtke, S. Reuter, K. Masur, K.D. Weltmann, Plasmas for medicine. Phys. Rep. **530**, 291–320 (2013)
65. M.G. Kong, G. Kroesen, G. Morfill, T. Nosenko, T. Shimizu, J. van Dijk, J.L. Zimmermann, Plasma medicine: an introductory review. New J. Phys. **11**, 115012 (2009)
66. G.E. Morfill, M.G. Kong, J.L. Zimmermann, Focus on plasma medicine. New J. Phys. **11**, 115011 (2009)
67. E.C. Neyts, M. Yusupov, C.C. Verlackt, A. Bogaerts, Computer simulations of plasma-biomolecule and plasma-tissue interactions for a better insight in plasma medicine. J. Phys. D. Appl. Phys. **47**(29), 293001 (2014)
68. X. Lu, Guest editorial: atmospheric pressure plasma jets and their applications. IEEE Trans. Plasma Sci. **43**, 701–702 (2015)
69. K. Fricke, H. Steffen, T. von Woedtke, K. Schroeder, K.-D. Weltmann, High rate etching of polymers by means of an atmospheric pressure plasma jet. Plasma Process. Polym. **8**, 51–58 (2011)
70. T.-C. Tsai, D. Staack, Low-temperature polymer deposition in ambient air using a floating-electrode dielectric barrier discharge jet. Plasma Process. Polym. **8**, 523–534 (2011)
71. D. Wang, Q. Yang, Y. Guo, X. Liu, J. Shi, J. Zhang, One step growth of TiO2 crystal trees by atmospheric pressure plasma jet. Mater. Lett. **65**, 2526–2529 (2011)
72. K. Kelly-Wintenberg, T.C. Montie, C. Brickman, J.R. Roth, A.K. Carr, K. Sorge, et al., Room temperature sterilization of surfaces and fabrics with a one atmosphere uniform glow discharge plasma. J. Ind. Microbiol. Biotechnol. **20**, 69–74 (1998)
73. O. Sakai, K. Tachibana, Plasmas as metamaterials: a review. Plasma Sources Sci. Technol. **21**(1), 013001 (2012)
74. K. Ostrikov, Colloquium: Reactive plasmas as a versatile nanofabrication tool. Rev. Mod. Phys. **77**, 489–511 (2005)

75. D. Kolacyak, J. Ihde, C. Merten, A. Hartwig, U. Lommatzsch, Fast functionalization of multi-walled carbon nanotubes by an atmospheric pressure plasma jet. J. Colloid Interface Sci. **359**, 311–317 (2011)
76. K. Ostrikov, E.C. Neyts, M. Meyyappan, Plasma nanoscience: from nano-solids in plasmas to nano-plasmas in solids. Adv. Phys. **62**, 113–224 (2013)
77. P. Bruggeman, J. Liu, J. Degroote, M.G. Kong, J. Vierendeels, C. Leys, Dc excited glow discharges in atmospheric pressure air in pin-to-water electrode systems. J. Phys. D. Appl. Phys. **41**, 215201 (2008)
78. P. Bruggeman, C. Leys, Non-thermal plasmas in and in contact with liquids. J. Phys. D. Appl. Phys. **42**, 053001 (2009)
79. X. Lu, S. Wu, On the active species concentrations of atmospheric pressure nonequilibrium plasma jets. IEEE Trans. Plasma Sci. **41**, 2313–2326 (2013)
80. G. Fridman, A.D. Brooks, M. Balasubramanian, A. Fridman, A. Gutsol, V.N. Vasilets, H. Ayan, G. Friedman, Comparison of direct and indirect effects of non-thermal atmospheric-pressure plasma on bacteria. Plasma Process. Polym. **4**, 370–375 (2007)
81. J.Y. Kim, Y. Wei, J. Li, S.-O. Kim, 15 um-sized single-cellular-level and cell-manipulatable microplasma jet in cancer therapies. Biosens. Bioelectron. **26**, 555–559 (2010)
82. S.E. Babayan, J.Y. Jeong, V.J. Tu, J. Park, G.S. Selwyn, R.F. Hicks, Deposition of silicon dioxide films with an atmospheric-pressure plasma jet. Plasma Sources Sci. Technol. **7**, 286–288 (1998)
83. J.Y. Jeong, S.E. Babayan, V.J. Tu, J. Park, I. Henins, R.F. Hicks, G.S. Selwyn, Etching materials with an atmospheric-pressure plasma jet. Plasma Sources Sci. Technol. **7**, 282–285 (1998)
84. J.-Y. Choi, N. Takano, K. Urabe, K. Tachibana, Measurement of electron density in atmospheric pressure small-scale plasmas using CO2-laser heterodyne interferometry. Plasma Sources Sci. Technol. **18**(3), 035013 (2009)
85. S.C. Snyder, D.M. Crawford, J.R. Fincke, Dependence on the scattering angle of the electron temperature and electron density in Thomson-scattering measurements on an atmospheric-pressure plasma jet (vol 61, pg 1920, 2000). Phys. Rev. E **61**, 7261–7261 (2000)
86. S.-Z. Li, D.-Z. Wang, W.-C. Zhu, Y.-K. Pu, Evaluations of electron density and temperature in atmospheric-pressure radio-frequency helium plasma jet. Jpn. J. Appl. Phys. Part 1 **45**, 9213–9215 (2006)
87. S. Hofmann, A.F.H. van Gessel, T. Verreycken, P. Bruggeman, Power dissipation, gas temperatures and electron densities of cold atmospheric pressure helium and argon RF plasma jets (vol 20, 065010, 2011). Plasma Sources Sci. Technol. **21**(6), 069501 (2012)
88. M. Qian, C. Ren, D. Wang, J. Zhang, G. Wei, Stark broadening measurement of the electron density in an atmospheric pressure argon plasma jet with double-power electrodes. J. Appl. Phys. **107**(6), 063303 (2010)
89. K.D. Weltmann, R. Brandenburg, T. von Woedtke, J. Ehlbeck, R. Foest, M. Stieber, E. Kindel, Antimicrobial treatment of heat sensitive products by miniaturized atmospheric pressure plasma jets (APPJs). J. Phys. D. Appl. Phys. **41**, 194008 (2008)
90. J.H. Liu, X.Y. Liu, K. Hu, D.W. Liu, X.P. Lu, F. Iza, M.G. Kong, Plasma plume propagation characteristics of pulsed radio frequency plasma jet. Appl. Phys. Lett. **98**, 151502 (2011)
91. Y. Hidaka, E.M. Choi, I. Mastovsky, M.A. Shapiro, J.R. Sirigiri, R.J. Temkin, Observation of large arrays of plasma filaments in air breakdown by 1.5-MW 110-GHz gyrotron pulses. Phys. Rev. Lett. **100**, 035003 (2008)
92. K.V. Aleksandrov, V.L. Bychkov, I.I. Esakov, L.P. Grachev, K.V. Khodataev, A.A. Ravaev, I.B. Matveev, Investigations of subcritical streamer microwave discharge in reverse-vortex combustion chamber. IEEE Trans. Plasma Sci. **37**, 2293–2297 (2009)
93. Q. Zhang, G. Zhang, L. Wang, X. Wang, S. Wang, Y. Chen, Measurement of the electron density in a microwave plasma torch at atmospheric pressure. Appl. Phys. Lett. **95**, 201502 (2009)
94. Z. Wang, G. Zhang, Q. Zhang, Z. Jia, A large-volume open-air microwave plasma based on parallel multislot rectangular waveguides. IEEE Trans. Plasma Sci. **40**, 1380–1385 (2012)

95. J. Hnilica, V. Kudrle, P. Vasina, J. Schaefer, V. Aubrecht, Characterization of a periodic instability in filamentary surface wave discharge at atmospheric pressure in argon. J. Phys. D. Appl. Phys. **45**, 055201 (2012)
96. S. Takamura, S. Amano, T. Kurata, H. Kasada, J. Yamamoto, M.A. Razzak, G. Kushida, N. Ohno, M. Kando, Formation and decay processes of Ar/He microwave plasma jet at atmospheric gas pressure. J. Appl. Phys. **110**, 043301 (2011)
97. Q. Wang, L. Hou, G. Zhang, B. Zhang, C. Liu, Z. Wang, J. Huang, Using indium tin oxide material to implement the imaging of microwave plasma ignition process. Appl. Phys. Lett. **104**, 074107 (2014)
98. Z. Chen, Z. Yin, M. Chen, L. Hong, G. Xia, Y. Hu, Y. Huang, M. Liu, A.A. Kudryavtsev, Self-consistent fluid modeling and simulation on a pulsed microwave atmospheric-pressure argon plasma jet. J. Appl. Phys. **116**, 153303 (2014)
99. H.W. Lee, S.K. Kang, I.H. Won, H.Y. Kim, H.C. Kwon, J.Y. Sim, J.K. Lee, Distinctive plume formation in atmospheric Ar and He plasmas in microwave frequency band and suitability for biomedical applications. Phys. Plasmas **20**, 123506 (2013)
100. C. Wu, A.R. Hoskinson, J. Hopwood, Stable linear plasma arrays at atmospheric pressure. Plasma Sources Sci. Technol. **20**, 045022 (2011)
101. A.R. Hoskinson, J. Hopwood, A two-dimensional array of microplasmas generated using microwave resonators. Plasma Sources Sci. Technol. **21**, 052002 (2012)
102. E. Carbone, N. Sadeghi, E. Vos, S. Hubner, E. van Veldhuizen, J. van Dijk, S. Nijdam, G. Kroesen, Spatio-temporal dynamics of a pulsed microwave argon plasma: ignition and afterglow. Plasma Sources Sci. Technol. **24**, 015015 (2015)
103. E.A.D. Carbone, S. Huebner, J.M. Palomares, J.J.A.M. van der Mullen, The radial contraction of argon microwave plasmas studied by Thomson scattering. J. Phys. D. Appl. Phys. **45**, 345203 (2012)
104. Z. Chen, G. Xia, C. Zou, P. Li, Y. Hu, Q. Ye, S. Eliseev, O. Stepanova, A.I. Saifutdinov, A.A. Kudryavtsev, M. Liu, Study on hairpin-shaped argon plasma jets resonantly excited by microwave pulses at atmospheric pressure. J. Appl. Phys. **118**, 023307 (2015)
105. F. Iza, J.A. Hopwood, Low-power microwave plasma source based on a microstrip split-ring resonator. IEEE Trans. Plasma Sci. **31**, 782 (2003)
106. J. Choi, F. Iza, H.J. Do, J.K. Lee, M.H. Cho, Microwave-excited atmospheric-pressure microplasmas based on a coaxial transmission line resonator. Plasma Sources Sci. Technol. **18**, 025029 (2009)
107. Y.S. Seo, H.W. Lee, H.C. Kwon, J. Choi, S.M. Lee, K.C. Woo, K.T. Kim, J.K. Lee, A study on characterization of atmospheric pressure plasma jets according to the driving frequency for biomedical applications. Thin Solid Films **519**, 7071–7078 (2011)
108. Z. Chen, G. Xia, C. Zou, X. Liu, D. Feng, P. Li, Y. Hu, O. Stepanova, A.A. Kudryavtsev, Bullet-shaped ionization front of plasma jet plumes driven by microwave pulses at atmospheric gas pressure. J. Appl. Phys. **122**, 093301 (2017)
109. J. Zhao, L. Nie, Five gaseous reactive oxygen and nitrogen species (RONS) density generated by microwave plasma jet. Phys. Plasmas **26**, 073503 (2019)
110. Q. Li, J.-T. Li, W.-C. Zhu, X.-M. Zhu, Y.-K. Pu, Effects of gas flow rate on the length of atmospheric pressure nonequilibrium plasma jets. Appl. Phys. Lett. **95**, 141502 (2009)
111. X. Lu, Z. Jiang, Q. Xiong, Z. Tang, X. Hu, Y. Pan, An 11 cm long atmospheric pressure cold plasma plume for applications of plasma medicine. Appl. Phys. Lett. **92**, 081502 (2008)
112. X. Lu, Z. Jiang, Q. Xiong, Z. Tang, Y. Pan, A single electrode room-temperature plasma jet device for biomedical applications. Appl. Phys. Lett. **92**, 151504 (2008)
113. V. Leveille, S. Coulombe, Design and preliminary characterization of a miniature pulsed RF APGD torch with downstream injection of the source of reactive species. Plasma Sources Sci. Technol. **14**, 467–476 (2005)
114. A. Shashurin, M.N. Shneider, A. Dogariu, R.B. Miles, M. Keidar, Temporal behavior of cold atmospheric plasma jet. Appl. Phys. Lett. **94**, 231504 (2009)
115. E. Stoffels, I.E. Kieft, R.E.J. Sladek, Superficial treatment of mammalian cells using plasma needle. J. Phys. D. Appl. Phys. **36**, 2908–2913 (2003)

116. X. Lu, Y.G. Cao, P. Yang, Q. Xiong, Z.L. Xiong, Y.B. Xian, Y. Pan, An RC plasma device for sterilization of root canal of teeth. IEEE Trans. Plasma Sci. **37**, 668–673 (2009)
117. R. Bussiahn, R. Brandenburg, T. Gerling, E. Kindel, H. Lange, N. Lembke, K.D. Weltmann, T. von Woedtke, T. Kocher, The hairline plasma: an intermittent negative dc-corona discharge at atmospheric pressure for plasma medical applications. Appl. Phys. Lett. **96**, 143701 (2010)
118. S. Wu, X. Lu, Z. Xiong, Y. Pan, A touchable pulsed air plasma plume driven by DC power supply. IEEE Trans. Plasma Sci. **38**, 3404–3408 (2010)
119. S. Wu, Z. Wang, Q. Huang, X. Lu, K. Ostrikov, Open-air direct current plasma jet: scaling up, uniformity, and cellular control. Phys. Plasmas **19**, 103503 (2012)
120. Z. Cao, J.L. Walsh, M.G. Kong, Atmospheric plasma jet array in parallel electric and gas flow fields for three-dimensional surface treatment. Appl. Phys. Lett. **94**, 21501 (2009)
121. X. Pei, Z. Wang, Q. Huang, S. Wu, X. Lu, Dynamics of a plasma jet Array. IEEE Trans. Plasma Sci. **39**, 2276–2277 (2011)
122. X. Pei, X. Lu, J. Liu, D. Liu, Y. Yang, K. Ostrikov, P.K. Chu, Y. Pan, Inactivation of a 25.5 um Enterococcus faecalis biofilm by a room-temperature, battery-operated, handheld air plasma jet. J. Phys. D. Appl. Phys. **45**(16), 165205 (2012)
123. J.F. Kolb, A.A.H. Mohamed, R.O. Price, R.J. Swanson, A. Bowman, R.L. Chiavarini, M. Stacey, K.H. Schoenbach, Cold atmospheric pressure air plasma jet for medical applications. Appl. Phys. Lett. **92**, 241501 (2008)
124. Y. Xian, X. Lu, S. Wu, P.K. Chu, Y. Pan, Are all atmospheric pressure cold plasma jets electrically driven? Appl. Phys. Lett. **100**, 123702 (2012)
125. X.L. Deng, A.Y. Nikiforov, P. Vanraes, C. Leys, Direct current plasma jet at atmospheric pressure operating in nitrogen and air. J. Appl. Phys. **113**, 23305 (2013)
126. Z. Cao, Q.Y. Nie, M.G. Kong, A cold atmospheric pressure plasma jet controlled with spatially separated dual-frequency excitations. J. Phys. D. Appl. Phys. **42**, 222003 (2009)
127. P.-S. Le, G. Li, S. Wang, H.-P. Li, C.-Y. Bao, Characteristics of kilohertz-ignited, radio-frequency atmospheric-pressure dielectric barrier discharges in argon. Appl. Phys. Lett. **95**(20), 201501 (2009)
128. P. Olszewski, E. Wagenaars, K. McKay, J.W. Bradley, J.L. Walsh, Measurement and control of the streamer head electric field in an atmospheric-pressure dielectric barrier plasma jet. Plasma Sources Sci. Technol. **23**, 015010 (2014)
129. G.V. Naidis, J.L. Walsh, The effects of an external electric field on the dynamics of cold plasma jets-experimental and computational studies. J. Phys. D. Appl. Phys. **46**(9), 095203 (2013)
130. J. Li, F. Wu, L. Nie, X. Lu, The effect of phase shift on the plasma driven by an AC voltage and a pulsed DC voltage. IEEE Trans. Plasma Sci. **47**, 1–9 (2019)

# Chapter 3
# Cold Atmospheric Plasma Sources for Cancer Applications and Their Diagnostics

**Eun Ha Choi**

## Contents



**Abstract** In general, the plasma used for the cancer therapy could be generated at the atmospheric pressure state under ambient air environment by injecting other kinds of gases such as He, Ar, nitrogen, water molecules, and their mixture for appropriate applications. This plasma should have following safety characteristics for cancer treatment such as nonthermal or cold temperature (less than 40 °C) characteristics, low electrical leakage currents to skin (less than 100 uA) for no electrical shock, and as low as amount of ozone generation (less than 0.05 ppm) for breathing safety. This kind of plasma is called as "nonthermal atmospheric pressure plasma (NAPP)," or "nonthermal biocompatible plasma (NBP)," or "cold atmospheric pressure plasma (CAP)." In addition, this plasma should be easy-handled for clinical test.

This chapter describes specific plasma devices for cancer application and their diagnostics.

E. H. Choi (✉)
Plasma Bioscience Research Center (PBRC), Kwangwoon University, Seoul, South Korea
e-mail: ehchoi@kw.ac.kr



M. Keidar (ed.), *Plasma Cancer Therapy*, Springer Series on Atomic, Optical, and Plasma Physics 115, https://doi.org/10.1007/978-3-030-49966-2_3

# 3.1 Plasma Sources

## *3.1.1 Soft Plasma Jet*

Plasma jets (see Chap. 2 for detailed description) are one of the simplest atmospheric plasma sources that could be created in the laboratory. Plasma jet is generated by using noble gas and has a benefit of producing long plasma plumes [1–4]. The soft plasma jet as shown in Fig. 3.1, it is consisting of an inner syringe needle acing as a high voltage (HV) electrode and gas inlet which covered and connected with quartz tube. Quartz tube serves as a guide for the plasma jet flow and is designed to induce dielectric discharge through the sheet current along the concave part at the end of the quartz tube between the grounded electrode and the needle. The distance between the needle tip and the grounded electrode is 2–3 mm and the plasma jet plume produced along 5 mm from nozzle. Soft plasma jet discharge operates using the HV inverter. Porous alumina with a diameter of 10–12 mm and a length of 17–20 mm could be machined and placed in tight contact between the two electrodes of the inner and outer electrodes according to the applications. The porous alumina if used in this work is approximately 30 vol % porosity with an average pore diameter of 80–100 um. Gas is injected into the injection needle and then ejected through the 1–2 mm hole in the outer electrode. Air or any other gases could be used as the feeding gas; the flow rate is controlled by an analog or mass flow controller. Once gas is introduced through the inner electrode and high-voltage ac power from inverter is applied, a discharge is fired into the porous alumina between the electrodes and a long plasma jet reaching lengths of up to 10 mm is ejected into the open air. Cells or biological targets were treated with jet plasma for appropriate exposure times such s 30 s, 60 s, 120 s, and 240 s under an electrical discharge power of less than about 2 W (Vpeak: 1–2 kV and Ip: 1–2 mA) from a duty-cycle controlled inverter voltage. The driving frequency is 20–50 kHz and flow rate of air gas is kept to 1–2 liter per minute (lpm). The working temperature of the plasma source is in the range of 26-36 °C at the time of treatment.

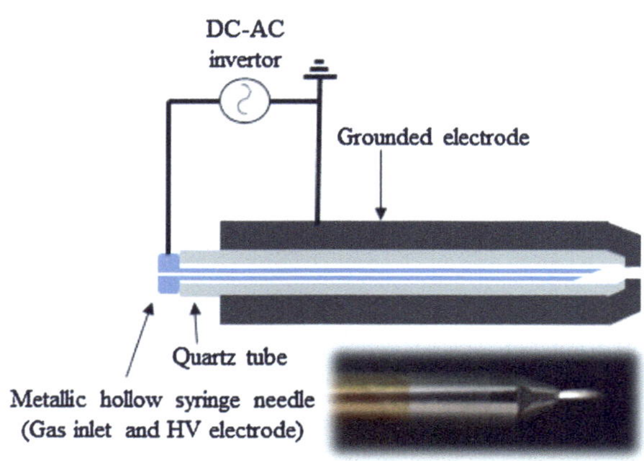


**Fig. 3.1** Schematic of soft plasma jet

### 3.1.2 μ-DBD (Micro-Dielectric Barrier Discharge) Plasma

Since the first study by Ernst Werner von Siemens in 1857, there have been various forms of dielectric barrier discharge (DBD) plasma source development. Figure 3.2 shows a nonthermal atmospheric pressure surface micro-dielectric barrier discharged (u-DBD) (S u-DBD) plasma for the coplanar electrodes [5] (a) and facing DBD (F u-DBD) electrodes [6] (b). Biological skins could be treated by these kinds of DBD plasma for 30 s—few minutes under electrical discharge power of about less than 3 W (1–2.2 kV, 1–2 mA, and phase angle 0.7 radian) with duty-cycle controlled high-voltage ac inverter or commercial high voltage used for neon light. The driving frequency of sinusoidal high voltage could be controlled to be 60 Hz to 100 kHz as well as its duty cycles 1–50%. The electrical discharge is sustained by alternating high voltage applied to the two electrodes covered by dielectric materials whose thickness is about 100 um. The dielectric materials could be chosen appropriately among glass, quartz, ceramics, and polymers for purposes. The distance between the electrodes can be varied from a distance of less than 0.1 mm to several centimeters. As shown in Fig. 3.2, typical u-DBD discharge occurs between two electrodes separated by an insulating dielectric barrier in surface u-DBD (S u-DBD) (a) and facing u-DBD (F u-DBD) (b). At S u-DBD micro-discharges occur on the coplanar dielectric surface, and plasma generation is more uniform than F u-DBD. The Fig. 3.3 shows typical fabrication processes for S u-DBD electrodes based on the semiconductor processing. The two parallel electrodes are made of Ag paste film patterned by screen printing method with drying and furnace treatment under 600 °C on the glass or ceramic substrate. Finally, aluminum oxide $Al_2O_3$ film has been deposited to be 1 um in thickness on the $SiO_2$ dielectric barrier by either e-beam evaporation or sputtering method for lowering the breakdown voltage in plasma generation.

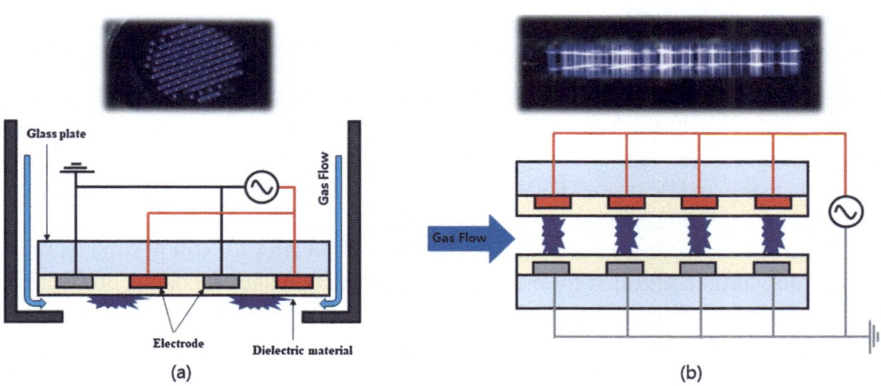


**Fig. 3.2** Schematic of u-DBD electrode structure (**a**) Surface u-DBD(S u-DBD), (**b**) Facing u-DBD(F u-DBD)

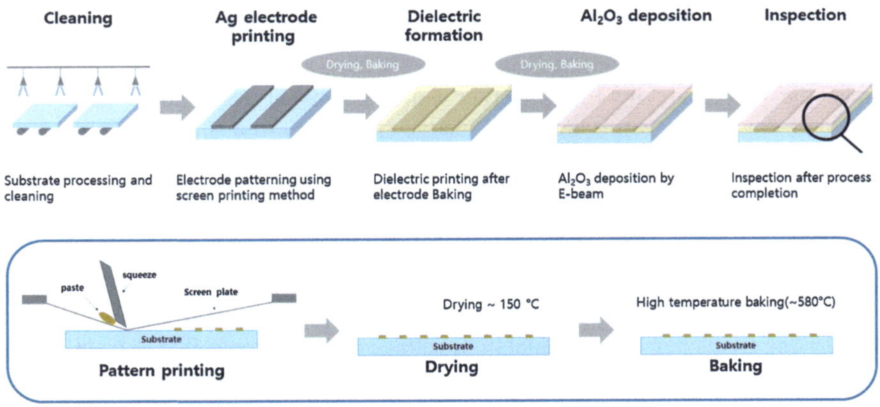


**Fig. 3.3** Surface u-DBD electrode manufacturing process

The separation of these coplanar two electrodes were kept to by 100–200 um, whose thickness are about 3–5 um, after that they are tightly covered with insulating $SiO_2$ thick film by screen printing method. The thickness of substrate is about 1.8 mm. The diameter of u-DBD surface plasma is 30–100 mm, which is designed for 20–100 mm petri-dishes. The discharge power is about less than 3 W whose peak voltage is about Vp ~ 1–2 kV and currents are less than 5 mA. Working temperature of plasma source is in the range of 24–32 °C at the time of treatment.

## 3.2 Electrical and Optical Characteristics of Plasma Sources for Cancer Applications

### *3.2.1 Electrical Characteristics*

An oscilloscope (DSO-X-3104T, Keysight) was used to measure the voltage and current of the discharge between the electrodes. As shown in Fig. 3.4, the voltage was measured between the HV electrode (syringe needle) and the grounded electrode using an HV probe (P6015A, Tektronix). The current was measured using a current probe (P6021, Tektronix).

Figure 3.5 shows an output voltage (a) of the HV inverter for soft plasma jet with voltage duty ratio controlled for minimization of heat damage and ozone production during plasma discharge. The discharge of the soft plasma jet has been generated by 2 kV and 50 kHz sinusoidal wave (b) under the nitrogen or any other gas flow rate of 1.5 lpm. The discharge on-time duration is fixed to be 25 ms, and the duty ratio could be adjusted by 9–14% by controlling the off-time duration in the driving voltage. In Fig. 3.6a, output voltage of the HV inverter for surface u-DBD has a voltage duty ratio to minimize the heat damage and ozone caused by plasma discharge like Fig.

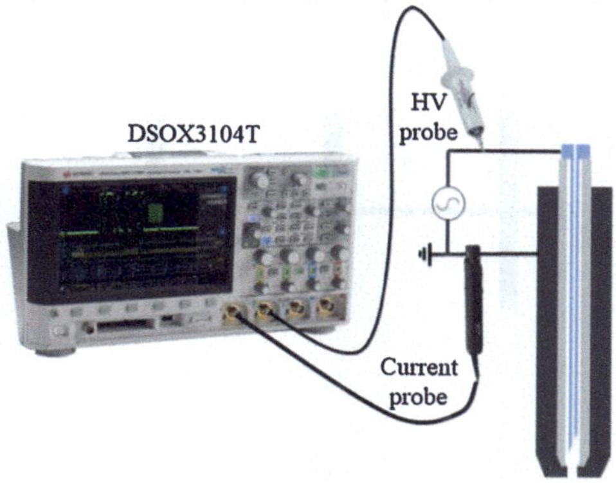


**Fig. 3.4** Electrical measurement of plasma source

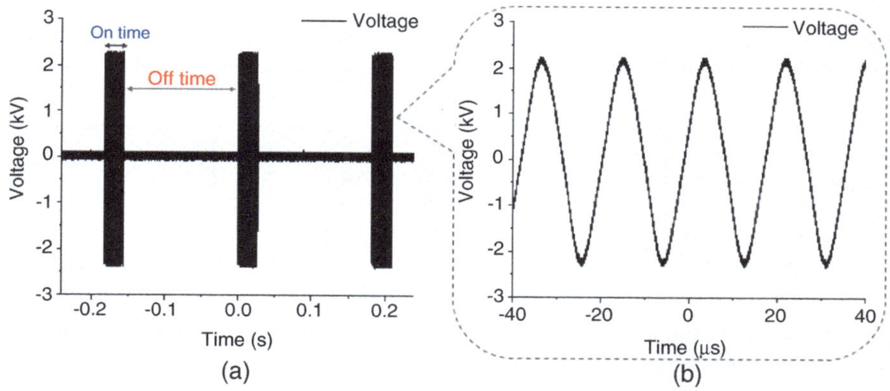


**Fig. 3.5** Voltage waveform from high voltage invertor for soft plasma jet (**a**) on-off time of high voltage invertor, (**b**) 2 kV, 50 kHz sinusoidal wave

3.5a. The discharge of the surface u-DBD plasma has been generated by 1.5 lpm of nitrogen gas and the sinusoidal driving voltage of 2 kV, 30 kHz as in Fig. 3.6b. The discharge on-time is fixed at 25 ms, and the duty ratio in driving voltage could be adjusted to be 9–14% by controlling the discharge off-time duration. The electrical energy per second could be obtained from the following equation with signal period T.

$$\text{Electrical energy per second} = \left(\frac{\text{On time}}{\text{On time} + \text{Off time}}\right) * \frac{1}{T}\int_0^T (V(t) * I(t))\,dt \tag{3.1}$$

The voltage and current waveforms used for the soft jet and surface u-DBD plasma are shown in the left-hand side and right-hand side of Fig. 3.7a, respectively. Positive discharge for soft jet and surface u-DBD plasma occur at voltage enhancement of (820 V, 880 V) with a discharge current of (40 mA, 10 mA), while a negative

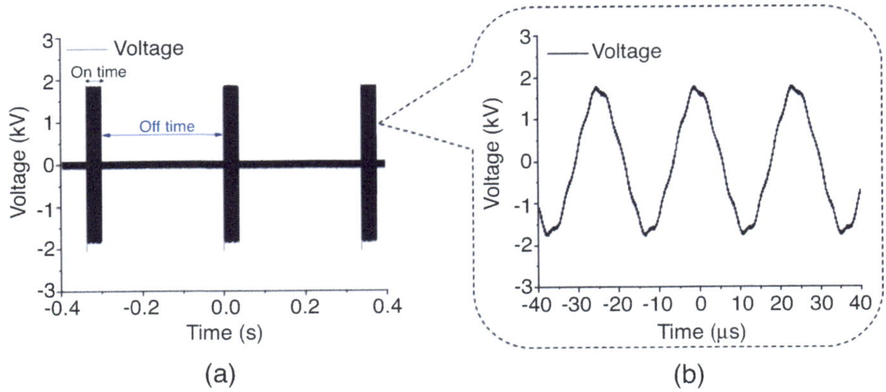


**Fig. 3.6** Voltage waveform from HV invertor for S u-DBD (**a**) on-off time of HV invertor, and (**b**) 2 kV, 30 kHz sinusoidal wave in on-time

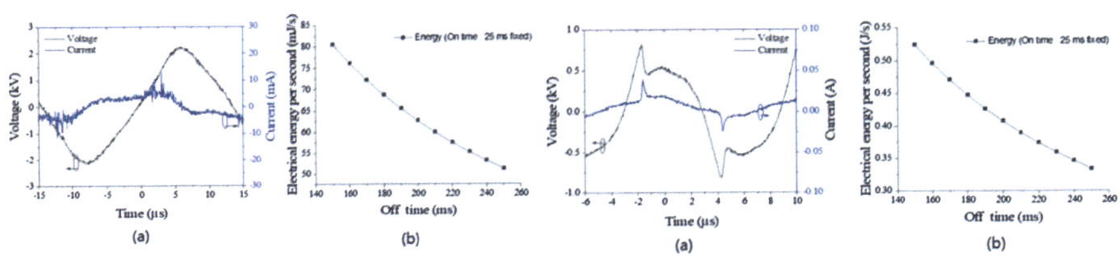


**Fig. 3.7** Electrical characteristics of soft plasma jet (left-hand side) and S u-DBD plasma (right-hand side), (**a**) voltage (black line) and current (blue line) curve in discharge, (**b**) Electrical energy per second with duty ratio, respectively

discharge occurs at voltage enhancement of (−820 V, −800 V) with a discharge current of (−25 mA, −10 mA), respectively, in which the voltage frequency for the soft jet has been increased to 85 kHz from 50 kHz, while it has not changed for surface u-DBD plasma. The electrical energy has been performed for several off-time durations. Electrical energy per unit time also has been measured to be 350–520 mJ/s and 50–80 mJ/s for the soft jet and surface u-DBD plasma, respectively, where they are linearly decreased in accordance with off-time durations as shown in Fig. 3.7b.

### *3.2.2 Physical Characterization Measurement Methods*

Ozone could be measured in accordance with the measurement distance from the nozzle of the soft plasma jet using commercial device (200 series, aeroqual), as shown in Fig. 3.8a. The gas temperature could also be measured by a fiber optic thermometer (FOT Lab kit, Luxtron), and this optical fiber for the measurement

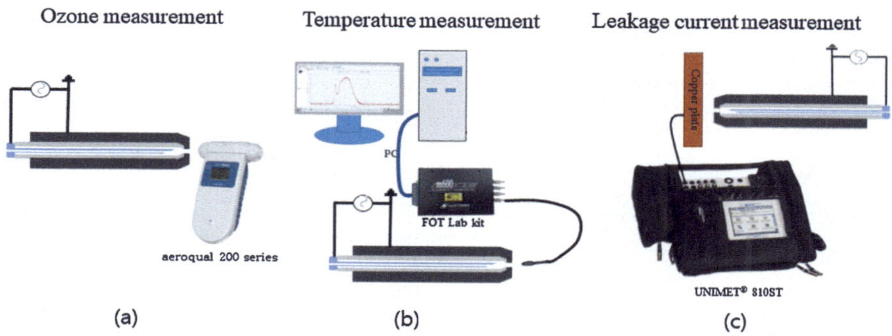


**Fig. 3.8** Medical requirement limitation measurement; (**a**) ozone measurement, (**b**) axial temperature measurement, (**c**) leakage current measurement

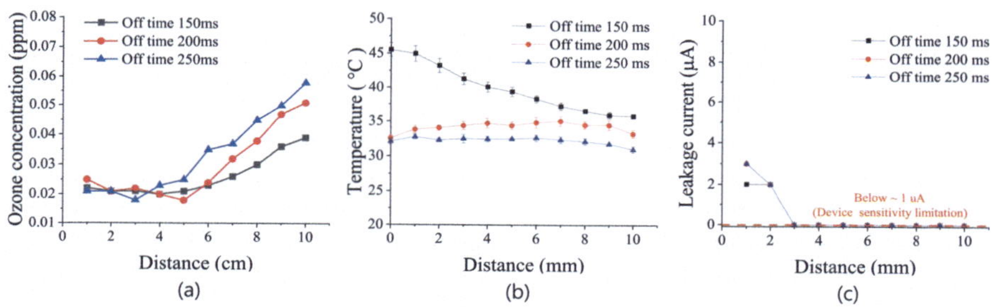


**Fig. 3.9** Ozone concentration (**a**), temperature (**b**), and leakage current (**c**) measurement versus distance under several off-time durations in discharge for soft plasma jet

could be moved along the distance away from the discharge, as shown in Fig. 3.8b. Measurement results for temperatures have been time-averaged over a minute with resolution time of 1 s. Leakage current for the soft jet could also be measured by a copper plate target (40 × 40 × 5 mm$^3$) connected to a test device (UNIMET® 800ST, BENDER), as shown in Fig. 3.8c, according to the axial distance [7]. Optical emission spectroscopy (OES) has also been measured using a CCD spectrometer (HR4000, Ocean optics) and this OES has been performed with a distance of 1 cm away from either the plasma nozzle in soft plasma jet or u-DBD discharged plasma surface.

In Fig. 3.9a, ozone production from the soft plasma jet has been measured as increasing with distance, and it has been a little bit increased in accordance with increasing in off-time duration. The temperature has been found to be decreased by increasing either position of measurement distance or cooling off-time duration, as shown in Fig. 3.9b. As shown in Fig. 3.9c, leakage current has been measured and detectable up to only 2 mm from the soft jet nozzle, and no more leakage current has been measured beyond 3 mm from the nozzle since the lower measurement limit of the instrument is 1 uA.

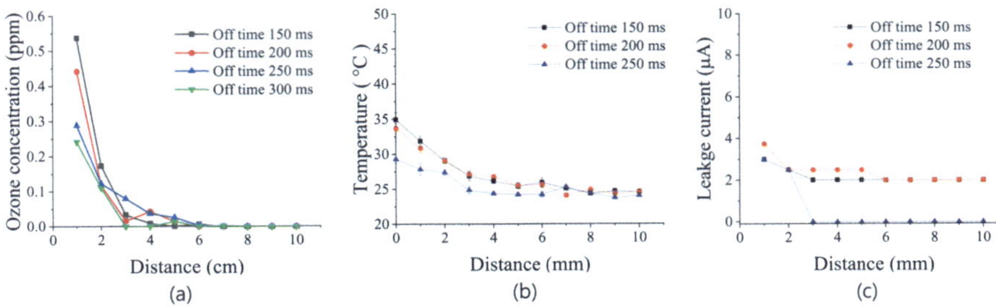


**Fig. 3.10** Ozone, temperature, and leakage current measurement versus distance from surface of discharged u-DBD plasma for several off-time durations, (**a**) ozone concentration, (**b**) temperature, and (**c**) leakage current for surface u-DBD plasma

Figure 3.10 shows ozone concentration (a) versus distance from the discharged surface from S u-DBD plasma. It has been measured to maximum 0.55 ppm at 1 mm distance and it decreases with increase in measurement distance. The ozone concentration has also been decreased by increasing the off-time duration. The temperature has been measured to be below 35 °C in S u-DBD, and it is also decreased by increasing the cooling off-time duration as shown in Fig. 3.10b. The leakage current has been decreased up to 3 mm, and saturated beyond 3 mm for off-time duration of 150–200 ms. For off-time duration of 250 ms the leakage current has been measured to be less than 1 uA, which is under the low limitation detectable value of device, as shown in Fig. 3.10c.

### 3.2.3 Electron Temperature Measurement

The collisional radiative model is based on the relative intensity of the nitrogen optical emission line and can be used to calculate the electron temperature of the plasma jet. The diagnosis of electron temperature could be performed based on the collisional radiative model (CRM) using the rate balancing equation between $N_2$ SPS ($N_2$ second positive system, $C^3\Pi_u \rightarrow B^3\Pi_g$) and $N_2$ FPS ($N_2$ first positive system, $B^3\Pi_g \rightarrow A^3\Sigma_u{}^+$) [8–11]. The rate equilibrium equation in the nitrogen molecule's transition model can be expressed through several reactions. First, the excitation reaction to the base states $X^1\Sigma_u{}^+$, $B^3\Pi_g$, and $C^3\Pi_u$ by electron excitation is the most important process, and the attenuation due to the collision, the wall deactivation reaction by the interaction with the surface, and the natural emission due to the transition to the lower energy level. The equilibrium equations for CRM are as follows [8].

$$n_e \sum_{\nu} n_{X,\nu} Q_{e-N_2}^{X,\nu\rightarrow A} + n_B \sum_{\nu} n_{X,\nu} Q_{N_2(B)-N_2(X)}^{B\rightarrow A} + A_B n_B \tag{3.2}$$

$$= n_A \sum_{5\le\nu\le 14} n_{X,\nu} Q_{N_2(A)-N_2(X)}^{A\rightarrow B} + 2n_A{}^2 \left( Q_{N_2(A)-N_2(A)}^{A,\ A\rightarrow B} + Q_{N_2(A)-N_2(A)}^{A,A\rightarrow C} \right) + K_{\text{wall}}^{A} n_A$$

$$n_e \sum_{\nu} n_{X,\nu} Q_{e-N_2}^{X,\nu\rightarrow B} + n_A \sum_{\nu} n_{X,\nu} Q_{N_2(A)-N_2(X)}^{A\rightarrow B} + n_A{}^2 Q_{N_2(A)-N_2(A)}^{A,\ A\rightarrow B} + \sum_{\nu''} A_{C,\nu''} n_{C,\nu''} \tag{3.3}$$

$$= n_B \sum_{\nu} n_{X,\nu} Q_{N_2(B)\rightarrow N_2(X)}^{X,\nu\rightarrow C,\nu''} + A_B n_B$$

$$n_e \sum_{\nu} n_{X,\nu} Q_{e-N_2}^{X,\nu\rightarrow V,\nu''} + n_A{}^2 Q_{N_2(A)-N_2(A)}^{A,\ A\rightarrow C,\nu''} = A_{C,\nu''} n_{C,\nu''} \tag{3.4}$$

In the equilibrium equation above, $n_e$ is the electron density, and $n_X$, $n_A$, $n_B$, and $n_C$ denote excited density of $X^1\Sigma_u{}^+$, $A^3\ \Sigma_u{}^+$, $B^3\Pi_g$, $C^3\Pi_u$ energy levels, respectively. A is Einstein coefficient for transition, ν is the vibrational quantum number, and $K_{\text{wall}}$ is the rate coefficient for wall deactivation [9, 10]. $Q$ is a reaction constant for electron excitation in gaseous state [9–11] under assumptions of Maxwell electron energy distribution. The densities $n_X$, $n_A$, $n_B$, and $n_C$ on each excited level could be obtained by using the above Eqs. (3.2), (3.3), and (3.4). The ratio of the emission line is obtained by using the emission intensities of $N_2$ SPS and $N_2$ FPS. The formula for the ratio $R_1$ of emission lines is as follows.

$$R_1 = \frac{A_C n_C (T_e)}{A_B n_B (T_e)} = \frac{I_C}{I_B} \tag{3.5}$$

Electron temperature could be calculated from cross point of theoretical value from Eq. (3.5) with electron temperature and experimental determined value of $R_1 = 28.88$, which is given by Eq. (3.5).

Figure 3.11 shows the OES of the soft plasma jet, where $N_2$ SPS (297–400 nm) and $N_2$ FPS (550–800 nm) are observed. The electron temperature has been measured to be 1.76 eV based on CRM model.

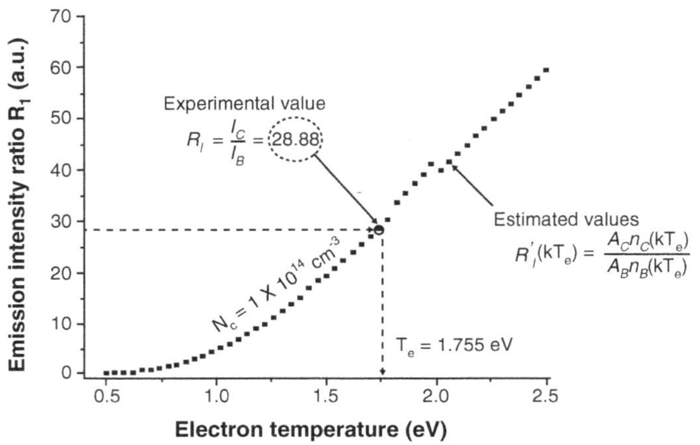


**Fig. 3.11** Cross point of experimental value $R_1$ and estimated values [9–11]

### *3.2.4 Plasma Electron Density*

Electron density in plasma is an important parameter to understand the formation of various reactive oxygen and nitrogen species in plasma jets. The electron density could be estimated using convective wave packet model from the jet current measurements [12]. Electron flux $\Gamma_e$ might be a little greater than ion flux $\Gamma_i$ due to electron has bigger diffusion coefficient, $D_e$ than that of ion diffusion coefficient, $D_i$ as well as $\upsilon_e \gg \upsilon_i$ where $\upsilon_e$ is electron mobility and $\upsilon_i$ is ion mobility, respectively. Therefore, the current density of plasma region could be expressed by

$$J = e(\Gamma_i - \Gamma_e) = e[(D_e - D_i)\nabla n + n(\upsilon_e + \upsilon_i)E] \approx e[D_e \nabla n + n\upsilon_e E_{\text{total}}] \approx e n_e \upsilon_e \tag{3.6}$$

where $e$ is the electron charge, $A$ is the cross-sectional area of the plasma, $v_e$ is the electron speed, and $n_e$ represents the electron density. From this background reason, the downstream current density is not zero, i.e., $J \neq 0$, whenever we measured it in this experiment. It could be interpreted that the current density might be expressed almost by electron flux only. In addition, the ambipolar electric field $E_{\text{polar}}$ causes ambipolar diffusion for $\Gamma_e \approx \Gamma_i$, the electron speed $v_e$ should be close to ambipolar diffusion velocity, $v_n$. Therefore, current density in downstream region could be expressed by [12–16]

$$J \approx e n_e \upsilon_e \approx e n_e \upsilon_n \tag{3.7}$$

which is dominant electron flux under conditions of $\Gamma_e \approx \Gamma_i$. Therefore, the electron density could be estimated by using Eq. (3.3). The ambipolar diffusion velocity has been estimated to be $3.30 \times 10^2$ m/s [12]. The peak value of downstream electron current has been measured to be about Ipeak $\approx$ 12 mA with plasma bullet radius of $r_b = 2.67 \times 10^{-4}$ m in this experiment. The electron density has been estimated to be $3 \times 10^{15}$ cm$^{-3}$ from Eq. (3.7) based on ambipolar diffusion model [12].

### *3.2.5 Rotational and Vibrational Temperature Measurement Method*

Determination of the molecular gas temperature is of fundamental importance for understanding the characteristics of the plasma jet, as it plays the role of plasma thermometer. In nitrogen containing plasmas, the rotational distribution of nitrogen quickly achieves thermodynamic equilibrium within the gas, due to the faster exchange of rotational energy with heavy particles than with electrons. The Boltzmann distribution of rotational levels can be used to estimate the rotational temperature in such plasmas [17]. Optical emission spectroscopy has been extensively used to estimate the rotational temperature in nonthermal plasma jets [13, 17, 18]. Bruggeman et al. showed that more accurate value of gas temperature could be obtained using the $N_2$ (C–B), so called the second positive system (SPS) emission band [19] rather than the first positive system (FPS), where their transition energy diagram and real emission spectra are shown in Fig. 3.12a and b, respectively. Figure 3.12b shows the OES of the soft plasma jet, where $N_2$ SPS (297–400 nm) and $N_2$ FPS (550–800 nm) are observed. The OES spectrum from the surface u-DDB plasma is similar to that of soft jet as in Fig. 3.12b, where NO-$\gamma$ band (213–258 nm), $N_2$ SPS, and $N_2$ FPS could also be observed [20]. The electron temperature for the soft jet and the surface u-DDB plasma have been measured to be 1.76 eV and be 0.6 eV based on CRM model and electron density for them have been estimated to be $3 \times 10^{15}$ cm$^{-3}$ and $5 \times 10^{14}$ cm$^{-3}$ from ambipolar diffusion model, respectively. Especially, optical emission spectroscopy (OES) at interfacial region of the biosolutions is extremely useful for bioplasma jet bombarding onto the surface since the informations for induced ROS and their spatial distribution characteristics could be obtained from these data.

In this work, we have selected rotational levels of the SPS belonging to the 0–0 vibrational band of $N_2$ ($C^3\,\Pi_u$) ($\nu = 0$) $-$ $B^3\Pi g$ ($\nu = 0$), as indicated by magnified spectrum around 337 nm, in Fig. 3.12c. The intensity of a particular rotational band in this SPS is given by: [17]

$$I_{\text{em}} = C\left(J' + J'' + 1\right)\exp\left(-\frac{B'_{\nu}hc}{k_B T_{\text{rot}}}J'\left(J' + 1\right)\right) \tag{3.8}$$

where, $C$ is a constant for a single band, $J'$ and $J''$ are the rotational quantum numbers for the transitions ($\upsilon' = 0, J'$) $\rightarrow$ ($\upsilon' = 0, J''$), $B'_{\upsilon}$ is the rotational term, $h$

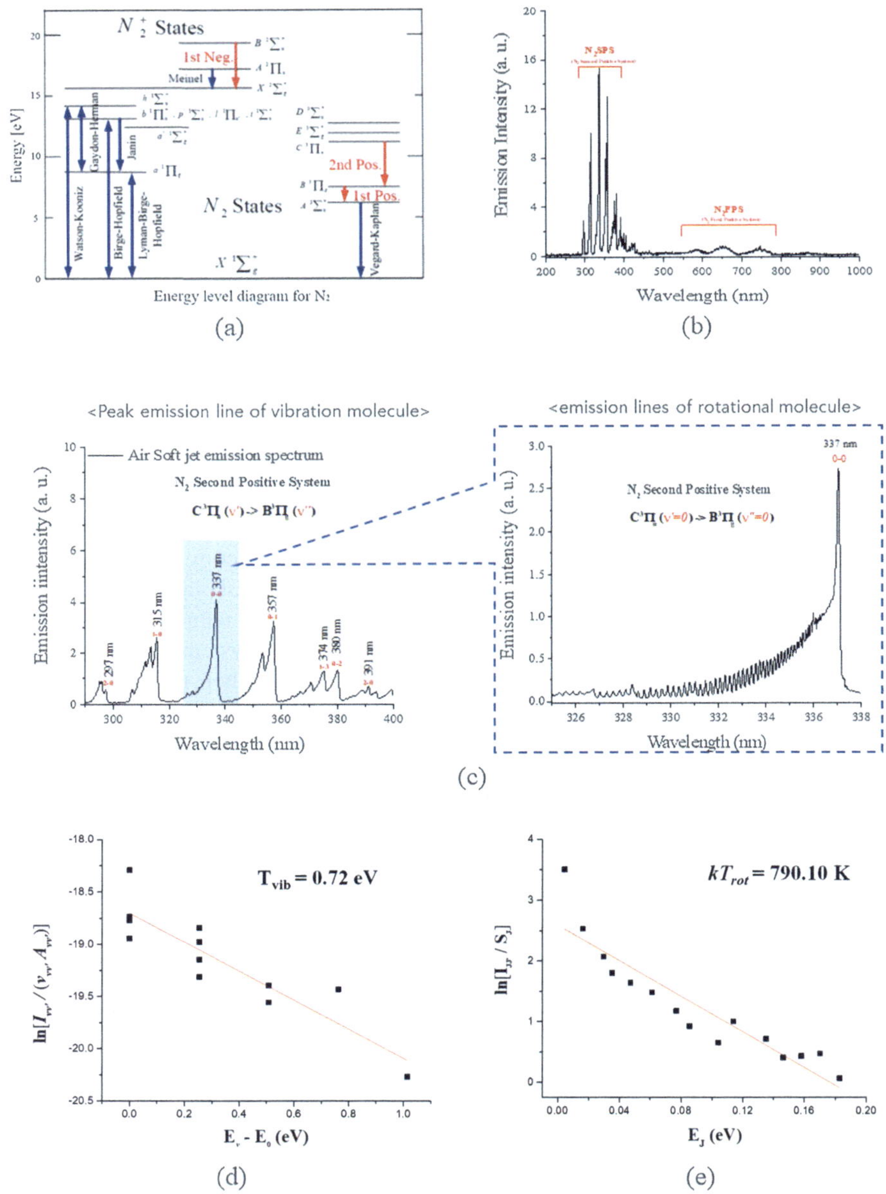


**Fig. 3.12** Transition energy diagram $N_2$ (297–400 nm), so called the second positive system (SPS), emission band and $N_2$ (550–800 nm), so called the first positive system (FPS) (**a**) and their real emission spectra (**b**), respectively. Vibrational spectra of the $N_2$ SPS and magnified rotational spectrum belonging to the 0–0 vibrational band around 337 nm of $N_2$ ($C^3$ $\Pi u$) (v = 0) − $B^3\Pi g$ ($v = 0$), as indicated by dotted box (**c**). Rotational temperature of 790.10 K (**d**), and vibrational temperature of 0.72 eV (**e**) for 1 mm irradiation position of plasma jet above the water surface, respectively

is Planck's constant, $c$ is the velocity of light, $k_B$ is the Boltzmann constant, and $T_{rot}$ is the rotational temperature. For the estimation of rotational temperature, Eq. (3.8) can be rearranged as:

$$\ln\left(-\frac{I_{em}}{J' + J'' + 1}\right) = \ln C - \frac{B'_v hc}{k_B T_{rot}} J'\left(J' + 1\right)\Big) \tag{3.9}$$

If the left-hand side is plotted against $J'(J' + 1)$, the slope of the resulting linear distribution will be equal to $B_\upsilon{}'$ hc/($k_B T_{rot}$), yielding the rotational temperature of the molecule. Figure 3.12d shows the rotational temperature calculated very close to the water surface for the 1 mm irradiation position of the plasma jet. From repeated experiments, the value is estimated to be 790.10 ± 2.43 °K. In this work, we have also selected vibrational levels of the SPS belonging to the ν–ν' vibrational band of $N_2$ ($C^3$ $\Pi_u$) (ν) – $B^3 \Pi g$ (ν'), as indicated by Fig. 3.12c. The intensity of a particular vibrational band in this SPS is given by [17],

$$\ln\left(\frac{I_{\nu\nu'}}{\nu_{\nu\nu'} A_{\nu\nu'}}\right) = \text{Constant} - \frac{E_\nu - E_0}{kT_{vib}} \tag{3.10}$$

where $E_\nu$ is given by

$$E_\nu(eV) = 1.2398 \times 10^{-4}\left(\frac{1}{2} + \nu\right)\left[\omega_e(\text{cm})^{-1}\right] \tag{3.11}$$

If the left-hand side of Eq. (3.10) is plotted against $E_\nu - E_o$, the slope of the resulting linear distribution will be equal *to*1/($kT_{vib}$), yielding the vibrational temperature of the gas molecules. Figure 3.12e shows the vibrational temperature calculated very close to the water surface for the 1 mm irradiation position of the plasma jet. From repeated experiments, the value is estimated to be 0.72 eV.

### 3.2.6 *Radical Species OH Measurement Methods*

For measurement of hydroxyl radical OH density, the nonthermal plasma jet has been in contact with the biosolution surface and its optical emission signal has been observed by charge-coupled device (CCD) spectrometer connected by the double slit and optical fiber with collimator lens [21]. Figure 3.13 shows the optical emission spectrum for Ar plasma jet measured by CCD spectrometer with optical fiber at the 2 mm above the deionized (DI) biosolution surface (a), and 2 mm below the DI biosolution surface (b). From the result of spectrum of Fig. 3.13a, the Ar emission lines between 700 nm and 900 nm are shown to be dominantly appeared and it is found that the 309 nm emitted from the hydroxyl OH radical [22–24] species, 224 nm from the nitric oxide NO [25], the emission lines 245 nm from the reactive species of superoxide anion $O_2{}^{*-}$ [26], and the emission lines from

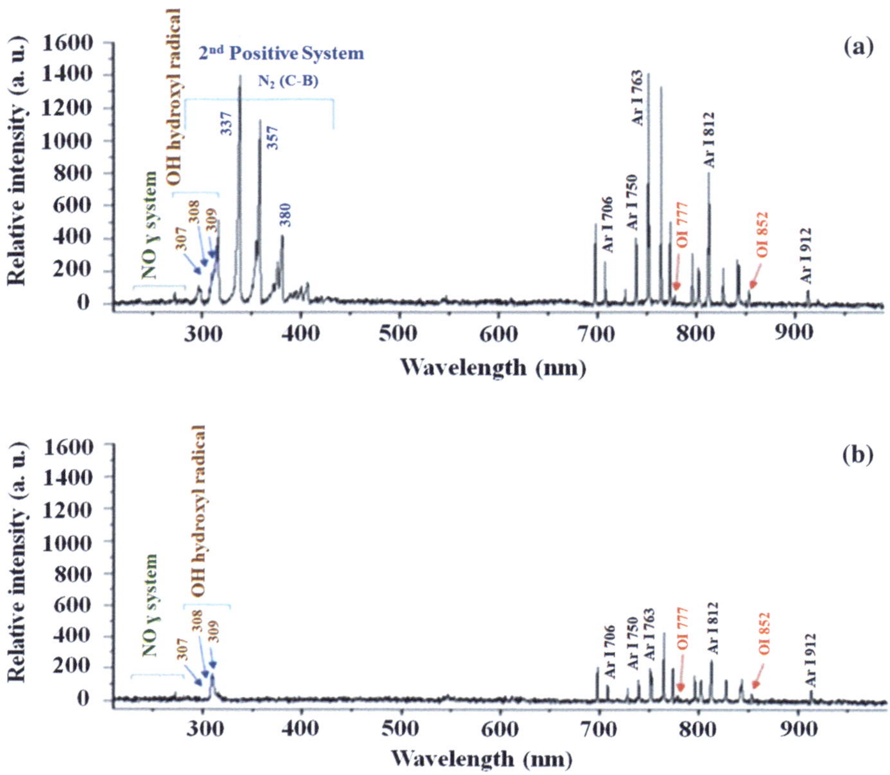


**Fig. 3.13** Optical emission spectrum measured by CCD spectrometer with optical fiber at 2 mm above DI water (**a**), and 2 mm below the deionized (DI) surface (**b**) [21]

$N_2$ second positive system ($C^3\Pi u$–$B^3\Pi g$) ranged from 320 nm to 380 nm strongly appeared just above the biosolution surface. While as seen in Fig. 3.13b, as moved 2 mm downward into the DI biosolution surface, the emission lines of 224–280 nm from the nitric oxide NO [25], the 245 nm from the reactive species of superoxide anion $O_2^{*-}$ [26], and $N_2$ second positive system ($C^3\Pi u$–$B^3\Pi g$) emission lines ranged from 320 nm to 380 nm are shown to be drastically disappeared. However, it is found that the 309 nm emitted from the hydroxyl OH radical species is getting weaker in the biosolution than that above the surface. It is also noted from the Figs. 3.13a and b that the both lines of 777 nm and 852 nm inside the DI biosolution, which are emitted from the $O_2$ first positive system, including the emission lines of Ar I and $N_2$ first positive system (B–A) are getting weaker than those above the solution.

Also we have measured the reactive oxygen species, especially, for the hydroxyl OH radicals by the optical emission spectroscopy as well as its absolute density inside or just interfacial region of the biosolutions by the ultraviolet (UV) absorption

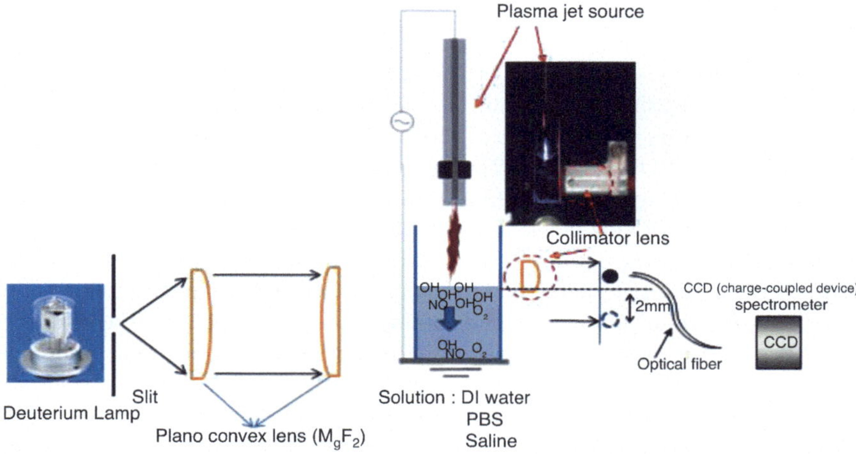


**Fig. 3.14** Schematic experimental setup of ultraviolet absorption spectroscopy inside the biosolutions [21]

spectroscopy when the nonthermal plasma has been bombarded onto the biosolution surface. Figure 3.14 shows the schematic experimental setup for the ultraviolet absorption spectroscopy inside the biosolutions. This system consists of deuterium UV lamp in this experiment, whose power is 30 W with spectral wavelength between 160 nm and 800 nm, plano-convex lens whose transmission wavelength ranges are from ultraviolet to infrared, double slit connected to the collimator lens, and CCD spectrometer with optical fiber whose spectral range is from 200 to 1100 nm. The light of UV lamp has been incident and focused by 3 mm plano-convex lens in front of detector system for the measurement of absorption profiles occurred at 309 nm, of the hydroxyl radical OH species [21, 22], as shown in the Fig. 3.14. This detector system consists of double slit in this experiment, whose slit width 100 um and their gap is 1 mm, and then collimator lens has been connected to the optical fiber whose diameter is 500 um. With this experimental situation, we observed no stray light from above the surface since the plasma jet has been bombarded onto the central surface region in the measurement [21].

The hydroxyl OH radical density inside the biosolution, which is generated by the argon nonthermal plasma jet interaction with DI water surface, could be obtained by ultraviolet absorption spectroscopy using the Lambert–Beer's law [22]. The incident UV light incident on the biosolution has the intensity $I_o$ and the transmitted light intensity is denoted by $I_v$ after passing through the OH existing region $x = 10$ mm in this experiment, inside the biosolution, in which the hydroxyl OH radical species are generated by the nonthermal plasma bombardment onto the solution surface. The density of hydroxyl radical species, OH, inside the biosolution, which is induced by the nonthermal plasma bombardment onto the biosolution is given by [22]

$$N = -\frac{1}{\sigma \cdot x} \ln\left(\frac{Iv}{I_0}\right) \tag{3.12}$$

where $N$ is the density or concentration for absorbing species of hydroxyl OH radicals, $\sigma$ is the cross-sectional area of about $1.2 \times 10^{-16}$ cm$^2$ for absorbing species of hydroxyl radical OH species [22]. The hydroxyl radical OH density inside the biosolution could be obtained Eq. (3.12), by measurement of the ratio $I_v/I_o$ of the transmitted intensity to the incident one during nonthermal plasma irradiation onto the biosolution. Here the optimized gas flow condition has been also investigated for maximizing the density of hydroxyl OH radical species inside the biosolution under the argon nonthermal plasma jet bombardment onto the solution. The hydroxyl radical density has been investigated at the region of 2 mm above the interfacial solution surface [23], where the UV passing length is 3 mm, as well as inside the biosolution whose spatial passing length is 10 mm, by the ultraviolet absorption spectroscopy by using the deuterium lamp in Ar gas flow ranged from 100 sccm to 350 sccm in argon bioplasma jet. To get an absorption spectrum caused by the hydroxyl radical species, two plano-convex lenses have been used for providing parallel UV lights produced from the deuterium lamp and then making a crossing beam with focused diameter of 200 um when transmission through a middle position of the biosolution along the passing way of 10 mm. The deuterium lamp is a gas-discharged light source, which operates with similar process to arc lamp and has a continuous spectrum from 200 nm to 500 nm in this experiment. After passing through the plasma region of 3 mm located 2 mm above the interfacial surface as well as inside the biosolutions whose passing length is 10 mm, the parallel UV lights whose beam diameter is around 0.5 mm are provided by the convex lenses for UV collection to the optical fiber with collimator lens. At this time the position of the optical lens system has been fixed during the measurement. Here, the spatial resolution of the system is enough for about 1 mm resolution since the output UV has beam diameter of around 0.5 mm. For the measurement of OH radical species at the different position inside the biosolution, only the detector of optical fiber could be moved from the region of 2 mm below water surface by spatial resolution 1 mm without moving of optical lens and collimation system.

Figure 3.15 shows the absorption profile around 309 nm of hydroxyl OH radical species (a) and density of hydroxyl OH radical species (b) at the region of 2 mm above the DI water, respectively, versus the Ar gas flow rates ranged from 100 sccm to the 350 sccm, under the low electrical power of 4.9 W and driving frequency of 35 kHz. It is noted that these measurement values are within 5% error ranges. For the gas flow rate of around 250 sccm, it is found that the density of hydroxyl OH radical species reaches the maximum value of $5.2 \times 10^{15}$ cm$^{-3}$. It is also noted that this density of OH radical species has been rapidly decreased to $3.1 \times 10^{15}$ cm$^{-3}$ as the gas flow rate is increased to 350 sccm in this experiment. These OH densities are in relatively good agreement with other results of $(0.3 \sim 7.5) \times 10^{15}$ cm$^{-3}$ reported by other groups under the low contents of water molecules less than 3% in their employed gases of He, $N_2$, and $N_2/O_2$ mixtures [27, 28]. They have used a microwave frequency of 2.45 GHz and RF frequency of 13.56 MHz [27, 28] with

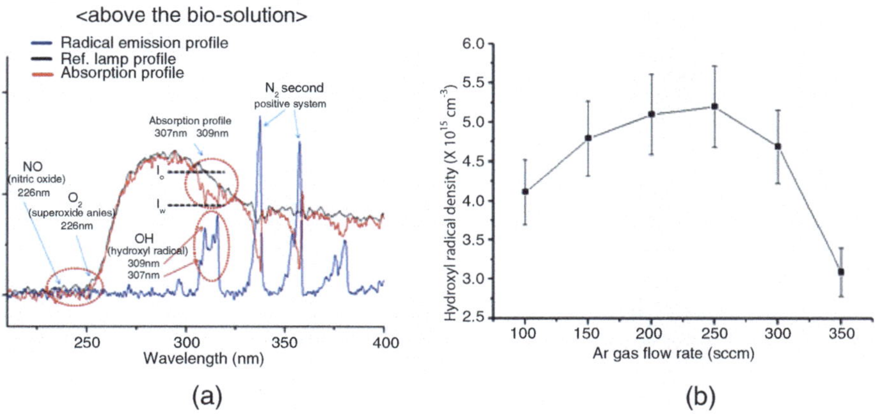


**Fig. 3.15** Absorption profile around 309 nm of hydroxyl OH radical species (**a**) and density of hydroxyl OH radical species (**b**) at the region of 2 mm above the DI water, respectively, versus the Ar gas flow rates ranged from 100 sccm to the 350 sccm [21]

relatively high power greater than 100 W. However, we have used the low driving frequency of 35 kHz and low electrical power of 4.9 W in this experiment for the production of OH radical species inside the DI biosolutions by bombardment of the atmospheric pressure nonthermal plasma onto the DI with Ar gas for various ranges of gas flow rates.

## 3.3 Other ROS Species in Water

Figure 3.16 shows a plot of the OH• (a) and $H_2O_2$ (b) densities, respectively, denoted by $\mu$Mole by $\mu$Mole, nitric acid compounds (c), and especially the $NO_3^-$ ion concentration (d) denoted by mMole, in deionized water activated by the $N_2$ plasma jet, versus $O_2$ mole fraction $\xi$ $=n_{O2}/n_0$, where $n_{O2}$ is $O_2$ density and $n_0$ is ambient neutral density in the atmospheric pressure at room temperature [29, 30]. The $N_2$ density $n_{N2}$ is then $(1 - \xi)\ n_0$. Most of the hydroxyl molecules are generated at water surface, and some of them can penetrate into water. The theoretical result obtained is least-squared-fitted to the experimental data. Typical error bar is shown in the data, where the size of the error bar is about 8% of its measurement value. We can see that the hydroxyl density is strong at a small $\xi$, but it quickly disappears as $\xi$ increases to 0.2, corresponding to the $O_2$ mole fraction of air. Dots in the figure are the experimental data from the averaged value of three times of measurements. The hydroxyl molecules are mostly generated at water surface, and some of them can penetrate into water. At a small mole fraction of oxygen, the hydroxyl density is strong; but as the oxygen mole fraction increases to 0.2, corresponding to the oxygen mole fraction of air, it quickly disappears. At a small value of oxygen mole

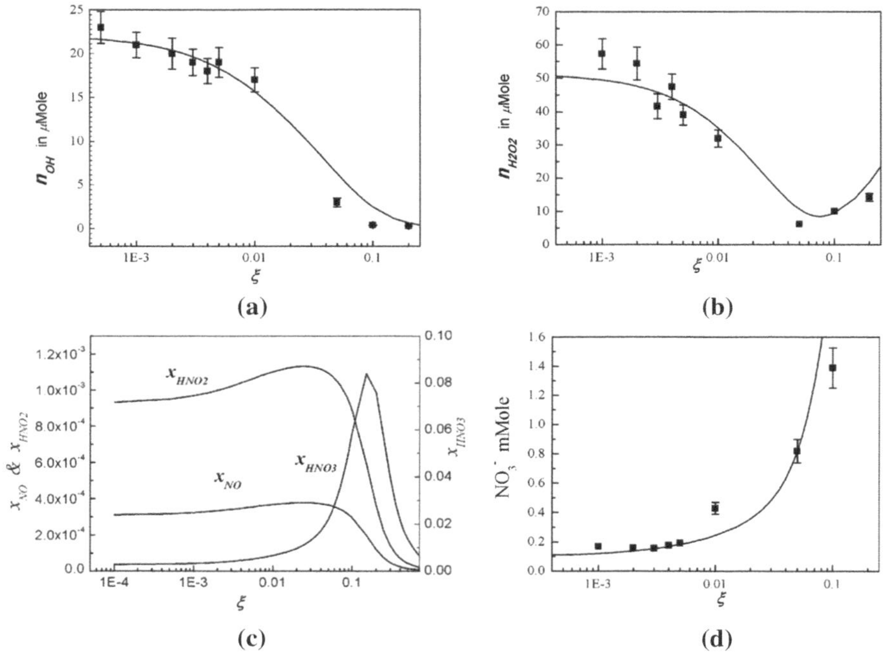


**Fig. 3.16** Densities of OH• (**a**), $H_2O_2$ (**b**) denoted by μMole, nitric acid compounds (**c**), and especially the $NO_3^-$ ion concentration (**d**) denoted by mMole, in deionized water activated by the $N_2$ plasma jet, versus $O_2$ mole fraction $\xi = n_{O2}/n_O$, where $n_{O2}$ is $O_2$ density and $n_O$ is ambient neutral density in the atmospheric pressure at room temperature [29]

fraction, the intensity of the hydrogen peroxide is very high, where a relatively high intensity of hydroxyl generates $H_2O_2$, as expected from Fig. 3.16b. But as the oxygen mole fraction increases, its intensity decreases, like the hydroxyl density. However, as the oxygen mole fraction increases to a large value, the hydrogen peroxide density increases again, due to the strong surge of hydrogen dioxide $HO_2$ molecules. Therefore, as the oxygen mole fraction increases from a small value to a large value, the hydrogen peroxide density decreases, reaches its minimum value at $\xi = 0.05$, and then increases again. The nitric acid compounds are also shown in Fig. 3.16c and d in terms of the $O_2$ mole fraction. The nitric acid density ($xHNO_3 = nHNO_3/n_O$) is low at a small mole fraction of oxygen, but its intensity increases to a peak value of around $\xi \approx 0.2$, and then decreases, as the $\xi$ increases. Meanwhile, the densities of $HNO_2$ and NO are at moderate levels in the range of $\xi \leq 0.25$. Note that $HNO_3$ is a strong acid; meanwhile, $HNO_2$ is a weak acid. We remind the reader that a moderate level of nitrogen monoxide (NO) exists in the entire range of the $O_2$ mole fraction. The experimental data of the $NO_3^-$ ion concentration in water is also represented in Fig. 3.16d, in terms of the $O_2$ mole fraction of $\xi$. It is shown that nitric acid increases as $\xi$ increases to 0.2; however, the nitric acid (HNO3) decreases drastically as the $\xi$ increases beyond 0.2 as seen in Fig. 3.16c. The *pH* value of the plasma activated water, which is slightly alkali without oxygen, decreases to three as $\xi$ increases to 0.25, which can be expected by the nitric acid in Fig. 3.16c, and was also confirmed by experiments. The $\xi$ of ambient air is about 0.2, so that the *pH* value of water activated by the air plasma is acidic, with $pH \approx 3$.

Solvation of acid species such as $HNO_2$ and $HNO_3$ results in the lowering of the water pH. The strong bactericidal activity of water is often attributed to the synergistic action of plasma-generated species in the aqueous phase coupled with low pH [31, 32]. In addition, the effect of alkane-like hydrocarbons, to mimic the chemistry of a biological target, was also investigated in the two-dimensional model study [33]. It was discovered that RNS are unaffected by alkane-like hydrocarbons, whereas ROS were consumed leaving alkyl radicals to reach the tissue target. The major mode in RONS delivery is through convection and diffusion of neutral species into the liquid resulting in overall lower RONS concentrations after plasma treatment. However, the RONS delivery could be synergistically done through plasma initiated UV photolysis [34] and diffusion processes during plasma exposure on liquid or tissues. Another important consideration for applications of air DBD plasmas jet in plasma medicine is that the plasma is usually operated at a frequency in the few tens of kHz—few MHz ranges and the treatment time can be in the order of seconds to minutes. It was found that the spatial location of the plasma plume strongly affects the spatial distribution of the solvated species such as OH and $H_2O_2$, which have high rates of solvation. Species with low rates of solvation, such as NO, are weakly affected by the location of the plasma plume because of their diffusion in air resulting in a uniform distribution directly above the liquid surface [35].

**Acknowledgements** This work was supported by the National Research Foundation of Korea (NRF) grant supported by Korea Government (MSIP) (NRF-2016K1A4A3914113 and NRF-2010-0027963) and in part by Kwangwoon University, Seoul, Korea.

## References

1. J.-H. Lee, E.-H. Choi, K.-M. Kim, K.-N. Kim, Corrigendum: Effect of non-thermal air atmospheric pressure plasma jet treatment on gingival wound healing. J. Phys. D: Appl. Phys. **49**, 249501 (2016)
2. P. Shaw, N. Kumar, H.S. Kwak, J.H. Park, H.S. Uhm, A. Bogaerts, E.H. Choi, P. Attri, Bacterial inactivation by plasma treated water enhanced by reactive nitrogen species. Sci. Rep. **8**, 11268 (2018)
3. J.-S. Kwon, Y.H. Kim, E.H. Choi, K.-N. Kim, The effects of non-thermal atmospheric pressure plasma jet on attachment of osteoblast. Curr. Appl. Phys. **13**, S42–S47 (2013)
4. K.Y. Baik, Y.H. Kim, Y.H. Ryu, H.S. Kwon, G. Park, H.S. Uhm, E.H. Choi, Feeding-gas effects of plasma jets on Escherichia coli in physiological solutions. Plasma Process. Polym. **10**, 235–242 (2013)
5. N.K. Kaushik, N. Kaushik, N.N. Linh, B. Ghimire, A. Pengkit, J. Sornsakdanuphap, S. Lee, E.H. Choi, Plasma and nanomaterials: fabrication and biomedical applications. Nano **9**, 98 (2019)

6. S.H. Ki, K.Y. Baik, E.H. Choi, Effects of humidity on room disinfection by dielectric barrier discharge plasma. J. Phys. D., accepted for publication **52**, 425204 (2019)
7. M.S. Mann, R. Tiede, K. Gavenis, G. Daeschein, R. Bussiahn, K.-D. Weltmann, S. Emmert, T. von Woedtke, R. Ahmed, Introduction to DIN-specification 91315 based on the characterization of the plasma jet kINPen® MED. Clin. Plasma Med. **4**, 35–45 (2016)
8. J.H. Ahn, Study on optical diagnostics of radical density in non-thermal atmospheric pressure plasma, MS thesis, Kwangwoon University, Korea, 2017
9. N. Kang, S. Oh, Determination of the absolute nitrogen atom density in an Ar-N2 ICP discharge. J. Phys. Soc. **59**(5), 3031–3036 (2011)
10. M. Capitelli, C.M. Ferreora, B.F. Gordiets, A.I. Osipov, *Plasma Kinetics in Atmospheric Gases* (Springer Science & Business Media, Berlin, 2013), p. 196
11. D. Mariotti, Y. Shimizu, T. Sasaki, N. Koshizaki, Method to determine argon metastable number density and plasma electron temperature from spectral emission originating from four 4p argon levels. Appl. Phys. Lett. **89**, 201502 (2006)
12. J. Sornsakdanuphap, P. Suanpoot, Y.J. Hong, B. Ghimire, G. Cho, H.S. Uhm, D. Kim, Y.J. Kim, E.H. Choi, Electron temperature and density of non-thermal atmospheric pressure argon plasma jet by convective wave packet model. J. Korean Phys. Soc. **70**(11), 979–989 (2017)
13. E. Karakas, M.A. Akman, M. Laroussi, The evolution of atmospheric-pressure low-temperature plasma jets: Jet current measurements. Plasma Sources Sci. Technol. **21**, 034016 (2012)
14. E.J. Baek, H.M. Joh, S.J. Kim, T.H. Chung, Effects of the electrical parameters and gas flow rate on the generation of reactive species in liquids exposed to atmospheric pressure plasma jets. Phys. Plasmas **23**, 073515 (2016)
15. Z. Xiong, X. Lu, Q. Xiong, Y. xian, C. Zou, J. Hu, W. Gong, J. Liu, Z. Jiang, Y. Pan, IEEE Trans. Plasma Sci. **38**, 1001 (2010)
16. E. Karakas, M. Laroussi, Experimental studies on the plasma bullet propagation and its inhibition. J. Appl. Phys. **108**, 063305 (2010)
17. S.B. Bayram, P.T. Arndt, Rotational spectra of N2+: an advanced undergraduate laboratory in atomic and molecular spectroscopy. Am. J. Phys. **83**, 867 (2015)
18. N.U. Rehman, M.A. Khan, M.Y. Naz, M. Shafiq, M. Zakaullah, Characterization of 13.56 MHz RF Ne–N2 mixture plasma using intrusive and non-intrusive diagnostic techniques. Phys. Scr. **88**, 045503 (2013)
19. P. Bruggeman, G. Cunge, N. Sadeghi, Plasma Sources Sci. Technol. **21**, 035019 (2012)
20. B. Ghimire, J. Sornsakdanuphap, Y.J. Hong, H.S. Uhm, K.-D. Weltmann, E.H. Choi, The effect of the gap distance between an atmospheric-pressure plasma jet nozzle and liquid surface on OH and N2 species concentrations. Phys. Plasmas **24**, 073502 (2017). https://doi.org/10.1063/1.4989735
21. Y.H. Kim, Y.J. Hong, K.Y. Baik, G.C. Kwon, J.J. Choi, G.S. Cho, S. Uhm, D.Y. Kim, E.H. Choi, Measurement of reactive hydroxyl radical species inside the biosolutions during non-thermal atmospheric pressure plasma jet bombardment onto the solution. Plasma Chem. Plasma Process. **34**, 457 (2014)
22. H.P. Dorn, R. Neuroth, A. Hofzumahaus, Investigation of OH absorption cross sections of rotational transitions in the A2S+, v0 = 0 X2Õ, v00 = 0 band under atmospheric conditions: Implications for tropospheric long-path absorption measurements. J. Geophys. Res. **100**, 7397 (1995)
23. Y.J. Hong, C.J. Nam, K.B. Song, G.S. Cho, H.S. Uhm, D.I. Choi, E.H. Choi, Measurement of hydroxyl radical density generated from the atmospheric pressure bioplasma jet. J. Instrum. **7**, C03046 (2012)
24. G.H. Dieke, H.M. Crosswhite, The ultraviolet bands of OH fundamental data. J. Quant. Spectrosc. Radiat. Transf. **2**, 97 (1961)
25. B. Benstaali, P. Boubert, B.G. Cheron, A. Addou, J.L. Brisset, Density and rotational temperature measurements of the OH and NO radicals produced by a gliding arc in humid air. Plasma Chem. Plasma Process. **22**, 553 (2002)

26. B.G. Kwon, J. Yoon, Superoxide anion radical: principle and application. J. Korean Ind. Eng. Chem. **20**, 593 (2009)
27. N. Srivastava, C. Wang, Effects of water addition on OH radical generation and plasma properties in an atmospheric argon microwave plasma jet. J. Appl. Phys. **110**, 053304 (2011)
28. M. Baeva, K. Rackow, M.M. Becker, J. Ehlbeck, D. Loffhagen, Characterization of atmospheric pressure microwave plasma in N2/O2/H2O gas mixtures. Topic C9, 30th ICPIG, Belfast, Northern Ireland UK, 2011
29. H.S. Uhm, S.H. Ki, K.Y. Baik, E.H. Choi, Influence of oxygen on generation of reactive chemicals from nitrogen plasma jet. Sci. Rep. **8**, 9318 (2018)
30. A.S. Blecker, R. Bansemer, S. Reuter, K.D. Weltmann, How to produce an NOx-instead of Ox-based chemistry with a cold atmospheric plasma jet. Plasma Process. Polym. **13**, 1120 (2016)
31. M. Naïtali, G. Kamgang-Youbi, J. Herry, M. Bellon-Fontaine, J. Brisset, Combined effects of long-living chemical species during microbial inactivation using atmospheric plasma-treated water. Appl. Environ. Microbiol. **76**, 7662–7664 (2010)
32. J. Julák, V. Scholtz, S. Kotúčová, O. Janoušková, The persistent microbicidal effect in water exposed to the corona discharge. Phys. Med. **28**, 230–239 (2012)
33. W. Tian, M. Kushner, Atmospheric pressure dielectric barrier discharges interacting with liquid covered tissue. J. Phys. D. Appl. Phys. **47**, 165201 (2014)
34. P. Attri, Y. Kim, D. Park, J. Park, Y. Hong, H. Uhm, K. Kim, A. Fridman, E.H. Choi, Generation mechanism of hydroxyl radical species and its lifetime prediction during the plasma-initiated ultraviolet (UV) photolysis. Sci. Rep. **5**, 9332 (2015)
35. A. Starikovskiy, Y. Yang, Y. Cho, A. Fridman, Non-equilibrium plasma in liquid water: Dynamics of generation and quenching. Plasma Sources Sci. Technol. **20**, 024003 (2011)

# Chapter 4
# Cancer Applications Overview

**Michael Keidar and Alexander Fridman**

## Contents



**Abstract** In this chapter, we overview the progress of CAP application for cancer therapy. This includes summary of in vitro results, in vivo treatment, as well as first clinical trials.

In recent years, cold atmospheric plasma (CAP) has emerged as a possible new modality for cancer treatment [1]. The novelty of CAP is in various forms and it is typically associated with reactive species produced in plasma and electric field formed when plasma hits the tissue. Outside of cancer applications, CAP's efficacy has been tested in various applications such as disinfection, wound healing, dentistry, and cancer therapy [2–7]. High-level hypothesis of CAP interaction with cells and tissue is based on the notion that chemical elements of the CAP are

M. Keidar (✉)
Mechanical and Aerospace Engineering, School of Engineering and Applied Science, The George Washington University, Washington, DC, USA
e-mail: keidar@gwu.edu

A. Fridman
Drexel University, Philadelphia, PA, USA



M. Keidar (ed.), *Plasma Cancer Therapy*, Springer Series on Atomic, Optical, and Plasma Physics 115, https://doi.org/10.1007/978-3-030-49966-2_4

potentially toxic, such as reactive oxygen species (ROS), which might promote a "plasma killing effect," while others such as reactive nitrogen species (RNS) could produce a "plasma healing" effect. Forming various combinations of these species might provide great potential for activation of specific signaling pathways in cells. CAP treatment possesses powerful lethal capabilities against tumor cells both in vitro and in vivo and just as importantly, the normal counterpart cells have been shown to be less sensitive to the same CAP treatment [8]. All these aforementioned effects are believed to be related to the plasma chemistry. While different species are produced as a result of plasma treatment, the role of other plasma effects such as charged particles and the electric field is still not fully understood.

The general understanding of the CAP action on living tissue is associated with the ROS and RNS formation and transport from the gaseous to the liquid phase and through the biological barriers. Many RONS generated by CAP are also active components in cell biology [9]. It has been argued that analogy between cellular and plasma-generated RONS represents the major logic of plasma application in medicine including cancer therapy and that many species produced directly or indirectly by plasma will function in cell biology as endogenous species [8]. RONS play a central role in 'redox' or oxidation–reduction biology [10].

The timeline of plasma interaction with cells is shown in Fig. 4.1. Plasma interaction with living tissue is essentially a multi-scale process spanning from the initial burst at the timescale of nano- and microseconds (depending on specific plasma device) to seconds, followed by the timescale of minutes, which is related to RONS formation and transport across the cellular membrane and finally triggering various cellular pathways at the timescale of hours and days [11].

It should be pointed out that there are many important functions associated with reactive oxygen species. Watson [12] suggested that ROS is "a positive force for

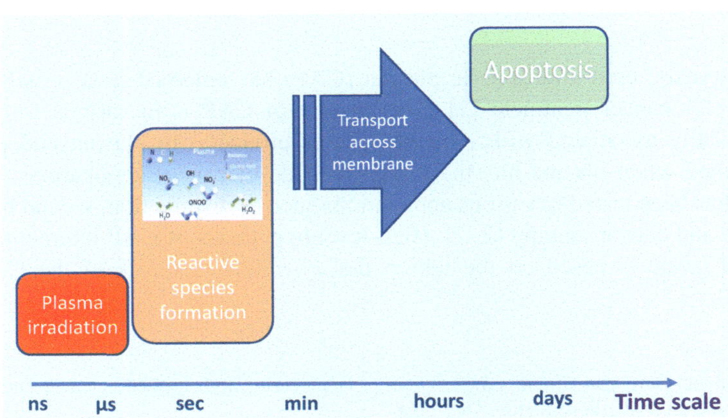


**Fig. 4.1** Schematics of multi-scale nature of plasma interaction with cells. Initial plasma impact leads to the formation of long-life species, transport of these species across the gas–liquid interphase, cell membrane, activation of various cellular pathways, and eventually to cell apoptosis or programmable death

life" due to their role in apoptosis—an internal program leading to cell death. At the same time, ROS are also well recognized "for their ability to irreversibly damage key proteins and nucleic acid molecules [e.g., DNA and RNA]." In a balanced cell, a normal level of ROS is maintained by the antioxidant system. Watson noted that "The vast majority of all agents used to directly kill cancer cells (ionizing radiation, most chemotherapeutic agents and some targeted therapies) work through either directly or indirectly generating reactive oxygen species that block key steps in the cell cycle." The general understanding is that the effect of ROS on cell development depends on the level of ROS [13]. Low level of ROS supports cell proliferation and helps to maintain cell functionality while a high level of ROS causes oxidative stress leading to cell death. Healthy cell function is preserved by an antioxidant system that maintains the ROS level at a tolerable level. Cancer cell abnormal metabolism causes an aberrant high level of ROS [11]. To survive, a cancer cell mutates to regulate ROS. However, rise in the intracellular ROS level might cause irreparable DNA damage [14, 15]. At the same time, the level of ROS in cancer cells is near the limit at which cell death occurs. On the other hand, the ROS level in the corresponding normal cells is generally lower [13]. Thus, selectivity toward tumor cells by pro-oxidative therapy is achieved when anticancer therapy produces ROS near the "threshold" between levels of ROS in normal cells and cancer cells.

It has been hypothesized that such an approach seems to be relevant to the CAP-based anticancer therapy. ROS produced by CAP might lead to cancer cells' death by damaging the function of intracellular regulatory factors [16, 17]. To that end, multiple studies demonstrated that biologically active neutral short- and long-living ROS molecules are produced by CAP including OH, O, O (1D), $O_2$ (1$\Delta$g), $O_3$, $HO_2$, and $H_2O_2$ [18]. RNS species such as $NO_2-$, $NO-$, and $NO+$ are generated directly during the discharge in a gas phase and in the plasma-activated media [7, 19].

Turning attention to reactive nitrogen species, nitrite oxide (NO) plays a very significant role in cellular processes. In particular, NO is an important factor in the electron transport chain. It is also known that NO might affect the electron transport system by attacking cytochrome oxidase [20]. In turn, the breaking of the electron transport will increase the generation of superoxide reacting with NO to form peroxynitrite [21].

It has been well established that one of the important species formed in the course of plasma interaction with cell culture media is hydrogen peroxide ($H_2O_2$) [22, 23]. To that end, a model based on aquaporin (AQP), which is a well-known $H_2O_2$ channel on the cell membrane, has been proposed [24, 25]. Multiple studies inform that in general cancer tissues express more AQP channels than corresponding normal tissues. According to the proposed hypothesis [22, 23], after plasma treatment, $H_2O_2$ transport across the cellular membrane of the cancer cell is higher than that of the normal cell. Such differential $H_2O_2$ consumption rate might be the possible mechanism of the selective anticancer action of CAP. As such, the rise of intracellular ROS as a result of ROS diffusion across cell membrane correlates with the rise of extracellularly plasma-originated ROS. The reaction of cells to ROS excess is well documented. Catalase is the major enzyme that controls concentrations of $H_2O_2$ in both cancer and normal cells [26]. For instance, a recent paper correlated the rate of $H_2O_2$ removal and activity of catalase for 15 cancer cell lines and 10 normal cell lines. This study showed that $H_2O_2$ produced from the oxidation of P-AscH- is a principle mediating factor in selective targeting of the cancer cells. More importantly, it was demonstrated that normal cells have a higher constant of $H_2O_2$ removal than that of cancer cells. This trend is illustrated schematically in Fig. 4.3. Granted, $H_2O_2$ is known to be a very strong oxidant but it has a slow reaction rate with the majority of biomolecules. As such, hydrogen peroxide accumulates in cells and the aforementioned process of $H_2O_2$ removal by catalase becomes a critical process for cell survival [24]. Detailed analysis of the biochemical process in cells caused by CAP is presented in Chap. 5.

Aforementioned processes suggest the possible role of plasmas in the biomedical application through effects on the cellular chemical balance by generating RONS. In that sense, what plasma creates is a reservoir for some critical species that can be utilized when the cell needs them. In addition, it should be pointed out that CAP is very different from chemotherapy and radiotherapy, though many chemotherapy and radiotherapy also increase intracellular RONS stress and further kill cancer cells. The primary difference is that CAP itself is the source of RONS including hydrogen peroxide, nitrite, nitrate, and nitric oxide.

The high-level model of plasma–cell interaction based on the aforementioned processes is shown schematically in Fig. 4.2. Current understanding of CAP action could be attributed to several cellular factors, i.e., higher expression of AQPs by the cancer cell membrane, the higher level of ROS in cancer cells, and lower expression of antioxidant enzyme (catalase) in cancer cells [22, 23].

## 4.1 Effect of CAP on Cell Cycle

According to conventional wisdom, a way to target cancer cells is to interfere with the cell cycle [27–29]. It stems from the fact that cancer cells proliferate at a faster rate than normal cells [30]. Several studies suggest that CAP interaction with cells, in particular, cancer cells, lead to modifications in the cell cycle. In fact, recent data confirmed that cancer cells are more susceptible to the effects of CAP because a greater percentage of cells are in the S phase.

Recall that one of the hallmarks of cancer is the approach to deregulate the mechanisms controlling cell cycle [31]. Note that the cell cycle or "life of cell" outlines the different stages taking place as a cell develops. In particular, the DNA synthesis occurs during the specific period of a cell cycle, which is called S-phase, the morphological changes in cell refer to the mitotic phase (M-phase), a phase of the cell division. The stages between M and S and between S and M phases are called G1 and G2 phases. It should be pointed out that non-proliferating cells are in the G1 phase of the cell cycle. Once cells proliferate, they move progressively from G1 to S (synthesis of DNA), G2, and then M (mitosis) phase where they divide. Cells in G1 can be either in quiescence (G0), terminally differentiate, or, in response to specific signals, be induced to proliferate. Most cells in normal adult tissues are in

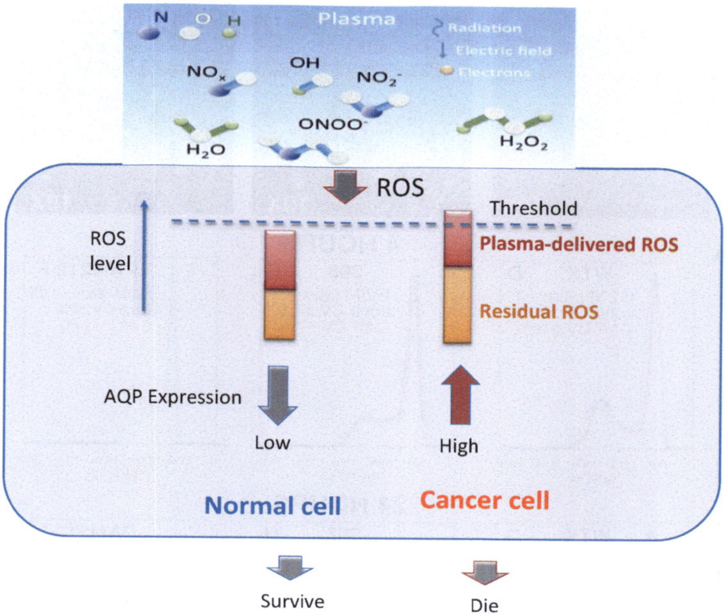


**Fig. 4.2** Schematically represented model of the plasma interaction with cancer and normal cells explains plasma selectivity

the quiescent G0 phase. Between S and G2 and between G2 and M phases there are so-called "checkpoints." Cell progression through the cell cycle is monitored and controlled by these checkpoints. Checkpoints verify whether the processes at each phase of the cell cycle have been accurately completed before moving into the next phase.

Initial analysis of the CAP effect on glioblastoma [32] suggested that the CAP treatment causes significant arrest in the G2/Mphase in U87MG cells. Vandamme et al. [22] found that in U87MG cells, plasma treatment led to a significant decrease in the number of cells in the G0/G1 phase with a significant increase of cells in the S-phase. It was found that this effect depends on the plasma dose.

The effect of CAP treatment on the cell cycle was studied for 3 different cell lines. Bright-field images with 10X magnification of wild-type keratinocytes, 308, and PAM212 cells morphology, respectively, are shown in Fig. 4.3. Figures 10 d–l are DNA content measurement of control (untreated) cells in blue, and cells treated with CAP for 60 s in red. The cells were investigated after 4 and 24 h after CAP treatment (60 s). One can see that no significant shifts in the G1–G2 peak positions were observed for all cell types considered. It can be seen that CAP-induced robust G2/M–cell cycle increases in both carcinoma and papilloma cells (it is about two to threefold increase in about 24 h after CAP treatment), whereas normal keratinocytes showed almost no variation. It should be pointed out that this change diminished in about 48 h.

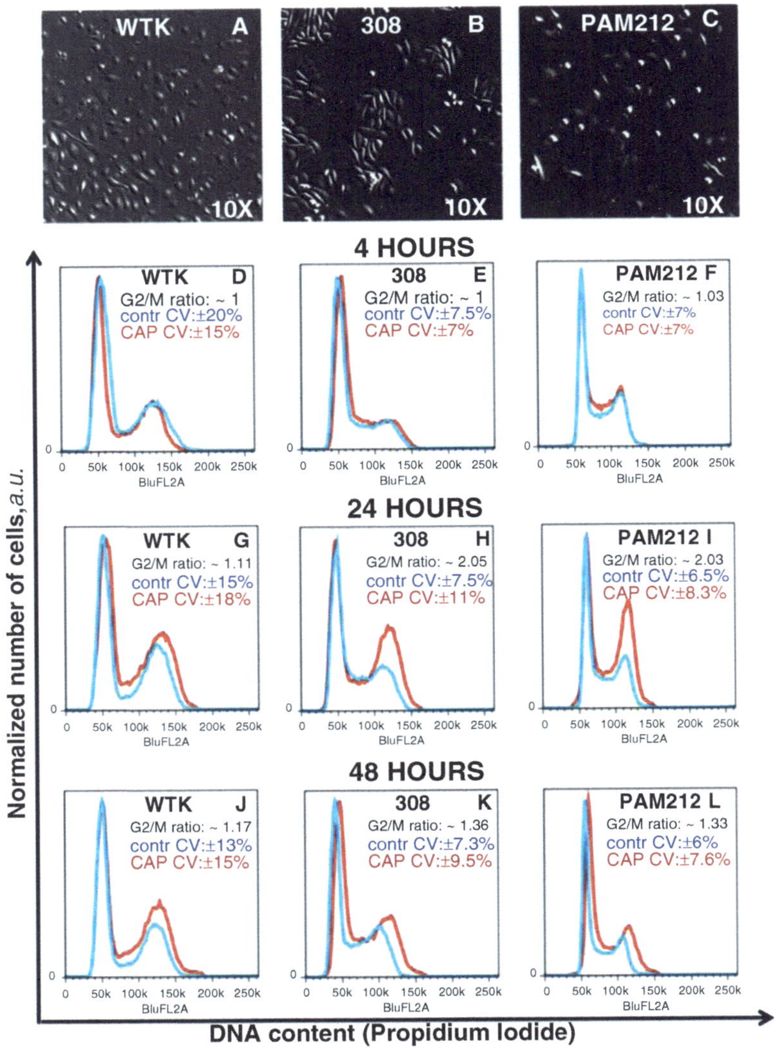


**Fig. 4.3** Identification of the cell cycle change in the G2/M-phase. (**a–c**) Bright-field images of wild-type keratinocytes (WTK), epidermal papilloma (308 cells), and epidermal carcinoma (PAM212 cells) cells are shown with a magnification ×10. (**d–l**) show cell cycle studies: Propidium Iodide content (horizontal axis) and normalized number of cells (vertical axis) are shown. The controls are shown in blue and cells CAP treated for 60 s are in red. The ratio of the number of cells (treated to untreated) in the G2/M-phase with coefficients of variation (CV, in percent) is shown in the right top corner of each figure. (**d–f**) shows cell cycle measurements in about 4 h; (**g–i**) in about 24 h; and Figure (**j–l**) in about 48 h after the CAP treatment for WTK, 308, and PAM 212 cells, respectively. Reprinted with permission from [26]. Copyright (2012) by Macmillan Publishers Ltd

### 4.1.1 In Vivo CAP Application. Subcutaneous Models

To date, a number of studies have been performed in order to access the anticancer effect of CAP treatment in vivo. To that end, CAP treatments were mainly performed by treating the skin above the tumor sites. In addition, in several studies, micro-size μ-CAP devices were applied to guide the tiny CAP jet to affect the exposed tumorous tissues underneath the skin or the skull [6, 12]. All reported studies reported a strong effect of CAP on the growth of tumors.

The earliest in vivo experiments were performed by Vandamme et al. [33, 34]. They used the glioblastoma U87MG tumor mouse xenograft model to test the anticancer effect of CAP treatment. This pioneering work demonstrated a significant tumor volume decrease of 56% on the treated mice after the micro-duration pulsed floating electrode dielectric barrier discharge (FE-DBD), which delivered about 0.75 W at 200 Hz on the mouse skin.

Corresponding mouse length of survival increased by about 60% after the FE-DBD treatment. Both the tumor volume measurement and bioluminescence imaging (BLI) have been used to assess the anticancer effect of FE-DBD treatment. In this study, the authors also discovered an important trend that the five consecutive plasma treatments will cause a much better anticancer effect than a long single plasma treatment. Such a fractionated treatment strategy has been used in many subsequent studies (Fig. 4.4).

In the same year, Keidar et al. performed the study in vivo on a bladder tumor xenograft model on mouse through the direct CAP jet treatment (Fig. 4.5) [35].

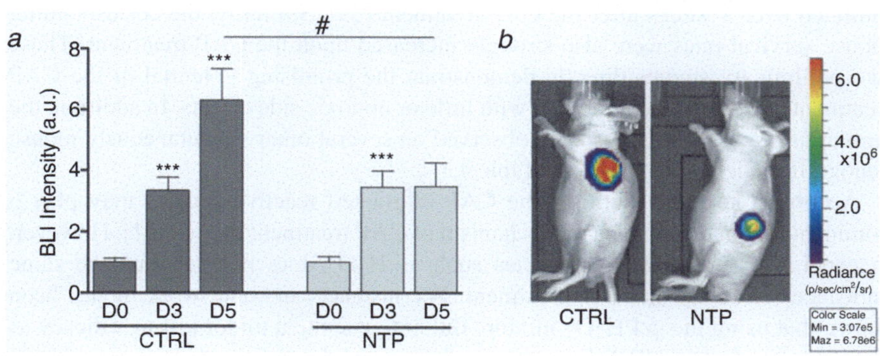


**Fig. 4.4** The early in vivo demonstration of the antitumor effect of CAP. When the tumor reached 150 ± 50 mm$^3$, mice were randomly assigned into two groups: control (CTRL) and plasma, eight mice per group. Plasma treatment was delivered each day for five consecutive days (6 min, 200 Hz). Mice in both groups were sacrificed 24 h after the last treatment. (**a**) BLI imaging was performed before the first treatment (Day 0), during treatment course (Day 3), and 24 h after the end of treatment protocol (Day 5). Tumor BLI was normalized to signal at Day 0. (**b**) Representative BLI imaging of CTRL and NTP treated mice at Day 5. Reproduced with permission from Vandamme et al., International Journal of Cancer, 130, 2185 (2012). Copyright 2012 John Wiley & Sons on behalf of the Union for International Cancer Control

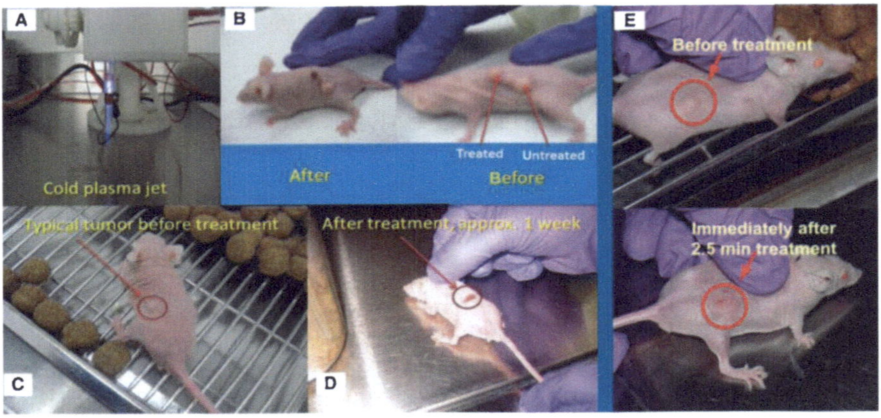


**Fig. 4.5** The anticancer effect on the subcutaneous tumor model. (**a**) CAP jet; (**b**) typical images of mice with three tumors before and after the CAP treatment (shown after 24 h); (**c** and **d**) typical images of mice with a single tumor before and approximately 1 week after CAP treatment; (**e**) tumor before and immediately after the 2.5 min of CAP treatment [29]. Reproduced with permission from Keidar et al., British Journal of Cancer, 105.9, 1295 (2011). Copyright 2011 Nature Publishing Group (a division of Macmillan Publishers Ltd)

After just 2 min of CAP treatment, the tumor significantly decreased its size and could not be observed via gross inspection after 24 h [13]. In addition, the authors have performed the similar experiments on a murine xenograft melanoma model and achieved a promising result that the tumor growth has been completely inhibited over 3 weeks after the CAP treatment [29]. Similarly, the corresponding mouse survival rates were also strongly increased upon the CAP treatment. These two preliminary studies directly demonstrate the promising potential of the CAP treatment as an anticancer method with little or no toxic side effects. In addition, the similar anticancer effect has been observed on several other subcutaneously mouse xenograft models as described in Table 4.1.

According to many authors, the CAP-originated reactive species may play a dominant role in the anticancer mechanism of CAP treatment in vivo [11]. However, in general, a single reactive species such as $H_2O_2$ does not generate the same anticancer effect as the CAP treatment. For instance, in some work, it has been shown that using the gel/$H_2O_2$ mixture did not generate a tumor killing efficacy as significant as that of DBD did on the melanoma in the mouse [11]. The gel/$H_2O_2$ mixture led to delayed tumor growth, but the tumor eventually progressed through the treatment [11]. Based on this result, it is hypothesized that other reactive species rather than $H_2O_2$ may be involved in the interaction between CAP and the cancerous tissue. As an alternative notion, it can be argued that the biological effect of $H_2O_2$ on tumor tissues may be quite different from what is seen in vitro in a petri dish or multi-wells plate as compared to the in vivo setting.

**Table 4.1** Summary of the anticancer CAP treatment in vivo

| Year | Tumor types | Models | Anticancer effect | Authors/References |
|---|---|---|---|---|
| 2011 | Xenografted glioblastoma | Mouse | Medium volume decrease | [27] |
| 2011 | Xenografted bladder cancer | Mouse | Strong volume decrease | [13] |
| 2011 | Melanoma | Mouse | Strong volume decrease | [13] |
| 2012 | Xenografted pancreatic carcinoma | Mouse | Strong volume decrease | [3] |
| 2012 | Xenografted glioblastoma | Mouse | Medium volume decrease | [2] |
| 2013 | Xenografted neuroblastoma | Mouse | Medium volume decrease | [5] |
| 2014 | Xenografted head and neck cancer | Mouse | Medium volume decrease | [4] |
| 2014 | Melanoma | Mouse | Slight volume decrease | [10] |
| 2015 | Melanoma | Mouse | Strong volume decrease | [11] |
| 2015 | Xenografted adenocarcinoma | Mouse | Little volume change | [7] |
| 2016 | Xenografted breast cancer | Mouse | Strong volume decrease | [6] |
| 2016 | Xenografted glioblastoma | Mouse | Medium volume decrease | [8] |
| 2017 | Melanoma | Mouse | Medium volume decrease | [9] |
| 2017 | Glioblastoma | Mouse | Strong volume decrease | [12] |
| 2017 | Head and neck cancer | Human | Partial remission | [14] |

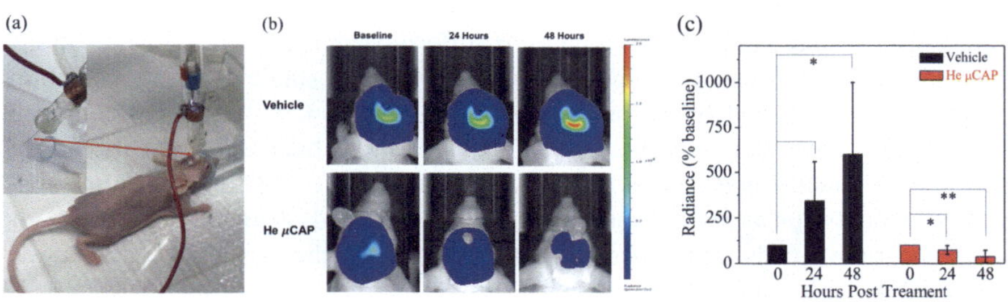


**Fig. 4.6** Treatment of brain tumor with micro-cold atmospheric plasma (μCAP). (**a**) μCAP photo showing plasma delivery through an intracranial endoscopic tube; (**b**) bioluminescence images of tumor volume; (**c**) summary data of radiance intensity showing both helium (no discharge) treatment (marked as "vehicle") and CAP treatment [Ref. 27]. Reprinted with permission from Cancers 9.6 (2017): 61

### 4.1.2 *In Vivo Treatment: Intracranial Model*

Recently, the first intracranial mouse model was utilized in CAP treatment. To study the effect of CAP on a brain tumor, a micro-CAP device has been developed. This device was directly applied to the glioblastoma tumor by utilizing an implanted endoscopic delivery system as shown in Fig. 4.6 [36]. Tumor volume was assessed by the bioluminescence imaging in real time (as shown in Fig. 4.6b). These experiments suggested that the tumor volume in the case of a control (i.e., helium only treatment) increased by about 600% after 2 days. At the same time, CAP treatment led to a decrease in the tumor volume by about 50% as shown in Fig. 4.6c.

Overall, by 2017, twenty-seven in vivo studies have been identified and concluded that there was a significant reduction in tumor size and an increase in survival rate [37]. To date, it should be pointed out that the in vivo applications were mostly performed on subcutaneous tumor xenografts in mice. Another study based on the use of a tumor chorioallantoic model was conducted [38]. Although the results corroborate the in vitro studies, the development of models closer to clinical situations is necessary. Only a single study was performed utilizing the intracranial model [27].

## 4.2 Clinical Studies

To date, several clinical studies have been performed. In Germany, two of such studies concerned with the application of CAP on patients with head and neck cancers as a palliative treatment or before tumor resection. The rationale of this therapeutic choice can be explained by the ability of CAP to decontaminate and to treat severely infected wounds or ulcerations [39] while inducing apoptosis in head and neck cancer cell lines. Summary of the first clinical studies was published by Metelmann et al. [33]. The trial enrolled 6 patients with locally advanced (pT4) squamous cell carcinoma of the oropharynx suffering from open infected ulcerations [33]. Patients were treated by a plasma jet in a cycle of 3 single applications within 1 week, each followed by an intermittence of 1 week. CAP treatment caused a reduction in odor and pain medication requirement, which improved the social function of patients and caused a positive emotional effect. The partial remission in two patients for at least 9 months has been observed. A moderate amount of apoptotic tumor cells and a desmoplastic reaction of the connective tissue was observed in the incisional biopsies [33].

In a most recent clinical follow-up, Metelmann and colleagues [40] investigated the effect of CAP on the surface of head and neck squamous cell carcinoma tumors. Overall, these results have shown an improvement in the quality of life of the patients, tumor reduction, and significant improvement in tumor decontamination.

In the USA, US Medical Innovation, LLC and GWU performed a clinical application by completing stage 4 colon cancer surgery to remove tumor and treat cancerous tissue remnants while sparing normal tissue at Baton Rouge General Medical Center in Baton Rouge, LA, in March 2015. Later in 2017, US Medical Innovation LLC used Canady Helios Cold Plasma and Hybrid Plasma Scalpels in the first clinical liver resection to remove and selectively kill liver cancer tumor cells [41]. More details of the clinical studies and perspectives of CAP application in health care are presented in Chap. 11.

## 4.3 Summary of Plasma Immunomodulation Studies

A recent new application of CAP in oncology involves plasma-based immunotherapy. There are two major reasons for recent significant interest in plasma immunomodulation studies in general and plasma immunotherapy of cancers in particular:

- First of all, most of the cancers are not superficial (like melanoma, or some types of carcinoma), but internal, that is located deep inside of the human body. Plasma species are in many cases simply unable to directly access the cancer tissue, even taking into account bystander and cell-to-cell communication effects. From this perspective, plasma stimulation of the immune system makes sense. Plasma species are able to stimulate major immune system factors either directly or through the intermediate contribution of bystander and cell-to-cell communication mechanisms; then the immune system suppresses malignancies wherever they are located in deep inside of the human body.
- The second aspect of interest in plasma immunomodulation studies is due to recent successes in cancer immunotherapy in general. Breakthroughs in cancer immunotherapies have demonstrated considerable success, though not without limitations mostly related to treatment side effects. Nonthermal plasma is highly multi-factor controllable substance, which provides an opportunity for extreme selectivity and suppression of side effects. It should be pointed out that CAP for cancer therapy has been emerging as a potential adjuvant treatment via induction of immunogenic cell death (ICD), which will be clarified below.

## 4.4 About Physical and Biochemical Mechanisms of Plasma Immunomodulation for Cancer Treatment

Cancer cells undergoing immunogenic cell death (ICD) stimulate a patient's immune system to mount an anticancer response. This is a major opportunity for plasma-based cancer immunotherapy. Underlying biochemical and biomedical mechanisms of NTP-induced ICD and relevant ICD consequences and their effect on cancer treatment are closely examined recently. Some novel experimental results in this direction (in particular, by Lin et al. [42]) are going to be shortly overviewed below.

The short-lived reactive oxygen and nitrogen species (e.g., hydroxyl radicals, atomic oxygen, nitric oxide) produced by plasma are the main effectors that elicit ICD, in particular, in melanoma. It has been demonstrated in vitro using dielectric barrier discharge DBD plasma systems, as well as validated in a vaccination assay in vivo. Plasma generation of reactive species appears to be dictated by many factors but especially by the total energy, which is a positive fact helping in defining effective dosimetry and standardization of plasma-medical treatment.

Recent advances in cancer immunotherapy have led to significant positive impacts on patient survival, especially in patients with cancers previously limited to first-line treatments. Cancer immunotherapy intends to assist a patient's natural cancer immunity cycle to fight cancer, and major success has been achieved with checkpoint inhibitors such as anti-PD-1/PDL-1 and anti-CTLA-4 therapies. However, the benefits of these treatments have been met with challenges including severe side effects and efficacy in only a subset of patients. Therefore, there is a considerable need to develop new treatment modalities that, in combination with current therapies, may help improve clinical outcomes by supporting different steps of the cancer immunity cycle. The goal of treatments should be to enable an effective, self-sustaining anticancer response in the patient.

## 4.5 Immunogenic Cell Death (ICD) as One of the Key Factors of Plasma-Stimulated Cancer Immunotherapy

There are several approaches to plasma immunomodulation for cancer treatment. Probably the most interesting approach to enhance the initial step of the cycle is to induce immunogenic cell death (ICD) in the tumor. ICD is a form of regulated cell death, characterized by the timely release of "danger signals" known as damage-associated molecular patterns (DAMPs).

Several damage-associated molecular patterns (DAMPs) have been linked to ICD (e.g., high-mobility group box 1 (HMGB1), adenosine triphosphate), the most critical and well-studied being surface-exposed calreticulin (CRT). CRT on the outer leaflet of the cell functions as an "eat-me" signal for uptake by dendritic cells. ICD inducers have been identified, and novel modalities continue to be explored. One such treatment is plasma generated at room temperature and atmospheric pressure, also known as nonthermal plasma (NTP). NTP treatment of mice has been shown to reduce tumor burden and extend survival in different cancer types.

The majority of these studies have used NTP for direct tumor cell killing or to induce cell senescence via reactive oxygen and nitrogen species (RONS)-mediated pathways. Small clinical studies with nonthermal plasma have only recently commenced for palliative and curative treatment of dermatological diseases, and to date, plasma has been effective with mild to no side effects [40, 43].

## 4.6 Immunogenic Cell Death: In Vitro, and In Vivo Studies, Understanding of Plasma-Chemical Mechanisms

Beginning in 2015, NTP has been investigated for its potential to induce ICD, and it has been reported to stimulate DAMP emission in multiple cancer cell lines in vitro [44, 45]. The first in vivo demonstration of NTP-induced ICD was performed on Balb/c mice bearing subcutaneous, syngeneic CT26 colorectal tumors [46]. Tumors treated with a dielectric barrier discharge (DBD) plasma resulted in higher expression of DAMPs (CRT and HMGB1) and increased recruitment of CD11c+ and CD45+ immune cells into the tumor environment. In combination with a therapeutic vaccine, DBD plasma treatment also enhanced cancer-specific T-cell responses [46].

However, the underlying mechanisms by which NTP elicits ICD are still not fully understood. When plasma is generated, a complex environment of reactive species, charged particles, neutral molecules, ultraviolet radiation, and electric fields is present and can interact with the biological target. Reactive species produced by the DBD plasma in the presence of oxygen were reported as the major contributors for ICD induction, not the physical components [45].

The exact chemical species that are responsible remain unclear. These include RONS with lifetimes ranging from fractions of a second (e.g., atomic species, radicals) to days and weeks (e.g., hydrogen peroxide, nitrite anion). Knowing which short-lived and persistent species are required for ICD induction will be critical to the understanding and development of NTP technology for cancer immunotherapy.

To address the above challenges in oncology and the underlying questions in plasma chemistry and biological interactions, one should delineate the RONS generated by DBD plasma that are responsible for plasma-induced ICD [42]. Knowing the species in plasma that are critical to ICD induction leads to the development of an optimized clinical device, more details on the subject can be found in [42].

Summarizing, the number of open questions in plasma immunotherapy is still very high. On the other hand, the addition of plasma systems into the existing arsenal of cancer immunotherapies opens the great possibility for new combination strategies for safer and more robust control of cancer.

## References

1. M. Keidar, D. Yan, J.H. Sherman, *Cold Plasma Cancer Therapy* (IOP Publishing, Bristol, 2018)
2. M. Laroussi, X. Lu, M. Keidar, Perspective: The physics, diagnostics, and applications of low temperature plasma sources used in plasma medicine. J. Appl. Phys. **122**, 020901 (2017)
3. M. Keidar, A. Shashurin, O. Volotskova, M.A. Stepp, P. Srinivasan, A. Sandler, B. Trink, Cold atmospheric plasma in cancer therapy. Phys. Plasmas **20**, 057101 (2013)
4. M. Laroussi, M. Kong, G. Morfill, W. Stolz (eds.), *Plasma Medicine* (Cambridge University Press, Cambridge, 2012)
5. A. Fridman, G. Friedman, *Plasma Medicine* (Wiley, New York, 2013)
6. M. Keidar, I.I. Beilis, *Plasma Engineering: Application in Aerospace, Nanotechnology and Bio-nanotechnology* (Elsevier, Oxford, 2013)
7. G. Fridman, G. Friedman, A. Gutsol, A.B. Shekhter, V.N. Vasilets, A. Fridman, Applied plasma medicine. Plasma Process. Polym. **5**, 503 (2008)
8. M. Keidar, Plasma for cancer treatment. Plasma Sources Sci. Technol. **24**, 033001 (2015)

9. D. Graves, Mechanisms of plasma medicine: coupling plasma physics, biochemistry, and biology. IEEE Trans. Radiat. Plasma Med. Sci. **1**, 281–292 (2017)
10. D. Graves, The emerging role of reactive oxygen and nitrogen species in redox biology and some implications for plasma applications to medicine and biology. J. Phys. D: Appl. Phys. **45**, 263001 (2012)
11. M. Keidar, A prospectus on innovations in the plasma treatment of cancer. Phys. Plasmas **25**, 083504 (2018)
12. J. Watson, Perspective: Oxidants, antioxidants and the current incurability of metastatic cancers. Open Biol. **3**, 120144 (2013)
13. R. Cairns, I. Harris, T.W. Mak, Regulation of cancer cell metabolism. Nat. Rev. Cancer **11**, 85–95 (2011)
14. H. Ahn, K. Kim, N. Hoan, C. Kim, E. Moon, K. Choi, S. Yang, J. Lee, Targeting cancer cells with reactive oxygen and nitrogen species generated by atmospheric-pressure air plasma. PLoS One **9**, e86173 (2014)
15. D. Trachootham, J. Alexandre, P. Huang, Targeting cancer cells by ROS-mediated mechanisms: a radical therapeutic approach? Nat. Rev. Drug Discov. **8**, 579–591 (2009)
16. M. Ishaq, M.D.M. Evans, K.K. Ostrikov, Atmospheric pressure gas plasma-induced colorectal cancer cell death is mediated by Nox2–ASK1 apoptosis pathways and oxidative stress is mitigated by Srx–Nrf2 anti-oxidant system. Biochim. Biophys. Acta Biomembr.-Mole. Cell Res. **1843**(12), 2827–2837 (2014)
17. E.A. Ratovitski, X. Cheng, D. Yan, J.H. Sherman, J. Canady, B. Trink, M. Keidar, Anti-cancer therapies of 21-st century. Novel approach to treat human cancers using cold atmospheric plasma. Plasma Process. Polym. **11**, 1128–1137 (2014)
18. G. Manda, M.T. Nechifor, T.-M. Neagu, Reactive oxygen species, cancer and anti-cancer therapies. Curr. Chem. Biol. **3**(1), 22–46 (2009)
19. P. Lukes, E. Dolezalova, I. Sisrova, M. Clupek, Aqueous-phase chemistry and bactericidal effects from an air discharge plasma in contact with water: evidence for the formation of peroxynitrite through a pseudo-second-order post-discharge reaction of H2O2and HNO2. Plasma Sources Sci. Technol. **23**, 015019 (2014)
20. S. Moncada, J. Erusalimsky, Does nitric oxide modulate mitochondrial energy generation and apoptosis? Nat. Rev. Mol. Cell. Biol. **3**, 214–220 (2002)
21. C. Szabo, H. Ischiropoulos, R. Radi, Peroxynitrite: biochemistry, pathophysiology and development of therapeutics. Nat. Rev. Drug Discov. **6**, 662–680 (2007)
22. D. Yan, A. Talbot, N. Nourmohammadi, X. Cheng, J. Canady, J. Sherman, M. Keidar, Principles of using cold atmospheric plasma stimulated media for cancer treatment. Sci. Rep. **5**, 18339 (2015)
23. N. Kurake, H. Tanaka, K. Ishikawa, T. Kondo, M. Sekine, K. Nakamura, H. Kajiyama, F. Kikkawa, M. Mizuno, M. Hori, Cell survival of glioblastoma grown in medium containing hydrogen peroxide and/or nitrite, or in plasma-activated medium. Arch. Biochem. Biophys. **605**, 102–108 (2016)
24. D. Yan, A. Talbot, N. Nourmohammadi, J. Sherman, X. Cheng, M. Keidar, Toward understanding the selectivanti-cancer capacity of cold atmospheric plasma—a model based on aquaporins. Biointerphases **10**, 040801 (2015)
25. D. Yan, H. Xiao, W. Zhu, N. Nourmohammadi, L.G. Zhang, K. Bian, M. Keidar, The role of aquaporins in the anti-glioblastoma capacity of the cold plasma-stimulated medium. J. Phys. D. Appl. Phys. **50**, 055401 (2017)
26. C.M. Doskey, V. Buranasudja, B.A. Wagner, J.G. Wilkes, J. Du, J.J. Cullen, G.R. Buettner, Tumor cells have decreased ability to metabolize H2O2: implications for pharmacological ascorbate in cancer therapy. Redox Biol. **10**, 274–284 (2016)
27. C.J. Lord, A. Ashworth, The DNA damage response and cancer therapy. Nature **481**, 287–294 (2012)
28. P.P. Connell, S. Hellman, Advances in radiotherapy and implications for the next century: a historical perspective. Cancer Res. **69**, 383–392 (2009)

29. T. Tsuruo et al., Molecular targeting therapy of cancer: drug resistance, apoptosis and survival signal. Cancer Sci. **94**, 15–21 (2003)
30. G.K. Schwartz, M.A. Shah, Targeting the cell cycle: a new approach to cancer therapy. J. Clin. Oncol. **23**, 9408–9421 (2005)
31. D.L. Longo, D. Longo, *Harrison's Hematology and Oncology* (McGraw-Hill Medical, New York, 2010), pp. 294–318
32. J. Koritzer, V. Boxhammer, A. Schafer, T. Shimizu, T.G. Klampfl, Y.-F. Li, C. Welz, S. Schwenk-Zieger, G.E. Morfill, J.L. Zimmermann, J. Schlegel, Restoration of sensitivity in chemo—resistant glioma cells by cold atmospheric plasma. PLoS One **8**(5), e64498 (2013)
33. M. Vandamme, E. Robert, S. Pesnel, E. Barbosa, S. Dozias, J. Sobilo, Antitumor effect of plasma treatment on U87 glioma xenografts: preliminary results. Plasma Process. Polym. **7**, 264–273 (2010)
34. M. Vandamme, E. Robert, S. Lerondel, V. Sarron, D. Ries, S. Dozias, ROS implication in a new antitumor strategy based on non-thermal plasma. Int. J. Cancer. **130**, 2185–2194 (2012)
35. M. Keidar, R. Walk, A. Shashurin, P. Srinivasan, A. Sandler, S. Dasgupta, Cold plasma selectivity and the possibility of a paradigm shift in cancer therapy. Br. J. Cancer **105**, 1295–1301 (2011)
36. Z. Chen, H. Simonyan, X. Cheng, L. Lin, J. Canady, J.H. Sherman, M. Keidar, A novel micro cold atmospheric plasma device for glioblastoma both in vitro and in vivo. Cancers **9**(6), 61 (2017)
37. A. Dubuc, P. Monsarrat, F. Virard, N. Merbahi, J.-P. Sarrette, S. Laurencin-Dalicieux, S. Cousty, Use of cold-atmospheric plasma in oncology: a concise systematic review. Ther. Adv. Med. Oncol. **10**, 1–12 (2018)
38. L.I. Partecke, K. Evert, J. Haugk, et al., Tissue tolerable plasma (TTP) induces apoptosis in pancreatic cancer cells in vitro and in vivo. BMC Cancer **12**, 473 (2012)
39. H.-R. Metelmann, D.S. Nedrelow, C. Seebauer, et al., Head and neck cancer treatment and physical plasma. Clin. Plasma Med. **3**, 17–23 (2015)
40. H.-R. Metelmann, C. Seebauer, V. Miller, A. Fridman, G. Bauer, D.B. Graves, J.-M. Pouvesle, R. Rutkowski, M. Schuster, S. Bekeschus, Clinical experience with cold plasma in the treatment of locally advanced head and neck cancer. Clin. Plasma Med. **9**, 6 (2018)
41. J. Canady, B. Trink, J. Sherman, M. Keidar, GWU-USMI plasma medicine research program, 5th International Workshop on Plasma for Cancer Treatment, Greifswald, Germany, March 20-21, 2018. Clin. Plasma Med. **9**, 3 (2018)
42. A. Lin et al., Non-thermal plasma as a unique delivery system of short-lived reactive oxygen and nitrogen species for immunogenic cell death in melanoma cells. Adv. Sci. **6**(6), 1802062 (2019)
43. P. Friedman, V. Miller, G. Fridman, A. Lin, A. Fridman, Successful treatment of actinic keratoses using nonthermal atmospheric pressure plasma: a case series. J. Am. Acad. Dermatol. **76**, 349 (2017)
44. A. Lin, B. Truong, A. Pappas, L. Kirifides, A. Oubarri, S. Chen, S. Lin, D. Dobrynin, G. Fridman, A. Fridman, Uniform nanosecond pulsed dielectric barrier discharge plasma enhances anti-tumor effects by induction of immunogenic cell death in tumors and stimulation of macrophages. Plasma Processes Polym. **12**, 1392 (2015)
45. A. Lin, B. Truong, S. Patel, N. Kaushik, E.H. Choi, G. Fridman, A. Fridman, V. Miller, Nanosecond-pulsed DBD plasma-generated reactive oxygen species trigger immunogenic cell death in A549 lung carcinoma cells through intracellular oxidative stress. Int. J. Mol. Sci. **18**, 966 (2017)
46. A.G. Lin, B. Xiang, D.J. Merlino, T.R. Baybutt, J. Sahu, A. Fridman, A.E. Snook, V. Miller, Non-thermal plasma induces immunogenic cell deathin vivoin murine CT26 colorectal tumors. OncoImmunology **7**, e1484978 (2018)

# Chapter 5
# Biochemistry of Plasma in Cancer Therapy

## Plasma-Derived ROS/RNS Cooperate with Redox Control Elements on the Surface of Tumor Cells During Induction of Selective Tumor Cell Death

**Georg Bauer and Sander Bekeschus**

**Contents**




**Abstract** This chapter summarizes the role of reactive oxygen/reactive nitrogen species (ROS/RNS) in multistep oncogenesis as well as a selection of central features of the composition and antitumor effects of cold atmospheric plasma (CAP)



G. Bauer (✉)
Institute of Virology, Medical Center - University of Freiburg, Freiburg, Germany
e-mail: georg.bauer@uniklinik-freiburg.de

S. Bekeschus
Leibniz Institute for Plasma Science and Technology, Greifswald, Germany





M. Keidar (ed.), *Plasma Cancer Therapy*, Springer Series on Atomic, Optical, and Plasma Physics 115, https://doi.org/10.1007/978-3-030-49966-2_5

and plasma-activated medium (PAM). It provides a compendium on important ROS/RNS sources and reactions in the context of CAP and PAM interactions with non-malignant as compared to malignant cells. As CAP- and PAM-derived ROS/RNS are directly interacting with the surface of targeted cells, the redox-related elements on the membrane of non-malignant cells, transformed cells (early stage of oncogenesis), and *bona fide* tumor cells (late stage of oncogenesis) are outlined in detail. Based on this information, the experimentally derived concept of primary singlet oxygen generation from CAP and PAM constituents and their triggering function for sustained and dynamic generation of secondary singlet oxygen by the tumor cells themselves is presented. These data allow presenting a scenario of CAP and PAM action during tumor therapy, in which CAP and PAM are the triggers, whereas the tumor cells themselves are the effectors for the induction of their cell death. Evidence for the selective damage in non-malignant over malignant cells is provided in such a context. This concept will be discussed in relation to existing concepts that are based on the assumption that CAP and PAM already contain the cell death-inducing compound at sufficient concentration to cause the observed effects.

## 5.1 Introduction

### *5.1.1 ROS/RNS in Multistep Oncogenesis*

Abrogation of senescence control, oncogene activation, and tumor suppressor gene inactivation are basic requirements for multistep oncogenesis [1–4]. These steps establish autocrine mechanisms that render the malignant cells independent of exogenous proliferation signals [5–7]. The survival and progression of the malignant cells to *bona fide* tumors strictly require achieving independence of control by neighboring cells [8–12], escape from immune surveillance [2, 3], release of prostaglandin E2 [13–17], resistance against hypoxia-induced p53-mediated cell death [18, 19], tumor angiogenesis [20], profound metabolic changes [21], and resistance towards intercellular ROS/RNS-mediated apoptosis-inducing signaling through expression of membrane-associated catalase and co-modulatory superoxide dismutase (SOD) [22–25].

ROS/RNS may act at many distinct and partially adverse steps during multistep oncogenesis [22, 23]. These are (1) mutagenic effects of ROS related to tumor initiation and tumor progression [26, 27], (2) ROS-related mechanisms during tumor promotion [27, 28], (3) proliferation control and maintenance of the transformed cell state through extracellular superoxide anion ($O_2^{\bullet -}$) generation by membrane-associated NADPH oxidase (NOX1) [22, 23, 29–51], (4) specific removal of transformed cells through intercellular apoptosis-inducing ROS/RNS signaling based on hypochlorous acid (HOCl)- and nitric acid/peroxynitrite ($^{\bullet}NO/ONOO^{-}$)-mediated apoptosis-inducing signaling pathways [22, 34, 36, 37, 39, 52–58]. ROS/RNS have been recently discussed as essential elements during the pro- and antioncogenic functions of neutrophils [59]. Interestingly, HOCl is not only the central molecule for the generation of apoptosis-inducing $^{\bullet}OH$ radicals through its reaction with $O_2^{\bullet -}$, as demonstrated for intercellular HOCl signaling [25] and discussed for antioncogenic effects of neutrophils [59]. HOCl-mediated modifications of proteins also seem to enhance the specific response of cytotoxic T cells towards tumor cells [60–65], thus enabling HOCl to act directly as well as indirectly towards tumor cells [25]. Hence, tumor development and progression can only proceed if antagonistic pro- and antioncogenic effects of ROS/RNS, as discussed in the preceding paragraph, are precisely balanced during oncogenesis. An essential step during tumor progression is to maintain the ROS-dependent proliferation stimulus, while efficient protection towards intercellular ROS/RNS signaling is established in parallel [13–17, 22, 23].

The classical tumor progression studies in vivo by Deichman and colleagues show that tumor cells acquire resistance towards exogenous $H_2O_2$ [13–17]. Later, the expression of membrane-associated catalase by tumor cells was recognized as the basis for such a "$H_2O_2$-catabolizing phenotype" of tumor cells [24, 25, 66–71]. Expression of membrane-associated catalase as a regular marker of tumor cells confers resistance towards intercellular apoptosis-inducing ROS/RNS signaling. It thus ensures the survival of these cells, as it efficiently interferes with the intercellular ROS/RNS-mediated apoptosis-inducing signaling pathways that are driven by the strong expression of NOX1 in the membrane of malignant cells [22, 23]. Due to its (frequently ignored) enzymatic versatility, membrane-associated catalase not only decomposes $H_2O_2$ but also protects towards RNS signaling through oxidation of $^{\bullet}NO$ and decomposition of $ONOO^-$ [22, 23, 68, 72]. Protection towards intercellular ROS/RNS signaling by membrane-associated catalase is a regular feature of *bona fide* human tumor cells derived from many organs [23]. These aspects are discussed in more detail in Sect. 5.2.2.

Initially, the presence of membrane-associated catalase on the surface of tumor cells and its biological consequence, i.e. the effective protection towards exogenous ROS/RNS, seemed to be in direct contradiction to classical findings: tumor cells had been found to contain less catalase than nontransformed cells [73–79], leading to the assumption that tumor cells should react more sensitive to exogenous ROS/RNS as compared to healthy tissue. Despite its unclear mechanistic basis, this view has been the starting point for many studies, also in plasma research, though it has been proven to be incorrect by direct measurements [68, 69].

What is the resolution of this paradoxical situation? Early studies [73–79] had quantified the overall concentration of catalase in extracts from tumor cells and non-malignant cells. Although the covalent attachment of catalase to the surface of tumor cells [68, 69] establishes a high local concentration of catalase, the total amount of catalase required to establish this effect is relatively low. Therefore, this amount of membrane-associated catalase does not change the *overall low catalase concentration* of tumor cells, as this is dominated by the low intracellular catalase concentration in combination with the relatively large volume of the cells.

The high local concentration of catalase on the membrane efficiently protects from exogenous ROS/RNS. It also drastically lowers the aquaporin-mediated influx of extracellular $H_2O_2$ and thus seems to represent an efficient gatekeeper for aquaporins. Therefore, apoptosis induction in tumor cells by a direct $H_2O_2$ effect (in the absence of HOCl signaling) requires relatively high concentrations of exogenous $H_2O_2$, inhibition of membrane-associated catalase and functional aquaporins (Bauer, unpublished result).

In order to maintain an influx of extracellular $H_2O_2$ (derived from $O_2^{\bullet -}$ generated by NOX) that is sufficiently high for proliferation stimulation, tumor cells need to express high concentrations of aquaporins, as will be discussed later. By contrast, the low catalase concentration inside the cells seems to be involved in the enhancement of ROS-mediated intracellular signaling and potentially also in enhanced genetic instability and cell damage by intracellular ROS/RNS. In line with this conclusion, overexpression of intracellular catalase in tumor cells decreases tumor cell proliferation in vitro and in vivo [80].

As catalase has one of the highest turnover numbers of all enzymes, it is not difficult to determine the activity of membrane-associated catalase of tumor cells by a functional assay. This assay is based on a challenge by exogenous $ONOO^-$ and is not affected by intracellular catalase [69, 81]. In contrast, the detection of the relatively low concentration of catalase protein on the cell surface by biophysical methods (i.e., determination of structure rather than function) requires a high level of sensitivity. Nevertheless, the structural demonstration of membrane-associated catalase located specifically on the surface of the human tumor cell lines MKN-45 (gastric carcinoma), CHP-134 (neuroblastoma), and IPC-238 (melanoma), but not on non-malignant cells, was provided using immunofluorescence microscopy [68]. Further confirmation of membrane-associated catalase was obtained through FACS analysis of MKN-45 gastric carcinoma cells and *ex vivo* isolates from intramuscular, as well as lymph node metastasis of human malignant melanoma [Bauer, unpublished result].

These aspects are of particular importance for the understanding of the selective apoptosis induction in tumor cells in vitro and in vivo by cold atmospheric plasma and plasma-activated medium, as these complex sources of ROS/RNS interact directly with redox control-related elements on the surface of non-malignant and malignant cells [24, 81–86].

### 5.1.2 Cold Atmospheric Plasma and Plasma-Activated Medium

The gaseous and liquid phases of cold atmospheric plasma (CAP) contain electrons, photons, and ROS/RNS such as atomic oxygen (O), superoxide anions ($O_2^{\bullet -}$), hydroperoxyl radicals ($HO_2^{\bullet}$), hydrogen peroxide ($H_2O_2$), hydroxyl radicals ($^{\bullet}OH$), singlet delta oxygen ($^1O_2$) (commonly called "singlet oxygen"), ozone ($O_3$), nitric oxide ($^{\bullet}NO$), nitrogen dioxide ($^{\bullet}NO_2$), peroxynitrite ($ONOO^-$), nitrite ($NO_2^-$), nitrate ($NO_3^-$), dichloride radicals ($Cl_2^{\bullet -}$), and hypochlorite anions ($OCl^-$), as summarized in [24, 82, 83, 87–89]. As will be shown in Sect. 5.2.1, these species have multiple reaction potentials, resulting in a scenario with ROS/RNS differing significantly in their free diffusion path lengths, life times, and biochemical potentials. The treatment of media with CAP results in the generation of plasma-activated medium (PAM), which seems to maintain many of the biological effects of CAP, although it only contains long-lived species from CAP [90–93]. The multiple effects of CAP- and PAM-derived ROS/RNS on biological systems triggered the establishment of the new field of plasma medicine [88, 94–107]. Thereby the promising antitumor effects of CAP and PAM in a wide variety of tumor systems in vitro and in vivo are particularly in the focus of ongoing research [99–102, 108–117]. Encouraging results after the first clinical application of CAP for tumor therapy have already been reported [103–106].

CAP and PAM were found to act selectively on malignant target cells in vitro and in vivo [91, 118–129]. However, it deserves the recognition that some investigators also reported on nonselective apoptosis-inducing effects of CAP or PAM [130–134]. This discrepancy may be explained by differences in the applied doses, as suggested by Bauer [81, 83] and discussed in more detail below. Whereas many CAP-derived species are characterized by a high reactivity and short live time, PAM (as well as other plasma-activated liquids) primarily contains only $H_2O_2$, $NO_2^-$, and $NO_3^-$ [90–93, 135]. Intriguingly, a ROS/RNS composition of such a low complexity can trigger profound antitumor effects in many tumor systems in vitro and in vivo [90–92, 119, 124, 127–129, 136–146]. Canal et al. [128] and Koensgen et al. [144] found that CAP and PAM caused equivalent antitumor effects in vitro, whereas Saadati et al. [147] reported that CAP induced a stronger antitumor effect in vivo when compared to PAM. Yan et al. [148] reported that CAP triggers $H_2O_2$ generation from tumor cells, whereas PAM did not. Selective effects of PAM towards tumor cells have been repeatedly reported [91, 124, 127, 128, 129]. Ikeda et al. [145] present evidence that PAM kills human cancer-initiating cells. PAM affects 3D multicellular tumor spheroids in vitro [149, 150] and is also effective in vivo [90, 137, 129, 147]. It seems that the ROS/RNS contained in PAM are responsible for apoptosis induction in tumor cells and that PAM treatment also elicits a ROS/RNS response of the target cells [90–92, 136, 138, 142].

As tumor cells express high concentrations of aquaporins in their membrane [151], Yan et al. [152, 153] proposed that an increased influx of $H_2O_2$ through aquaporins determines the selective effect of PAM on tumor cells. Girard et al. and Kurake et al. [91, 92] reported that a synergistic *effect* between PAM-contained $H_2O_2$ and $NO_2^-$ is required for the selective antitumor action of PAM. Reconstitution experiments with defined concentrations of $H_2O_2$ and $NO_2^-$ confirmed the requirement for a synergistic interaction between $H_2O_2$ and $NO_2^-$ in order to cause apoptosis induction selectively in tumor cells [81, 84]. The interaction between $H_2O_2$ and $NO_2^-$ comprises the reaction between a multitude of reactive species to eventually generate primary $^1O_2$, followed by local inactivation of membrane-associated catalase on tumor cells [81, 84]. This results in auto-amplificatory $^1O_2$ generation by the tumor cells and inactivation of their membrane-associated catalase. The inactivation of membrane-associated catalase has two major consequences for the survival of the tumor cells: (1) the influx of extracellular $H_2O_2$ through aquaporins is now possible and leads to the depletion of intracellular glutathione and (2) and intercellular apoptosis-inducing ROS/RNS signaling is activated [81, 84, 85]. Kinetic inhibitor studies showed that the influx of $H_2O_2$ through aquaporins after inactivation of membrane-associated catalase and the subsequent depletion of intracellular glutathione were not sufficient for apoptosis induction per se, but were strictly required as an intracellular sensitization step of the cells for apoptosis induction through extracellular ROS/RNS signaling [81, 84, 85]. Whereas $H_2O_2$ was directly responsible for the sensitization step, it was not sufficient for the apoptosis-inducing step under the conditions of selective apoptosis induction by CAP or PAM in tumor cells. Rather, peroxidase-mediated conversion of $H_2O_2$ to HOCl, followed by HOCl/ $O_2^{\bullet-}$ interaction was necessary. This finding was established through the inhibition profile of the process [81, 84, 85]. However, the aquaporin-dependent, $H_2O_2$-mediated sensitization step is in perfect agreement with the finding by Yan et al. [152, 153], who have shown that inhibition of aquaporins has a dominant inhibitory effect on CAP- and PAM-mediated apoptosis induction. In contrast to Yan‘s model, our experimental findings show the requirement for catalase inactivation and for intercellular ROS/RNS signaling for efficient CAP- and PAM-mediated apoptosis induction in tumor cells. However, we strongly agree with Yan et al., on the strict requirement for an aquaporine-mediated, $H_2O_2$-driven step, which according to our measurements, represents a $H_2O_2$-mediated sensitization step, based on intracellular glutathione depletion. This aquaporine-mediated, $H_2O_2$-driven sensitization step is not only required for the action of CAP and PAM [81, 84–86], but also for ROS/RNS-mediated apoptosis induction in tumor cells through inhibition or inactivation of membrane-associated catalase by catalase inhibitors, exogenous singlet oxygen generated by an illuminated photosensitizer, or singlet oxygen generated by tumor cells after modulation of their endogenous $^{\bullet}NO$ level [Bauer, unpublished result].

Freund et al. [150] presented strong evidence that plasma-treated saline generated with a plasma jet promotes an immunogenic phenotype of colon cancer cells. In contrast, Lin et al. [154] concluded from their study using a DBD that PAM was not sufficient to induce immunogenic cell death. Under DBD conditions, short-lived species in CAP seemed to be required for the induction of immunogenic cell death. The discrepancy between these two studies is possibly explained by the low degree of induction of cell death by PAM in the study by Lin et al. [154], and by the difference in plasma sources and cell types used.

Using a portable air plasma “corona pen” plasma source [155], Bauer et al. [85, 86] have shown that long-lived species were sufficient to trigger selective apoptosis induction in tumor cells after CAP and PAM treatment. (In the case of CAP treatment, but not after PAM treatment, an initial minor role of short-lived species was also seen [85, 86]). The responsible long-lived species were $NO_2^-$ and $H_2O_2$, which is in line with the results published by Girard et al. and Kurake et al. [91, 92]. The nonlinear, auto-amplificatory effects of CAP and PAM were established in a subsequent study [86]. One exciting result from this study [85] and from the confirming reconstitution experiments with pure $NO_2^-$ and $H_2O_2$ [81, 84] is the finding that the interaction between the long-lived species $H_2O_2$ and $NO_2^-$, in conjunction with the contributions of the specific redox system of tumor cells, leads to the site-specific generation of defined short-lived species that are instrumental for the overall success of CAP- or PAM-dependent selective antitumor effects. The precise biochemical steps of this novel mechanism [81, 84, 85, 86] are outlined in the next sections in detail.

## 5.2 Main Part

### *5.2.1 Basic Lesson on ROS/RNS Generation and Interactions*

During the last years, the view on ROS/RNS has been changing from the very narrow perception of merely damaging agents to a more balanced view of a complex biochemical entity. This is spanning from signaling to control functions, and finally, to potentially harmful effects. There are excellent reviews on the chemistry and chemical biology of ROS/RNS [87, 114, 156–162]. The understanding of the role of ROS/RNS for therapeutic approaches, based on treatment with CAP, requires the notion of essential modes of ROS/RNS generation and interactions. For instance, site-specific interactions of ROS/RNS are of particular interest, as many determining reactions need to occur in the close vicinity to the membrane of target cells.

Figure 5.1 is starting with the generation of $O_2^{\bullet-}$ by NADPH oxidase-1 (NOX1), which is typical for cells after malignant transformation. $O_2^{\bullet-}$ may also be generated due to the dysfunction of the mitochondrial respiratory chain—an aspect that is especially important during execution of the mitochondrial pathway of apoptosis. $O_2^{\bullet-}$ are also essential elements in the phagosome of neutrophils [163] and are contained in CAP. In addition to their potential for multiple interactions with other ROS/RNS, $O_2^{\bullet-}$ can inhibit catalase [25]. Spontaneous or SOD-catalyzed dismutation of $O_2^{\bullet-}$ leads to the generation of $H_2O_2$, one of the most versatile members of the ROS family [164, 165]. SOD-catalyzed dismutation of NOX1-derived $O_2^{\bullet-}$ also prevents $O_2^{\bullet-}$-dependent inhibition of membrane-associated catalase and thus ensures the protection of tumor cells towards exogenous ROS/RNS.

$H_2O_2$ has the potential for multiple specific signaling functions at low concentrations [164, 165]. Extracellular $H_2O_2$ generated by malignant cells enters the cells through aquaporins and drives their proliferation [22]. $H_2O_2$ is efficiently decomposed by catalase, as well as by glutathione peroxidase, in the presence of glutathione. The interaction of $H_2O_2$ with ferrous irons is known as the Fenton reaction. It results in the generation of highly reactive $^{\bullet}OH$. As $H_2O_2$ is a far-ranging species, the damaging effect of the Fenton reaction is not site-specific and therefore relatively weak. The interaction of $H_2O_2$ with peroxidases like MPO, peroxidasins, or Duox (Dual oxidase) leads to HOCl (or other hypohalous acids) [25]. HOCl may undergo Fenton reaction much more efficiently than $H_2O_2$ [166]. Importantly, HOCl reacts with $O_2^{\bullet-}$ to yield $^{\bullet}OH$ [167]. This reaction is the basis for site-specific $^{\bullet}OH$ generation in close vicinity to the membrane of $O_2^{\bullet-}$-producing malignant cells. This site-specific generation of $O_2^{\bullet-}$ causes much stronger effects than the diffuse

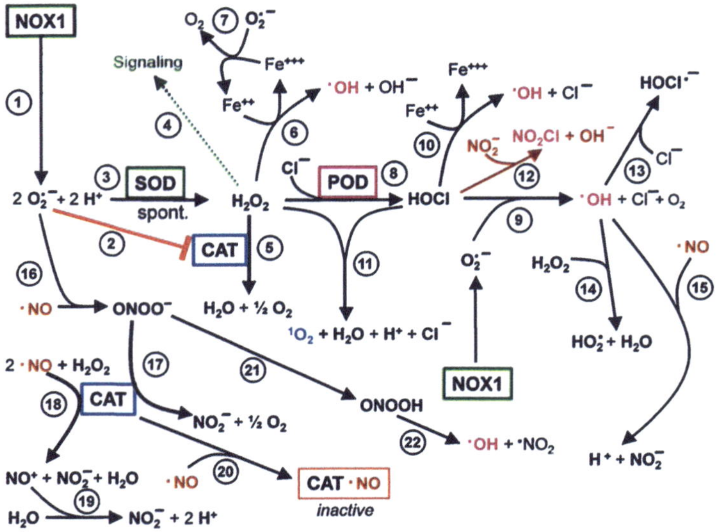

**Fig. 5.1** Central ROS/RNS reactions relevant to oncogenesis. Membrane-associated NADPH oxidase 1 (NOX1) generates extracellular $O_2^{\bullet-}$ (#1), which may inhibit catalase, a central regulatory enzyme (#2). Dismutation of $O_2^{\bullet-}$ by SOD (#3) leads to $H_2O_2$ and prevents $O_2^{\bullet-}$-dependent inhibition of catalase. In the absence of SOD, $O_2^{\bullet-}$ may dismutate spontaneously, though at a much slower rate. $H_2O_2$ exerts many signaling functions at low concentrations [164, 165] (#4). The level of $H_2O_2$ can be efficiently controlled by catalase (#5). Iron-catalyzed Fenton reaction yields damaging $^{\bullet}OH$ (#6, 7). $H_2O_2$ is used by certain peroxidases (MPO, peroxidasin, DUOX) as a substrate for the generation of HOCl (#8). HOCl can generate $^{\bullet}OH$ site-specifically after interaction with $O_2^{\bullet-}$ (#9), undergo Fenton chemistry much more efficiently than $H_2O_2$ (#10) or react with $H_2O_2$, yielding $^1O_2$ (#11). The interaction between HOCl and $NO_2^-$ yields nitrosyl chloride ($NO_2Cl$), an apoptosis-inducing and nitrating species (#12). $^{\bullet}OH$ derived from HOCl/$O_2^{\bullet-}$ interaction or Fenton chemistry can be scavenged by chloride (#13), $H_2O_2$ (#14), or $^{\bullet}NO$ (#15). The interaction between $O_2^{\bullet-}$ and $^{\bullet}NO$ yields $ONOO^-$ (#16). $ONOO^-$ can be decomposed by catalase (#17). Catalase also oxidizes $^{\bullet}NO$ (#18, 19) and thus prevents the formation of $ONOO^-$. This reaction is endangered by $^{\bullet}NO$-mediated inhibition of catalase (#20). In the vicinity to proton pumps, $ONOO^-$ is protonated and the resultant peroxynitrous acid yields $^{\bullet}OH$ (#21–22). This figure has been reprinted from Bauer [25] with permission from Elsevier

Fenton chemistry. Alternatively, HOCl may react with $NO_2^-$ to form nitryl chloride ($NO_2Cl$), a highly reactive compound with apoptosis-inducing potential [168]. The interaction between HOCl and $H_2O_2$ yields $^1O_2$ [169]. Therefore, in excess of the substrate $H_2O_2$, HOCl generation by peroxidase is biologically ineffective, as the reaction product HOCl is consumed by residual $H_2O_2$ [25, 67, 68]. $^{\bullet}OH$ generated through HOCl/$O_2^{\bullet-}$ interaction may induce lipid peroxidation and thus trigger the mitochondrial pathway of apoptosis [22, 25, 34]. Alternatively, they may be scavenged by $Cl^-$, $^{\bullet}NO$, or $H_2O_2$.

The interaction between $O_2^{\bullet-}$ and $^{\bullet}NO$ is one of the fastest reactions in chemical biology and leads to the formation of $ONOO^-$ [39, 170–174]. If $ONOO^-$ is generated in close vicinity to membrane-associated proton pumps, peroxynitrous acid is formed [72]. This compound readily decomposes spontaneously into $^{\bullet}OH$

and $^{\bullet}NO_2$ [171, 175]. In biological systems, and especially on the surface of tumor cells, $ONOO^-$ as well as $^{\bullet}NO$ are under control of catalase, which may oxidize $^{\bullet}NO$ [72, 176, 177] and decompose $ONOO^-$ [68, 178, 179]. There is a critical balance between oxidation of $^{\bullet}NO$ by catalase and reversible inhibition of catalase by $^{\bullet}NO$ [176, 180]. This balance is determined (1) by the concentration of $^{\bullet}NO$ and (2) by the availability of compound I of catalase, which is formed through the reaction between catalase and $H_2O_2$ or $ONOO^-$. $^{\bullet}NO$-mediated inhibition of catalase can be used for antitumor effects that utilize the modulation of the endogenous $^{\bullet}NO$ concentration of tumor cells [72].

NOX1-derived $O_2^{\bullet-}$ play a determinative role in the generation of $H_2O_2$ and $ONOO^-$ (Fig. 5.2). The concentration of $^{\bullet}NO$ in cellular systems is determined by the availability of arginine (the substrate for NO synthase (NOS)) and the concentration of arginase [181]. The activity of NOS determines the flux of generated NO, which is diminished by $^{\bullet}NO$ dioxygenase (NOD) [182–184], an enzyme that converts $^{\bullet}NO$ into $NO_3^-$ and that can be inhibited by many secondary

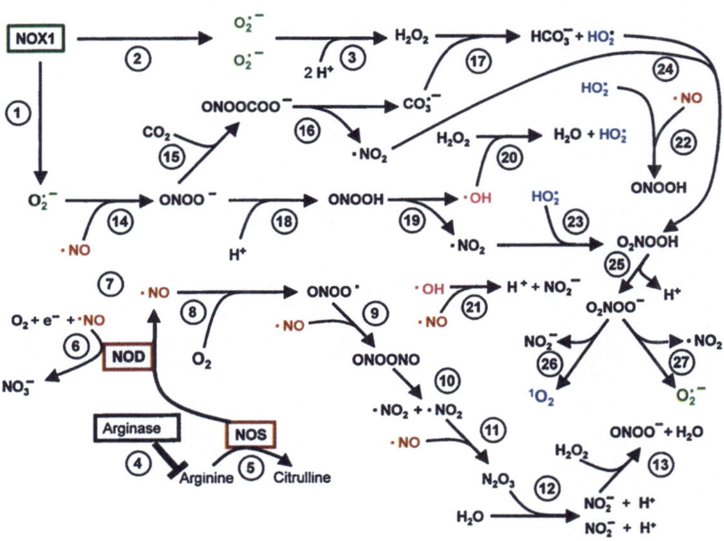


Fig. 5.2 Central reactions of $^{\bullet}NO$ and its derivatives. NOX1-derived $O_2^{\bullet-}$ are essential for the pathway leading to $ONOO^-$ (#1) or $H_2O_2$ (#2, 3). The concentration of free $^{\bullet}NO$ is controlled by arginase, arginine, $^{\bullet}NO$ synthase (NOS) (#4, 5), consumption by $^{\bullet}NO$ dioxygenase (NOD) (#6) and oxidation by molecular oxygen (#7–12). $ONOO^-$, formed through $^{\bullet}NO/O_2^{\bullet-}$ interaction (#14), has a high chance of reacting with $CO_2$, leading to the formation of nitrosoperoxycarboxylate (#15). Decomposition of nitrosoperoxycarboxylate yields $^{\bullet}NO_2$ and carbonate radicals (#16), which may interact with $H_2O_2$ to yield hydroperoxyl radicals ($HO_2^{\bullet}$) (#17). Protonation of $ONOO^-$ yields peroxynitrous acid (#18), which may decompose into $^{\bullet}OH$ and $^{\bullet}NO_2$ (#19). $^{\bullet}OH$ may induce lipid peroxidation and apoptosis (not shown in the figure), or react with $H_2O_2$ to yield hydroperoxyl radicals (#20) or with $^{\bullet}NO$ to yield $NO_2^-$ (#21). The interaction between $^{\bullet}NO$ and hydroperoxyl radical ($HO_2^{\bullet}$) yields peroxynitrous acid (ONOOH) (#22), whereas the interaction between $^{\bullet}NO_2$ and $HO_2^{\bullet}$leads to the formation of peroxynitric acid ($O_2NOOH$) (#23, 24). Deprotonation of peroxynitric acid ($O_2NOOH$) yields peroxynitrate ($O_2NOO^-$) (#25) which may decompose into $NO_2^-$ and $^1O_2$ (#26) or $^{\bullet}NO_2$ and $O_2^{\bullet-}$ (#27)

plant compounds and by azoles [72]. The concentration of $^{\bullet}NO$ is efficiently reduced through oxidation of $^{\bullet}NO$ by molecular oxygen ($O_2$), eventually resulting in $NO_2^-$. Interestingly, there is also evidence for the generation of $^{\bullet}NO$ by $NO_2^-$ [185]. Whereas the interaction of $^{\bullet}NO$ with molecular oxygen ($O_2$) requires several steps to yield $NO_2^-$ [186], the reaction between $^{\bullet}NO$ and $O_2^{\bullet-}$ to yield $ONOO^-$ is fast and comprises only one step [170–174]. In biological systems, the reaction between $ONOO^-$ and $CO_2$ is regarded as the favored reaction [187–189]. The reaction product nitrosoperoxycarboxylate ($ONOOCOO^-$) decomposes into $^{\bullet}NO_2$ and carbonate radicals ($CO_3^{\bullet-}$), whose interaction with $H_2O_2$ yields hydroperoxyl radicals ($HO_2^{\bullet}$). In the vicinity of proton pumps, protonation of $ONOO^-$ can outcompete its reaction with $CO_2$ [72]. The resultant peroxynitrous acid then decomposes into $^{\bullet}NO_2$ and $^{\bullet}OH$. These may cause lipid peroxidation and subsequent apoptosis induction [68, 71, 72]. Alternatively, these $^{\bullet}OH$ may be scavenged by $^{\bullet}NO$ or $Cl^-$, react with $H_2O_2$ to yield $HO_2^{\bullet}$, or react with other suitable molecules within a range of 40 nm. The reaction between $HO_2^{\bullet}$ and $^{\bullet}NO_2$ leads to the formation of peroxynitric acid ($O_2NOOH$) [190]. As soon as this compound is deprotonated to peroxynitrate ($O_2NOO^-$), spontaneous decomposition yields either $O_2^{\bullet-}$ and $^{\bullet}NO_2$, or $^1O_2$ and $NO_2^-$. It is important to keep in mind that $ONOO^-$ is much more stable than peroxynitrous acid (ONOOH), whereas in the case of peroxynitrate ($O_2NOO^-$) and peroxynitric acid ($O_2NOOH$), the acidic form is more stable [190].

If $^1O_2$ is generated in close vicinity to catalase or SOD, it can inactivate these enzymes through reaction with histidine at their active centers [191, 192]. This reaction seems to be crucial for several antitumor effects, including those triggered by CAP and PAM [24, 81, 82, 84–86], singlet oxygen generation by illuminated photosynthesizers [193–195] or modulation of the endogenous $^{\bullet}NO$ concentration [72, 193]. Alternatively, $^1O_2$ can activate the Fas receptor [72] which then activates caspase-8. As a consequence in malignant cells, this results in caspase-8-dependent enhancement of NOX1 and NOS. This has a substantial positive impact on the generation of secondary $^1O_2$ and ROS/RNS signaling [24, 71, 72, 81, 85]. It usually leads to autoamplificatory inactivation of membrane-associated catalase of tumor cells, reactivation of intercellular apoptosis-inducing ROS/RNS signaling and apoptotic cell death. This mechanism seems to be sufficiently efficient even if the concentration of FAS receptor is relatively low, as shown for SKN-MC Ewing sarcoma cells in the supplement to [24]. In the case of tumor cells with high FAS receptor expression, like SHEP neuroblastoma cells, treatment with high concentrations of $^1O_2$ derived from illuminated photofrin caused $^1O_2$-mediated activation of the canonical FAS receptor-dependent pathway of cell death [24].

### 5.2.2 Redox Elements on the Membranes of Non-malignant and Malignant Cells

Non-malignant cells carry few redox-relevant elements on their surface (Fig. 5.3A). Aquaporins allow the entry of exogenous $H_2O_2$ [196, 197]. Non-malignant cells

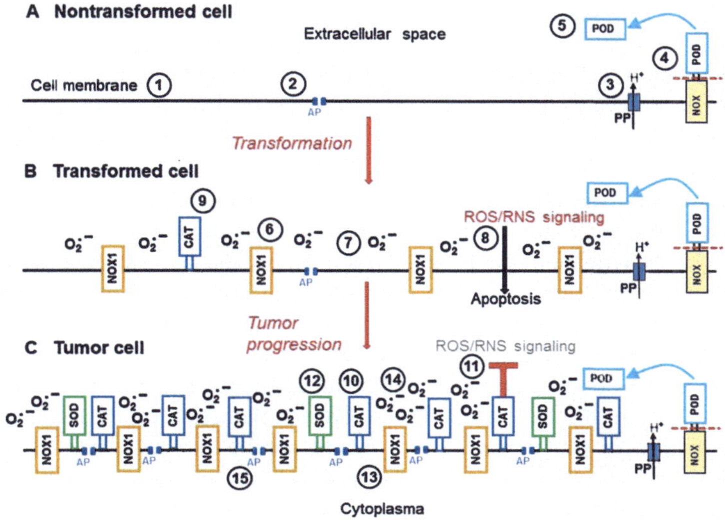


**Fig. 5.3** Redox-relevant membrane-changes during transformation and tumor progression. (**A**) The cell membrane (#) of nontransformed cells carries aquaporins (AP) (#2), proton pumps (#3), and Dual oxidase (DUOX) (#4) that is composed of a NOX domain and a peroxidase (POD) domain. The POD can be released (#5) from DUOX through matrix metalloproteases (MMP). (**B**) After malignant transformation, membrane-associated NADPH oxidase (NOX1) (#6) is expressed and generates extracellular $O_2^{\bullet-}$ (#7) in a sustained mode. Their dismutation product $H_2O_2$ enters the cells through aquaporins and drives their proliferation (not shown in the figure). NOX1-derived $O_2^{\bullet-}$ also control two intercellular ROS/RNS-mediated apoptosis-inducing signaling pathways (#8) that cause apoptosis induction in transformed cells. The expression of low concentrations of catalase (#9) on the surface of transformed cells is not sufficient to protect the cells towards apoptosis induction by their own ROS/RNS. (**C**) Cells from the late stage of tumor progression, i.e. *bona fide* tumor cells, express membrane-associated catalase at a high local concentration (#10) and therefore are protected towards apoptosis-inducing ROS/RNS signaling (#11). Co-modulatory SOD (#12) prevents $O_2^{\bullet-}$-mediated inhibition of catalase and thus warrants the full protection of the tumor cells. Tumor cells show increased NOX1 expression (#13) and enhanced production of extracellular $O_2^{\bullet-}$ (#14). Compared to non-malignant and transformed cells, the concentration of aquaporins (#15) in the membrane of tumor cells is strongly enhanced and their intracellular concentration of catalase is lowered (not shown in the figure). These two features allow sufficient intracellular $H_2O_2$-dependent proliferation stimulus despite the presence of high concentrations of membrane-associated catalase

from different lineages express Dual oxidase (DUOX), which is composed of a NOX and a peroxidase (POD) domain [25]. DUOX-associated NOX does not generate $O_2^{\bullet-}$, but only deficient concentrations of $H_2O_2$. The presence of active TGFβ increases the expression of DUOX and causes the release of the peroxidase domain through the activity of matrix metalloproteases [58, 198, 199]. As non-malignant cells do not generate substantial concentrations of exogenous $O_2^{\bullet-}$ and $H_2O_2$, the presence of released peroxidase has no adverse effect on their survival. Peroxidase released from non-malignant cells can supplement distant transformed cells with peroxidase, thus allowing or enhancing their extracellular HOCl synthesis. It thus contributes to the selective elimination of malignant cells [58]. The presence of HOCl-generating peroxidase is essential to avoid damaging effects of $H_2O_2$-generating lactobacilli on non-malignant cells while establishing a very efficient control system towards malignant cells at the same time [200]. In line with the hypothesis of Klebanoff, this mechanism has been suggested to explain the tumor-preventive effect of vaginal lactobacilli [200].

Transformation by many different agents regularly leads to the expression of NOX1 and subsequent sustained generation of extracellular $O_2^{\bullet-}$ (Fig. 5.3B) [22, 23, 29–51]. Their dismutation product $H_2O_2$ enters the cells through aquaporins and drives their proliferation. This principle is a striking example of autocrine growth control of malignant cells. On the flip side of the coin, NOX1-derived $O_2^{\bullet-}$ drive the HOCl and the $^{\bullet}NO/ONOO^-$ signaling pathways that selectively cause apoptosis in NOX1-expressing malignant cells but not in their nontransformed neighbors [22, 34, 36, 37, 39, 52–58]. Apoptosis induction by these intercellular signaling pathways might represent a biochemically defined, ROS/RNS-driven control system to prevent tumor formation. It is an interesting open question whether this primarily nonimmunological control system activates specific T cell-mediated control through immunogenic cell death (ICD). In order to escape this control, effective tumor formation requires the malignant cells to maintain their ROS-based autocrine proliferation stimulation, while in parallel efficiently interfering with intercellular apoptosis-inducing signaling. Galina Deichman has shown that the transition from a $H_2O_2$-sensitive phenotype to a $H_2O_2$-resistant phenotype is a regular phenomenon in experimental tumor progression in vivo [13–17]. This correlates with a substantial increase in tumorigenicity of the cells in a new round of experiments. The resistance defined by Deichman et al. seems to be selected for during tumor progression. This is explained by the expression of membrane-associated catalase [68] (Fig. 5.3C). As catalase is inhibited by free $O_2^{\bullet-}$ [25], catalase on the surface of NOX1 expressing tumor cells would not be functional. Expression of co-modulatory SOD on the membrane of tumor cells ensures the functionality of catalase by lowering the local concentration of $O_2^{\bullet-}$ [23, 201]. However, whereas the control of ROS/RNS signaling by membrane-associated catalase is tight, the concentration of SOD on the surface is only sufficient to reduce the $O_2^{\bullet-}$ concentration to a level where it is no longer inhibitory for adjacent catalase [201]. Therefore, despite the presence of SOD on their surface, tumor cells still provide sufficient $O_2^{\bullet-}$ to allow interaction with HOCl, resulting in the formation of $^{\bullet}OH$ radicals [201]. As membrane-associated catalase is decomposing $H_2O_2$ very efficiently, tumor cells would pay for their protection by catalase with a decrease in the concentration of their self-generated proliferation stimulus $H_2O_2$. The increased $O_2^{\bullet-}$ generation of tumor cells compared to transformed cells, as well as the increase of aquaporins in the membrane of tumor cells, may reflect the adequate response of the tumor cells to partially compensate for the decrease in available $H_2O_2$ through catalase action. The lower intracellular catalase level inside tumor cells compared to non-malignant cells may act along the same lines. This leads to the paradoxical situation in which the major part of the tumor cells contains lower concentrations of catalase than non-malignant cells, whereas the membrane of tumor cells carries a barrier of locally highly concentrated catalase. Therefore, *bona fide* tumor cells may react more sensitive to intracellular ROS signaling and also may encounter mutagenic effects by intracellular ROS more efficiently than non-malignant cells. This results in genetic instability that provides advantages for the survival, adaptation, and expansion of tumor cells. However, the response of tumor cells towards extracellular ROS/RNS is determined by their membrane-associated catalase and SOD that are firmly associated with the surface of the tumor cells and effectively protects them towards exogenous ROS/RNS attacks [13–17, 22–25, 66–71]. Membrane-associated catalase and SOD on tumor cells are especially relevant for the interaction between CAP or PAM-derived ROS/RNS and tumor cells, as this represents a very direct interaction between a complex extracellular source of ROS/RNS and the surface of tumor cells.

### Details of Intercellular ROS/RNS-Mediated Intercellular Signaling

In malignant cells of early stages, i.e. transformed cells, or in tumor cells with inactivated or inhibited catalase, NOX1-derived $O_2^{\bullet -}$ drive the HOCl or the $NO/ONOO^-$ signaling pathway (Fig. 5.4). Both signaling pathways depend on $O_2^{\bullet -}$ and are finalized by lipid peroxidation through $^{\bullet}OH$. These trigger the mitochondrial pathway of apoptosis [202]. The cell death-inducing potential of both pathways is explicitly directed towards the membrane of the NOX1-expressing malignant cells, as $O_2^{\bullet -}$ have a relatively short free diffusion path length of few μm. In addition, the function of the $^{\bullet}NO/ONOO^-$ signaling pathway is dependent on membrane-associated proton pumps. Both pathways depend on the density of the cells. It has been recognized that these two pathways seem to act mutually exclusive. This finding can be explained by a complex network of iron-catalyzed interactions [25, 203]. These interactions suppress $^{\bullet}NO$-dependent processes in the presence of abundant $H_2O_2$ and vice versa. In addition, several side reactions following some of the reactions shown in Figs. 5.1 and 5.2 are slowing down the intercellular apoptosis-inducing pathways. Interfering activities are oxidation of $^{\bullet}NO$ by molecular oxygen, consumption of $ONOO^-$ by $CO_2$, scavenging of $^{\bullet}OH$ by $^{\bullet}NO$, $Cl^-$, and $H_2O_2$, the interaction between $H_2O_2$ and HOCl, and others (please find more details in [25, 203]).

### Details of Tumor Cell Protection Towards ROS/RNS Signaling

The protection of *bona fide* tumor cells from intercellular apoptosis-inducing signaling is based on the expression of membrane-associated catalase and co-modulatory SOD (Fig. 5.5) [22–25, 66–69, 72]. The attachment of catalase on the membrane generates kinetic advantages for protection, as it provides a high local concentration of the enzyme at the site where it is needed. This effect is particularly relevant for protection towards $^{\bullet}NO/ONOO^-$ signaling, where all relevant processes like (1) formation of $ONOO^-$ through the interaction between tumor cell-derived $^{\bullet}NO$ and

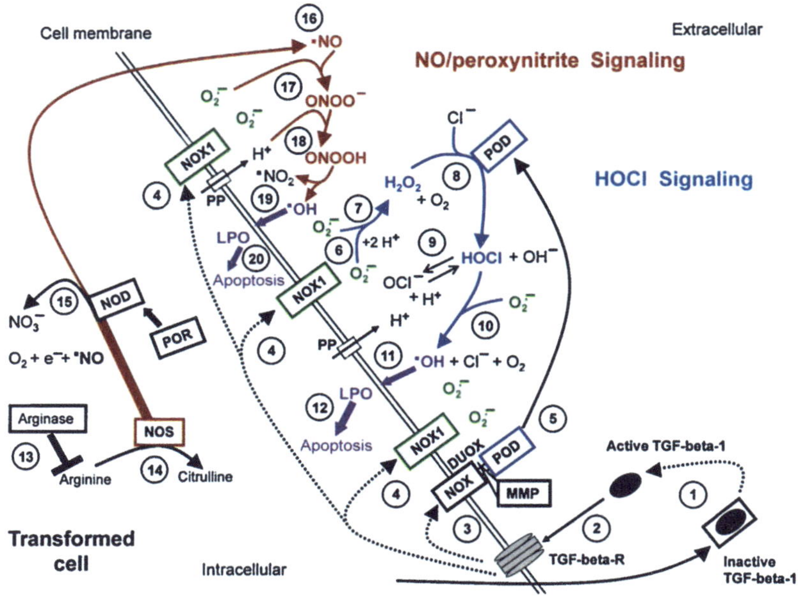


**Fig. 5.4** Intercellular ROS/RNS-mediated signaling through the HOCl and the NO/ONOO$^-$ signaling pathway. The HOCl and the $^{\bullet}$NO/ONOO$^-$ signaling pathway are established by transformed cells or by tumor cells with inhibited or inactivated catalase. Active TGF-beta controls the activity of NOX1 (#1–4). Matrix metalloprotease (MMP) causes the release of the peroxidase domain (POD) (#5) from DUOX (composed of NOX and POD). Generation of extracellular $O_2^{\bullet-}$ (#6), their dismutation to $H_2O_2$ (#7), and POD-mediated generation of HOCl (#8) are the basis for the HOCl signaling pathway. Proton pump-derived protons help to maintain a high local level of HOCl compared to hypochlorite anions (#9). The reaction between HOCl and $O_2^{\bullet-}$ yields $^{\bullet}$OH (#10) which induce lipid peroxidation (#11) and apoptosis induction by the mitochondrial pathway of apoptosis (#12). The concentration of extracellular $^{\bullet}$NO is controlled by steps #13–16. $O_2^{\bullet-}$ interact with $^{\bullet}$NO, resulting in the formation of ONOO$^-$ (#17). Close to the membrane, protonation of ONOO$^-$ is favored (#18) compared to the competing interaction between ONOO$^-$ and $CO_2$ (not shown). The decomposition of peroxynitrous acids yields apoptosis-inducing $^{\bullet}$OH (#19, 20). The selectivity of HOCl and $^{\bullet}$NO/ONOO$^-$ signaling is assured through the short free diffusion path length of $O_2^{\bullet-}$ and $^{\bullet}$OH and the presence of proton pumps. This figure has been reprinted from Bauer [25] with permission from Elsevier

$O_2^{\bullet-}$, (2) protonation of ONOO$^-$ through membrane-associated proton pumps, (3) decomposition of the resulting ONOOH into $^{\bullet}$OH and $^{\bullet}NO_2$, and (4) lipid peroxidation by $^{\bullet}$OH is occurring directly in close vicinity to the cell membrane. Protection towards this process can only be achieved by catalase that is located closely to the membrane and that oxidizes $^{\bullet}$NO in competition with $^{\bullet}NO/O_2^{\bullet-}$ interaction and that decomposes ONOO$^-$ in competition with the generation and decomposition of ONOOH. Experimental analysis shows that it only requires a minor reduction in the concentration of membrane-attached catalase by siRNA-mediated knockdown or by catalase inhibition in order to allow reactivation specifically of $^{\bullet}$NO/ONOO$^-$

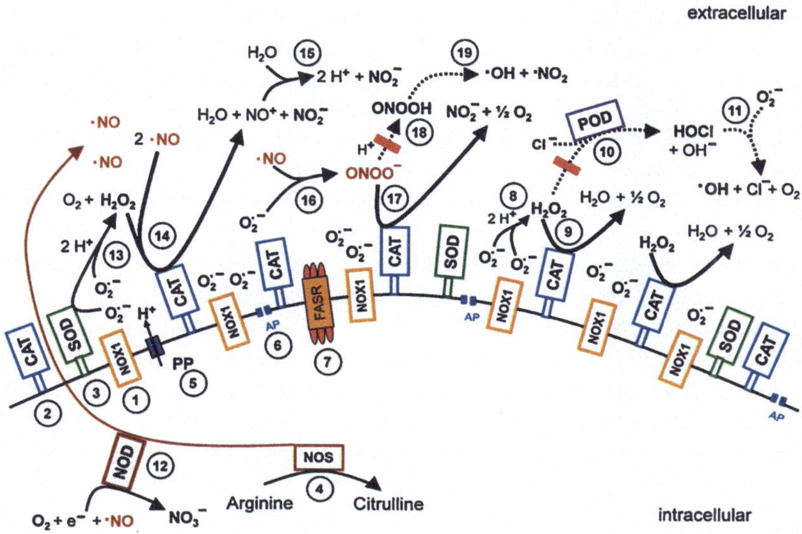


**Fig. 5.5** Protection of tumor cells towards ROS/RNS signaling. Tumor cells express membrane-associated NOX1, catalase, and SOD (#1–3). Intracellular •NO synthase (NOS) generates •NO that passes the membrane (#4). Further important elements in the membrane are proton pumps (#5), aquaporins (#6), and the Fas receptor (#7). NOX1-derived extracellular $O_2^{\bullet-}$ dismutate to $H_2O_2$ (#8). The decomposition of $H_2O_2$ by catalase (#9) prevents apoptosis induction through the HOCl pathway (#10, 11). NOD consumes •NO (#12). The dismutation of $O_2^{\bullet-}$ by SOD (#13) prevents $O_2^{\bullet-}$-mediated inhibition of catalase and yields $H_2O_2$ that is required for the oxidation of •NO by catalase (#14, 15). The reaction between •NO and $O_2^{\bullet-}$yields $ONOO^-$ (#16). The decomposition of $ONOO^-$ by catalase prevents apoptosis induction through •NO/$ONOO^-$ signaling (#18, 19)

signaling [204]. However, reverting the effect of minor siRNA-mediated knockdown of membrane-associated catalase on •NO/$ONOO^-$ signaling through addition of soluble catalase requires to add very high concentrations of catalase. These high concentrations of soluble catalase substituted for the missing catalase molecules in close vicinity to the membrane, whereas the vast excess of soluble catalase distant of the membrane seemed to be without relevance. Reactivation of HOCl signaling required higher concentrations of catalase inhibitors or a higher degree of siRNA-mediated knockdown of membrane-associated catalase than reactivation of •NO/$ONOO^-$ signaling, but the reversal of this process was achieved by lower concentrations of added soluble catalase than inhibition of •NO/$ONOO^-$ signaling [204]. This paradoxically seeming effect on HOCl signaling is explained by the facts that HOCl synthesis after catalase inhibition (1) is taking place within the diffusion range of the long-lived and wide-ranging species $H_2O_2$ by the soluble peroxidase domain derived from DUOX after the action of matrix metalloproteases, (2) that the resultant HOCl is wide-ranging as well, and (3) that only the generation of •OH through interaction between HOCl and $O_2^{\bullet-}$ is required to occur close to the membrane. Therefore, this process can be experimentally inhibited by lower concentrations of soluble catalase than $^{\bullet}NO/ONOO^{-}$ signaling, though intact tumor cells are utilizing membrane-associated catalase rather than soluble catalase for inhibition of both signaling pathways. Therefore, attempts to analyze the protective function of membrane-associated catalase by mathematical modeling of the effects of soluble catalase (Bengtson and Bogaerts, submitted) are not likely to lead to a result that is congruent with reality, as they neglect the necessity for site specificity of the biological system that has been experimentally established and confirmed [204].

Attachment of catalase to the membrane (or extracellular matrix) is mediated by transglutaminase [69]. When transglutaminase was inhibited, soluble catalase was found in the supernatant of tumor cells [69], indicating that catalase is not anchored in the lipid bilayer of the membrane itself, but seems to be attached to membrane-associated proteins by transglutaminase. The exact binding partner for catalase on the membrane has not yet been determined. The concentration of membrane-associated catalase is controlled by the tumor cells through monitoring the concentration of $H_2O_2$. Therefore, experimental decomposition of $H_2O_2$ or inhibition of $O_2^{\bullet-}$ generation (thus preventing $H_2O_2$ generation through dismutation) over 12–14 h causes complete depletion of membrane-associated catalase [69]. In line with these conclusions, scavenging of $O_2^{\bullet-}$ by SOD did not cause depletion of membrane-associated catalase, as this process caused the generation of $H_2O_2$ [69]. The latter finding confirms that $H_2O_2$ is the controlling factor, whereas $O_2^{\bullet-}$ are the precursers for $H_2O_2$, but are not directly involved in the monitoring process.

Catalase interferes efficiently with HOCl and $^{\bullet}NO/ONOO^{-}$ signaling. This requires several distinct activities of this crucial enzyme [68, 24, 25]. As shown in Fig. 5.6, enzymatic decomposition of $H_2O_2$ by catalase in a two-step reaction prevents HOCl synthesis and in this way blunts HOCl signaling. Catalase interferes with $^{\bullet}NO/ONOO^{-}$ signaling through two distinct activities. $^{\bullet}NO$ is oxidized by compound I of catalase in a two-step reaction with compound II as intermediate. This prevents (1) inhibition of catalase by $^{\bullet}NO$ and (2) $^{\bullet}NO/O_2^{\bullet-}$ interaction leading to the formation of $ONOO^{-}$. Eventually formed $ONOO^{-}$ is then decomposed by catalase by the same enzymatic mechanism as decomposition of $H_2O_2$, with compound I as active intermediate [68]. The enzymatic identity of these two decomposition steps has been elucidated through the determination of the active compound I during decomposition of $H_2O_2$ as well as the decomposition of $ONOO^{-}$ [68] This was achieved by utilizing the potential of compound I to preferentially oxidize methanol, leading to formaldehyde that was quantitatively determined by its reaction with purpald reagent (4-amino-3-hydrazino-5-mercapto-1,2,4-triazole). $H_2O_2$ and $ONOO^{-}$ were found to generate compound I with nearly equal efficiency [68].

The utilization of several substrates and the performance of multiple activities is not restricted to catalase. It has also been shown for other heme-containing enzymes, such as MPO [25]. These findings show the remarkable flexibility of redox-relevant enzymes, which allows their adequate responses under varying conditions. It is time to bring this information about the plasticity of enzyme activities into the center

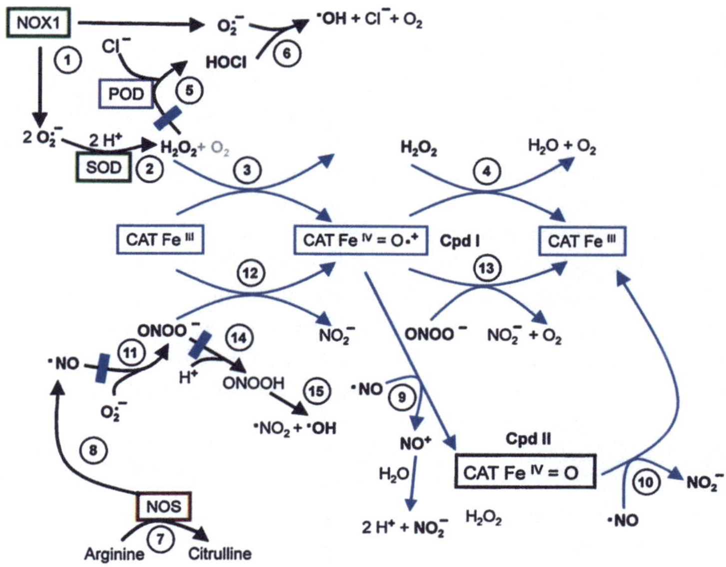


**Fig. 5.6** Multiple enzymatic functions of catalase. SOD dismutates NOX1-derived $O_2^{\bullet-}$ (#1) to $H_2O_2$ (#2). This reaction prevents the inhibition of catalase by $O_2^{\bullet-}$. Decomposition of $H_2O_2$ by catalase, with compound I (Cpd I) as intermediate (#3, 4), prevents HOCl synthesis and the subsequent generation of $^{\bullet}$OH (#5, 6). $^{\bullet}$NO is generated by NOS (#7, 8) and is oxidized by compound I of catalase in a two-step reaction involving compound II (Cpd II) (#9. 10). Eventually, formed $ONOO^-$ is decomposed by catalase in a two-step reaction with CpdI as intermediate (#12. 13). This prevents the protonation of $ONOO^-$ and $^{\bullet}$OH generation through the decomposition of peroxynitrous acid (#14, 15). This figure has been reprinted from Bauer [25] with permission from Elsevier

of our conceptional considerations, rather than to stick to outdated conventional concepts that strictly assign only one activity to one enzyme. This historically explained view on the activity of certain enzymes, especially on catalase, has often hampered our conceptional progress in the past.

The extracellular localization of catalase on the membrane of tumor cells has been established by (a) a specific immunofluorescence staining of living tumor cells, but not of non-malignant cells, using antibodies directed towards catalase on the cell surface, (b) a substantial decrease of catalase-mediated protection of tumor cells towards exogenous ROS/RNS after trypsinization, (c) sensitization of tumor cells for $H_2O_2$ or $ONOO^-$-dependent apoptosis induction as well as for autocrine ROS-mediated apoptosis induction through inhibition of catalase with blocking antibodies directed against catalase, (d) efficient protection of tumor cells against apoptosis induction by exogenous $ONOO^-$ [68, 69] and abrogation of this protection by blocking antibodies directed towards catalase, (e) release of catalase into the extracellular space after inhibition of transglutaminase, and (f) inactivation of protective catalase by extracellular $^1O_2$. The strength of these arguments is particularly based on the established (and experimentally confirmed) finding that antibodies are not penetrating into intact cells. The site-specific interaction between membrane-associated catalase and exogenous $ONOO^-$ has been exploited for the establishment of a test system in vitro that allows quantifying the degree of inhibition or inactivation of membrane-associated catalase [69, 193]. This system was crucial for the elucidation of the effects of CAP and PAM on tumor cells.

The high local concentration of catalase on the membrane of tumor cells, and the relatively low concentration of the enzyme within the tumor cell perfectly meet the three ROS-related requirements for tumorigenesis: (1) The membrane-associated extracellular catalase protects the tumor cells against the apoptosis-inducing effects of their own apoptosis-inducing ROS signaling. (2) The low intracellular catalase facilitates the carcinogenic effects of ROS, i.e. the induction of proliferation and genomic instability. (3) As the compartment with relatively high local concentration of catalase in close vicinity of the cell membrane represents a minor part of the total amount of catalase of the tumor cell, the finding of high concentration of catalase on the outside of the membrane is consistent with the established finding that the average concentration of catalase in tumor cells is usually lower than in non-malignant cells [73–79]. Importantly, tumor cells are more affected by endogenous ROS/RNS than non-malignant cells due to their low intracellular catalase concentration. However, they are more protected towards exogenous ROS/RNS than non-malignant cells, due to membrane-associated catalase.

## Inhibition and Inactivation of Tumor Cell-Protective Catalase Induces a Nonlinear, Auto-Amplificatory, Dynamic Process

Inhibition of membrane-associated catalase by pharmaceutical inhibitors such as 3-aminotriazole (3-AT) or salicylic acid, or by blocking antibodies, Fab fragments, and single-domain antibodies, causes reactivation of intercellular apoptosis-inducing ROS/RNS signaling [68, 69, 193, 205]. The same effect can be achieved by the siRNA-mediated knockdown of catalase or knockdown of transglutaminase that is required for the attachment of catalase to the membrane [69, 204]. Furthermore, inhibition of SOD by blocking antibodies leads to indirect inhibition of catalase through free NOX1-derived $O_2^{\bullet -}$ [205]. Under all these conditions, the mode of inhibition allows to affect many catalase molecules on the membrane and therefore causes a more or less synchronized beginning of intercellular signaling.

It was observed that a lower degree of overall inhibition of membrane-associated catalase first allows the reactivation of $^{\bullet}NO/ONOO^-$ signaling. Further increase in catalase inhibition causes an increase of the $H_2O_2$ concentration and subsequent HOCl signaling, but at the same time, interfered with $NO/ONOO^-$ signaling [25, 203]. Interestingly, some tumor cell systems like neuroblastoma, ovarian carcinoma, mammary carcinoma, and small lung cell carcinoma never reach HOCl signaling even at maximal catalase inhibition. They died from $^{\bullet}NO/ONOO^-$ signaling at all concentrations of the catalase inhibitor and did not establish HOCl signaling. This finding is explained by the generation of relatively high concentrations of $^{\bullet}NO$ compared to $H_2O_2$.

When relatively limited or localized effects inhibit membrane-associated catalase of tumor cells, for instance, a local increase of $^{\bullet}NO$ or $O_2^{\bullet-}$, a subsequent and direct reactivation of intercellular apoptosis-inducing ROS/RNS signaling is not possible. The concentration of $^{\bullet}NO$ can be enhanced through inhibition of NOD, inhibition of arginase, induction of NOS expression by interferons or the Fas receptor [72], whereas a high local concentration of $O_2^{\bullet-}$ can be achieved through local inhibition of SOD by excess $H_2O_2$. Under these rare conditions of $NO^-$ or $O_2^{\bullet-}$-mediated catalase inhibition, a $^1O_2$-mediated mechanism is required for propagation of catalase inactivation (Fig. 5.7A). The local inhibition of a few catalase molecules by $^{\bullet}NO$ or $O_2^{\bullet-}$ allows for the local generation of $^1O_2$ through a chain of reactions that starts with the complex interaction between free $ONOO^-$ and free $H_2O_2$ at the site of inactivated catalase (see Eqs. (5.1), (5.2) [171, 206], (5.3) [207], (5.4) [188, 190], (5.5) and (5.6) [190, 208]):

$$ONOO^- + H^+ \rightarrow ONOOH \tag{5.1}$$

$$ONOOH \rightarrow {}^{\bullet}NO_2 + {}^{\bullet}OH \tag{5.2}$$

$${}^{\bullet}OH + H_2O_2 \rightarrow H_2O + HO_2^{\bullet} \tag{5.3}$$

$${}^{\bullet}NO_2 + HO_2^{\bullet} \rightarrow O_2NOOH \tag{5.4}$$

$$O_2NOOH \rightarrow O_2NOO^- \tag{5.5}$$

$$O_2NOO^- \rightarrow {}^1O_2 + NO_2^- \tag{5.6}$$

$$O_2NOO^- \rightarrow O_2^{\bullet-} + {}^{\bullet}NO_2 \tag{5.7}$$

$^1O_2$ generated according to reaction (5.6) may then inactivate a neighboring catalase molecule and thus start a new round of $^1O_2$ generation. This auto-amplificatory process is further enhanced by $^1O_2$-dependent activation of the Fas receptor, which activates caspase-8 and subsequently NOX1 and NOS. Finally, a substantial number of catalase molecules on the surface of the tumor cells are inactivated and subsequently, ROS/RNS-mediated apoptosis-inducing signaling is initiated. This process is induced by NOD inhibitors like anthocyanidins, diallyl disulfide, certain azoles, by interferons, activation of the Fas receptor, and inhibition of arginase [72, 193, 194]. Tumor cell treatment with the direct application of a $^1O_2$-generating photosensitizer [194] does not allow a substantial degree of direct catalase reactivation because of the short free diffusion path length of the generated (primary) $^1O_2$. Also, under these conditions, the subsequent generation

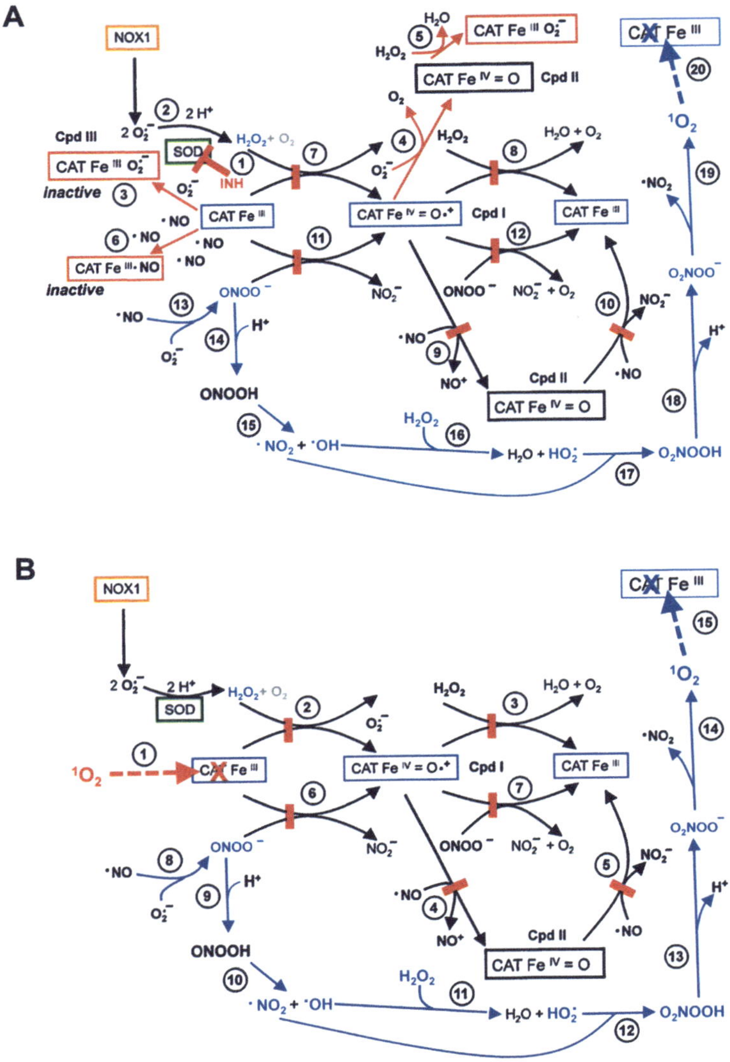


**Fig. 5.7** Local inhibition of catalase of tumor cells leads to the generation of $^1O_2$. A. High concentrations of exogenous $H_2O_2$ cause local inactivation of SOD (#1) on the surface of tumor cells [71]. This leads to an increase in the concentration of free $O_2^{\bullet-}$ (#2). Dismutation of $O_2^{\bullet-}$ to $H_2O_2$ is restricted to spontaneous dismutation under these conditions. Free $O_2^{\bullet-}$ inhibit catalase through the direct formation of compound III (CpdIII) (#3) or reduction of compound I to compound II (#4), followed by conversion of CpdII to the terminally inactive CpdIII (#5). The local increase in the concentration of $^{\bullet}NO$ causes reversible inhibition of catalase (#6). Inhibition of catalase by $O_2^{\bullet-}$ or by $^{\bullet}NO$ interferes with catalase-specific reaction steps #7–12. As a result, $ONOO^-$ formation (#13) is favored due to higher concentrations of available $^{\bullet}NO$. $ONOO^-$ is not decomposed by catalase but gets protonated (#14). The resultant $^{\bullet}OH$ and $^{\bullet}NO_2$ (#15) establish the generation of $^1O_2$ through reactions # 16-19. $^1O_2$ inactivates neighboring catalase molecules on the membrane of the tumor cells and thus leads to auto-amplification of $^1O_2$

of secondary $^1O_2$ by the tumor cells is the driving force for the substantial catalase inactivation (Fig. 5.7B). Then, intercellular ROS/RNS-mediated apoptosis signaling is reactivated and causes substantial apoptotic cell death. The dissection of these experimental systems shows that the dynamics of auto-amplification of $^1O_2$ generation are not restricted to the cell that had been originally targeted, but rather spreads in a bystander effect-like signaling cascade within the cell population [82, 84, 86].

### *5.2.3 The Interaction Between CAP and PAM-Derived ROS/RNS with the Surface of Non-malignant and Malignant Cells*

Our studies on the biological effects of defined ROS/RNS towards non-malignant and malignant cells [22, 23, 34, 35, 37, 39, 69, 72, 194], model experiments applying defined long-lived species of CAP and PAM on tumor cells and controls [81, 84], and the analysis of the biological ROS/RNS-related effects of a Corona plasma pen [85, 86], in combination with data on the free diffusion path length of CAP and PAM-derived species [209–213] allow to draw experimentally grounded conclusions on the interaction between CAP and PAM-derived ROS/RNS on non-malignant cells and tumor cells.

The strength of these experiments is the application of functional assays, in which defined species were removed from the system through scavengers or inhibitors of their source, and the effect of this treatment on the reaction studied was determined. This type of analysis was combined with kinetic analysis and the definition of distinct steps, which were studied separately. Finally, reconstitution experiments with defined compounds were used to verify or falsify the previously drawn conclusions. These reconstitution experiments were counter-controlled by inhibitor/scavenger studies to ensure their validity.

As shown in Fig. 5.8, non-malignant cells covered by cell culture medium in vitro, or covered by biological materials in vivo, are not likely to be reached by highly reactive, but short-lived species like $^{\bullet}OH$, $O_2^{\bullet -}$, $^1O_2$, or $O_3$. In contrast, $H_2O_2$, a long-lived and prominent species both in CAP and PAM, has a good chance to reach the cells and intrude into the cytoplasm through aquaporins. Several intracellular defense mechanisms would then counteract the intruding $H_2O_2$, with

**Fig. 5.7** (continued) generation and catalase inactivation. B. Direct inactivation of catalase by exogenous $^1O_2$ leads to the generation of secondary $^1O_2$. Direct inactivation of tumor cell-associated catalase by $^1O_2$ derived from an illuminated photosensitizer (#1) interferes with reaction steps # 2–7. This allows for the generation of secondary $^1O_2$ through reaction steps #10–14, followed by inactivation of further catalase molecules (#15) and the onset of auto-amplificatory $^1O_2$ generation

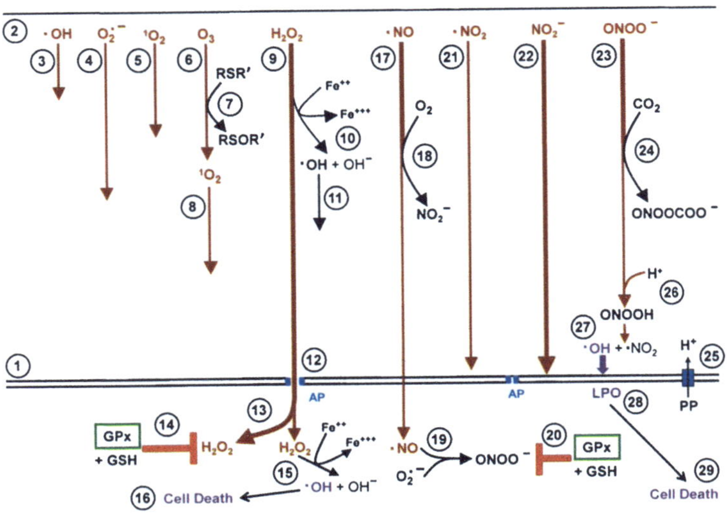

**Fig. 5.8** Interaction of CAP with non-malignant cells. The figure shows the cell membrane of non-malignant cells (#1) and the border of overlaying medium or biological material (#2). CAP-derived species like $^{\bullet}OH$ (#3), $O_2^{\bullet-}$ (#4), or $^1O_2$ (#5) do not reach the membrane of the cell due to their high reactivity and therefore short free diffusion path length. $O_3$ (#6) also does not reach the cells. Its reaction with specific aminoacids (#7) may yield $^1O_2$ (#8) that is too far to reach the membrane. $H_2O_2$, a long-lasting species from CAP and PAM (#9), may undergo a certain degree of Fenton chemistry (#10) which does not affect the cells due to the long distance between the resultant $^{\bullet}OH$ (#11) and the cells. $H_2O_2$ reaches the membrane and intrudes into the cells through aquaporins (#12). Inside the cells, $H_2O_2$ is either decomposed by glutathione peroxidase and glutathione (#13, 14), intracellular catalase (not shown) or causes Fenton chemistry and cell death (#15, 16). $^{\bullet}NO$ (#17) is efficiently oxidized by molecular oxygen (#18) in a multistep reaction, finally leading to $NO_2^-$. The surviving $^{\bullet}NO$ passes through the membrane and may generate intracellular $ONOO^-$ (#19), which is decomposed by glutathione peroxidase/glutathione. $^{\bullet}NO_2$ (#21), $NO_2^-$ (#22), and $NO_3^-$ (not shown) reach the cell membrane and cause no consequences. The concentration of $ONOO^-$ (#23) is lowered through its reaction with $CO_2$ (#24). Protonation of residual $ONOO^-$ by proton pumps (#25) at the membrane yields peroxynitrous acid (ONOOH) (#26) and subsequently apoptosis-inducing $^{\bullet}OH$ (#28, 29)

glutathione peroxidase and reduced glutathione the probably most effective one. This would diminish the concentration of $H_2O_2$ that yields cell death-inducing $^{\bullet}OH$ after iron-catalyzed Fenton reaction. The concentration of $H_2O_2$ applied would, therefore, determine between survival or cell death.

Though $^{\bullet}NO$ is not particularly long-lived, it is conceivable that the population of $^{\bullet}NO$ that is not oxidized by molecular oxygen on its move towards the cells enters the cells through passing the membrane. As nontransformed cells do not generate substantial concentrations of extracellular $O_2^{\bullet-}$ (in contrast to their transformed counterparts), $ONOO^-$ formation through the interaction between $^{\bullet}NO$ and $O_2^{\bullet-}$ cannot occur on the outside of the cell membrane but will be restricted to the intra-cellular space, possibly determined by mitochondria-derived $O_2^{\bullet-}$. The resultant $ONOO^-$ will be readily decomposed by glutathione peroxidase and glutathione

[214]. Therefore, $^{\bullet}NO$ over a wide concentration range is not expected to induce cell death in non-malignant cells, in line with experimental analysis [39, 215]. However, higher concentrations of $^{\bullet}NO$ in the submillimolar concentration range most probably have a negative impact on the mitochondrial function of the cells.

$NO_2^-$ and $NO_3^-$ have a good chance to reach non-malignant cells but remain without significant biological effect. The generation of $^{\bullet}NO$ through $NO_2^-$ acidification by proton pumps is not likely to reach critical levels for the survival of the cells.

$ONOO^-$ derived from CAP and PAM will be partially consumed through reaction with $CO_2$ or by oxidizing biological materials but has a chance to reach the surface of the cells. This assumption is based on the experimental finding of successful apoptosis induction in non-malignant cells by exogenous $ONOO^-$ in vitro [39]. Protonation of $ONOO^-$ close to the surface will lead to the formation of unstable peroxynitrous acid, followed by its decomposition into $^{\bullet}OH$ and $^{\bullet}NO_2$. Lipid peroxidation by $^{\bullet}OH$ then triggers the onset of the mitochondrial pathway of apoptosis that is finalized by the action of caspase-3.

Therefore, as soon as an apoptosis-inducing level of $H_2O_2$ (and to a minor degree of $ONOO^-$) is reached by CAP or PAM treatment, non-malignant cells will readily undergo apoptosis, potentially followed by secondary necrosis. Further increase in dosage can be predicted to induce necrosis directly.

- Therefore, any regimes to induce selective antitumor action of CAP or PAM have to ensure that the critical concentrations of $H_2O_2$ (and potentially also $ONOO^-$) are not reached during treatment. It is rational to assume that the few reports on plasma treatment of tumors that did not detect selective action of plasma towards tumor cells might have been performed in a dose range where $H_2O_2$ or $ONOO^-$ concentrations were sufficient to induce cell death in non-malignant cells.

Analysis of the application of CAP or PAM-derived ROS/RNS to tumor cells shows on the first sight, that species like $H_2O_2$, $ONOO^-$, and $^{\bullet}NO$ that are potentially endangering the survival of non-malignant cells are efficiently decomposed or oxidized by membrane-associated catalase on the membrane of tumor cells (Fig. 5.9) [68, 69].

- *Therefore, direct action of $H_2O_2$ cannot explain selective CAP or PAM action directed towards tumor cells.* This conclusion was verified in experiments where i) tumor cells were shown to be much less sensitive towards the action of $H_2O_2$ and $ONOO^-$ than non-malignant cells, and that ii) inhibition of membrane-associated catalase by non-cell-permeable, blocking antibodies sensitized the tumor cells towards the apoptosis-inducing potential of these compounds [68, 69].

$^1O_2$ and $O_3$ have the potential to inactivate catalase [191, 192, 216, 217], but due to their high reactivity, $^1O_2$ and $O_3$ from CAP have no realistic chance to reach the cells under usual experimental conditions. Lowering the layer of cell culture medium above the cells or direct application on tumor material during surgery might change this situation and allow $^1O_2$ or $O_3$ interaction with the protective catalase.

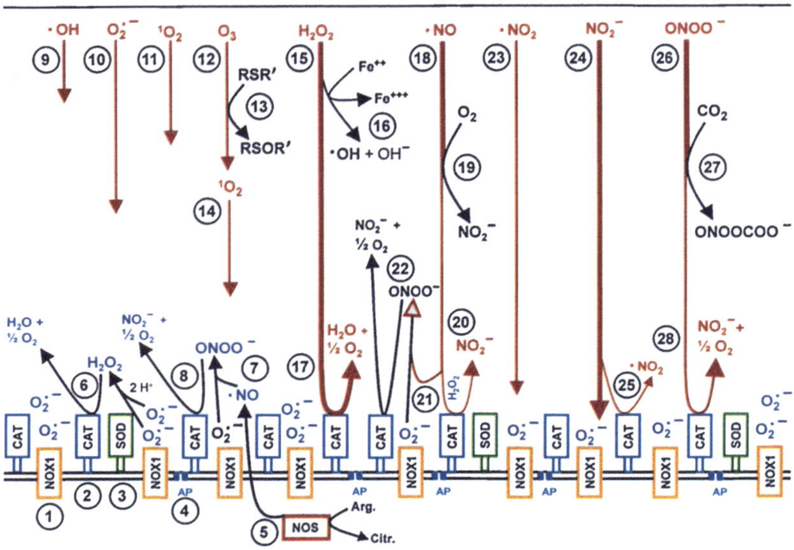

**Fig. 5.9** Interaction of CAP with tumor cells. *Bona fide* tumor cells are characterized by high expression of NOX1, leading to high local concentrations of extracellular $O_2^{\bullet -}$ (#1), and by the expression of membrane-associated catalase (#2) that protects the tumor cells towards damaging effects of their own ROS/RNS. Co-modulatory SOD (#3) prevents $O_2^{\bullet -}$-mediated inhibition of catalase. Tumor cells show higher expression of aquaporins (AP) in their membrane (#4) than non-malignant cells. NOS generates $^{\bullet}NO$ that may pass the membrane (#5). $H_2O_2$ derived from dismutation of $O_2^{\bullet -}$ is decomposed by catalase (#6). Catalase also oxidizes $^{\bullet}NO$ (not shown in the Figure). $ONOO^-$ eventually formed through the reaction between $^{\bullet}NO$ and $O_2^{\bullet -}$ (#7) is readily decomposed by catalase (#8). Thus, tumor cell-derived ROS/RNS (edited in blue color) are under stringent control by catalase/SOD. Under experimental conditions in vitro or during tumor treatment in vivo, short-ranging species like $^{\bullet}OH$, $O_2^{\bullet -}$, $^1O_2$, and $O_3$ most likely have no chance to reach tumor cells. Also, wide-ranging but more or less unreactive species like $^{\bullet}NO_2$, $NO_2^-$, and $NO_3^-$ are probably not affecting the tumor cells. Concerning these two groups of species, there is no noticeable difference when the interaction with tumor cells is compared to that of non-malignant cells. However, species like $H_2O_2$ (#15), $^{\bullet}NO$ (#18), and $ONOO^-$ (#26) that potentially induce apoptosis in non-malignant cells (Fig. 5.8) can be expected to have no negative impact on tumor cells over a wide range of concentrations, as they are decomposed (like $H_2O_2$ and $ONOO^-$, #17, # 28) or oxidized (like $^{\bullet}NO$, #20)

## 5.2.4 *Selective Antitumor Effects Through Synergistic Mechanisms*

How, then, can CAP and PAM elicit selective antitumor activity under experimental conditions and during tumor treatment? The solution to this central problem was initiated by the findings by Kurake et al. and Girard et al. [91, 92]. Both groups presented evidence that PAM can induce selective antitumor effects analogous to CAP, though it only contains long-lived species like $H_2O_2$, $NO_2^-$, and $NO_3^-$. Furthermore, both groups demonstrated that the antitumor action by PAM was necessarily dependent on a synergistic effect between $H_2O_2$ and $NO_2^-$. This synergistic effect is the key to understand selective action of CAP and PAM towards tumor cells, as it is functional at concentrations of $H_2O_2$ that (1) are neither damaging nontransformed nor tumor cells, but (2) only affects tumor cells, based on their specific redox elements on their surface.

The finding by Kurake et al. and Girard et al. are contrasted by the report by Van Boxem et al. [218], who concluded that $H_2O_2$ alone has the dominant function for antitumor action of PAM. The conclusions by Van Boxem et al. [218] are not supported by the experimental findings on the stronger effects of $H_2O_2$ on non-malignant compared to tumor cells [69], whereas their model predicts a more effective killing of tumor cells by $H_2O_2$, which is not the case. Kurake et al. and Girard et al. [91, 92] suggested that $ONOO^-$ generated through the reaction between $NO_2^-$ and $H_2O_2$ [219] might play a role for the observed antitumor effects. Figure 5.10 extends this suggestion and shows that despite the decomposition of $H_2O_2$ at the surface of the tumor cells through catalase, $ONOO^-$ formation through $H_2O_2/NO_2^-$ interaction should be possibly distant from the membrane. As long as the newly generated $ONOO^-$ is out of reach of membrane-associated catalase, it has a chance to react with residual $H_2O_2$, leading to the generation of $^1O_2$. This is the key for the understanding of the synergistic effect between $H_2O_2$ and $NO_2^-$.

DiMascio et al. [220] have originally demonstrated the formation of $^1O_2$ through the interaction between $H_2O_2$ and $ONOO^-$ and hypothesized that $^1O_2$ is generated via the direct Russell reaction between the two compounds according to the formula

$$H_2O_2 + ONOO^- \rightarrow {}^1O_2 + H_2O + NO_2^-$$

Later it was shown that the interaction is more complex and requires several ROS/RNS-driven steps [24, 71, 81, 84–86]. These will be discussed in more detail in the next figure. $^1O_2$ generated through the reaction between $H_2O_2$ and $ONOO^-$ will not affect the tumor cells if it is generated distant of the membrane (Fig. 5.10). However, if $ONOO^-/H_2O_2$ interaction occurs just out of reach of bound catalase, but close enough to the membrane, the resultant $^1O_2$ (# 16 in Fig. 5.10) has a chance to reach and inactivate a catalase molecule. It is evident that decomposition of $H_2O_2$ and $ONOO^-$ by catalase will be dominant and therefore minimize the generation and effects of $^1O_2$. Therefore, it is doubtful that $^1O_2$ generated in this way can cause direct damage or cell death to the cells. However, as at the site of inactivated catalase, $H_2O_2$ and $ONOO^-$ that are produced continuously by tumor cells are no longer decomposed, the rare local effect of primary $^1O_2$ generated by CAP or PAM components is translated into the sustained generation of secondary $^1O_2$. Secondary $^1O_2$ then inactivates neighboring catalase molecules on the same tumor cell and neighboring tumor cells in an auto-amplificatory mode and with the characteristics of bystander signaling. This process is highly dynamic and is fostered by cells that express NOX1 and membrane-associated catalase. It does not affect non-malignant cells without active NOX, nor is it further propagated by these cells. This scenario

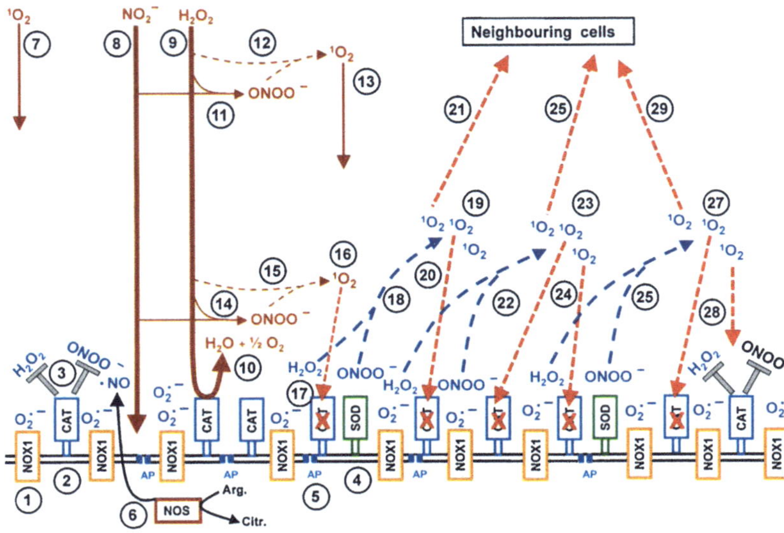


**Fig. 5.10** The impact of the synergistic effect between $NO_2^-$ and $H_2O_2$ on tumor cells. Membrane-associated NADPH oxidase-1 (NOX1), catalase, SOD, aquaporins, and intracellular NOS are essential redox-related elements of tumor cells (#1, 2, 4, 5). Catalase protects the cells towards damaging effects of their own ROS/RNS (#3). $^1O_2$ from the gaseous phase of CAP has no realistic chance to reach the tumor cells (#7). However, despite the efficient decomposition of CAP-derived $H_2O_2$ by tumor cell catalase (#10), the interaction between $NO_2^-$ (#8) and residual $H_2O_2$ (#9) may lead to the generation of $ONOO^-$ (# 11, #14). A multistep reaction between $ONOO^-$ and residual $H_2O_2$ (simplified in this figure) (#12, #15) yields to the generation of $^1O_2$ (#13, #16). If the generation of $^1O_2$ is too distant from the cell (#13), it will be without effect. If it is close enough to reach membrane-associated catalase (#16, 17), it may inactivate the catalase molecule. As a result, tumor cell-generated $H_2O_2$ and $ONOO^-$ are not decomposed locally. They can interact and generate more $^1O_2$ (#18, 19). As NOX1 and NOS constantly resupply $H_2O_2$ and $ONOO^-$, this secondary $^1O_2$ generation will be sustained (#20-28), yet initially strict locally. Through inactivation of neighboring catalase molecules on the same cell and on neighboring tumor cells, $^1O_2$ generation and catalase inactivation will reach an auto-amplificatory potential. Finally, a substantial population of membrane-associated catalase will be inactivated, resulting in the reactivation of intercellular apoptosis-inducing ROS/RNS signaling

therefore establishes a highly selective and effective mode of catalase inactivation of tumor cells.

These findings show that CAP and PAM only supply a highly specific triggering molecule, at concentrations that do not damage the cells per se. However, due to the biochemical composition of the membrane of tumor cells (containing NOX1, catalase, and SOD), its drives tumor cells to maximally contribute to their cell death, in a process that mirrors the triggering effect concerning the chemical biology of $^1O_2$ generation, but is far more efficient and sustained with respect of performance.

### 5.2.5 *Biochemical Details of Primary and Secondary $^1O_2$ Generation*

The biochemical details of the generation of primary $^1O_2$ generation after initial interaction between $NO_2^-$ and $H_2O_2$, as well as the subsequent generation of secondary $^1O_2$, are outlined in Fig. 5.11, following recently published experimental work [81, 84–86]. It was shown that the generation of primary $^1O_2$ is starting with $NO_2^-/H_2O_2$ interaction leading to $ONOO^-$ formation, whereas secondary

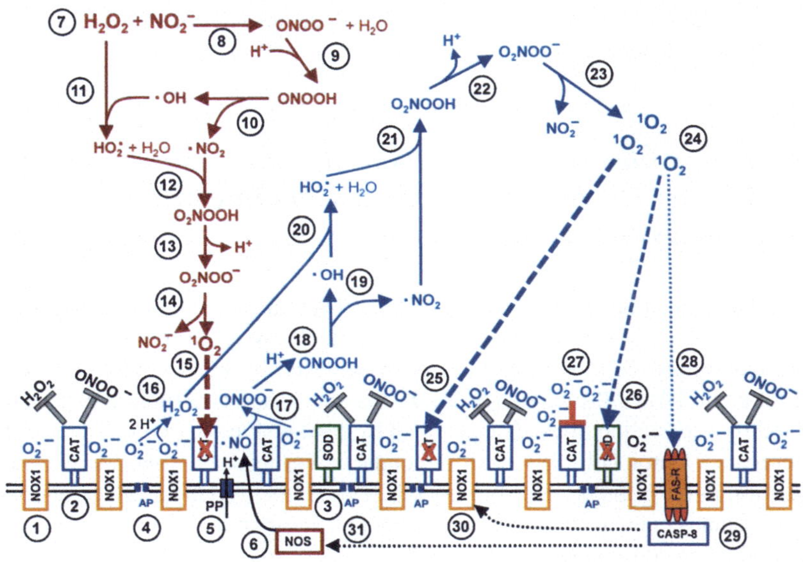


**Fig. 5.11** Biochemical details of the generation of primary and secondary $^1O_2$. In this chapter, $^1O_2$ derived from CAP or generated by CAP constituents is called "primary singlet oxygen," whereas $^1O_2$ generated by tumor cells as a response to a triggering effect by primary $^1O_2$ is called "secondary $^1O_2$ ." The membrane of tumor cells carries NOX1, catalase, SOD, aquaporins, and proton pumps (PP) (#1–5). Proton pumps are especially important for the protonation of $ONOO^-$. Intracellular NOS generates NO (#6). Long-lived species $NO_2^-$ and $H_2O_2$, both derived from CAP and maintained in PAM (#7) form $ONOO^-$ (#8). The reaction between $ONOO^-$ and residual $H_2O_2$ through reactions # 9–13 yields peroxynitrate ($O_2NOO^-$), which may decompose into $^{\bullet}NO_2$ and primary $^1O_2$ (#15). Primary $^1O_2$ locally inactivates catalase. This results in the local appearance of free $H_2O_2$ (#16) and $ONOO^-$ (#17) which are (1) constantly resupplied by NOX1 and NOS and (2) can interact and form secondary $^1O_2$ (#18-24). Secondary $^1O_2$ may inactivate catalase (#25) and thus promote auto-amplificatory $^1O_2$ generation, or inactivate SOD (#26), which will lead to local $O_2^{\bullet -}$-mediated inhibition of catalase (#27), followed by secondary $^1O_2$ generation (not shown for simplicity). $^1O_2$ can also activate the Fas receptor (#28) in a ligand-independent mode, resulting in caspase-8-mediated enhancement of NOS and NOX1 (#29–31), which contributes to the efficiency of the production of secondary $^1O_2$

$^1O_2$ generation is directly starting with $ONOO^-/H_2O_2$ interaction. Otherwise, both pathways are identical. Whereas the generation of primary $^1O_2$ according to this scheme is unique for CAP and PAM, secondary $^1O_2$ through the interaction between $H_2O_2$ and $ONOO^-$ at the site of inactivated membrane-associated catalase of tumor cells has also been demonstrated after direct application of $^1O_2$-generating photosensitizer [194], after application of 4-hydroxynonenal [195], after modulation of the endogenous $^{\bullet}NO$ level through inhibitors of NOD [72, 193], enhancers of NOS expression or inhibitors of arginase [72], and after inhibition of SOD by rather high concentrations of $H_2O_2$ [71].

- Secondary $^1O_2$ generation, therefore, seems to be a system that is utilized in many antitumor treatments which are based on interference with tumor cell-specific ROS/RNS control.

### *5.2.6 Finalization of CAP and PAM-Mediated Antitumor Cell Effects: The Role of Aquaporins and Intercellular Apoptosis-Inducing Signaling*

The preceding data have shown that CAP and PAM initially supply a trigger that induces the generation of secondary $^1O_2$ selectively by tumor cells. The strength and conclusiveness of these data have been achieved by kinetic inhibitor studies, using CAP and PAM [85], as well as $H_2O_2$ and $NO_2^-$ in reconstitution experiments [81], and by verification through analysis of bystander signaling between pretreated and untreated tumor cells [84, 86]. Secondary $^1O_2$ leads to substantial catalase inactivation in the population of tumor cells. Due to the bystander effect analogous scenario, catalase inactivation can be expected at sites that had not been initially hit by CAP or PAM components.

Based on previous concepts, one would predict that the inactivation of catalase would then allow for immediate intercellular ROS/RNS signaling and apoptosis induction. Kinetic analysis shows, however, that the onset of intercellular ROS signaling after catalase inactivation only has an apoptosis-inducing effect, if a second control step has been overcome. Figure 5.12A demonstrates that lipid peroxidation by ROS signaling derived $^{\bullet}OH$ remains initially ineffective, as it is efficiently counteracted by glutathione peroxidase 4 and reduced glutathione. However, the high number of aquaporins in the membrane of tumor cells, no longer being gated by catalase that prevented $H_2O_2$ influx under conditions of intact catalase, now allows for the rapid influx of cell-derived and CAP/PAM-derived $H_2O_2$ into the cells. This leads to the depletion of glutathione and abrogates the inhibitory action of glutathione peroxidase/glutathione towards the apoptosis-inducing effect of lipid peroxidation [221]. Kinetic measurements showed that the aquaporin-mediated effect required about 20 min more time than inactivation of catalase, in line with the conclusion that this step is a consequence of catalase inactivation [81, 85]. The aquaporin-mediated effect was abrogated when the tumor cells were pretreated with inhibitors of glutathione synthesis and thus their glutathione level was lowered to a level where cell death was not induced in control cells without CAP or PAM treatment, but cell death through ROS signaling started immediately after catalase inactivation and was no longer prevented by inhibitors of aquaporin function [85]. Further experiments have shown that $H_2O_2$ from CAP or PAM not necessarily have to be present for glutathione depletion, as $H_2O_2$ derived from tumor cells is generated in sufficient concentrations to perform the required task.

- These data indicate that the dominant controlling function of aquaporins, as defined by Keidar's group [152, 153], is of central importance for tumor cell death, but has to cooperate with $^1O_2$-mediated catalase inactivation. As catalase is gating the aquaporins, catalase inactivation is required to be initiated first. This mode of interaction is manifested in the experimentally determined kinetics of these processes [81, 85].

As shown in Fig. 5.12B, tumor cells after inactivation of catalase and depletion of intracellular glutathione (the two primary key events that control their cell death) establish ROS-mediated signaling that is immediately transformed into the activation of the mitochondrial pathway of apoptosis. The knowledge about the kinetics and biochemistry of these four key events ($^1O_2$-mediated catalase inactivation, GSH depletion through the concerted action of aquaporins and $H_2O_2$, intercellular ROS signaling, and mitochondrial apoptosis signaling) should enable us to optimize this system further through the application of conceivable synergistically acting compounds.

### 5.2.7 CAP and PAM Action: The Window of Selectivity

The preceding sections have shown that within defined concentration ranges, the synergistic effect between $NO_2^-$ and $H_2O_2$, but not the effect of $H_2O_2$ alone, can trigger an auto-amplificatory response specifically of tumor cells, which leads to their selective cell death. On the other hand, non-malignant cells are more sensitive to $H_2O_2$-mediated apoptosis induction than tumor cells [69]. This difference is due to the protective function of membrane-associated catalase of tumor cells, which is missing in non-malignant cells. Therefore, selective apoptosis induction by CAP and PAM in tumor cells requires to avoid concentrations of $H_2O_2$ that might damage non-malignant cells, but are still sufficiently high to allow the generation of $^1O_2$. The biochemical basis for selective antitumor cell effects of CAP and PAM has been addressed by Bauer [81] through the use of the long-lived species $NO_2^-$ and $H_2O_2$ that determine the effects of PAM, in combination with the target cells (non-malignant versus tumor cells) that were engineered to be genetically identical, but to differ with respect to the expression of the tumor cell markers NOX1 and membrane-associated catalase. This was achieved through siRNA-mediated knockdown of NOX1 and the resultant downmodulation of membrane-associated

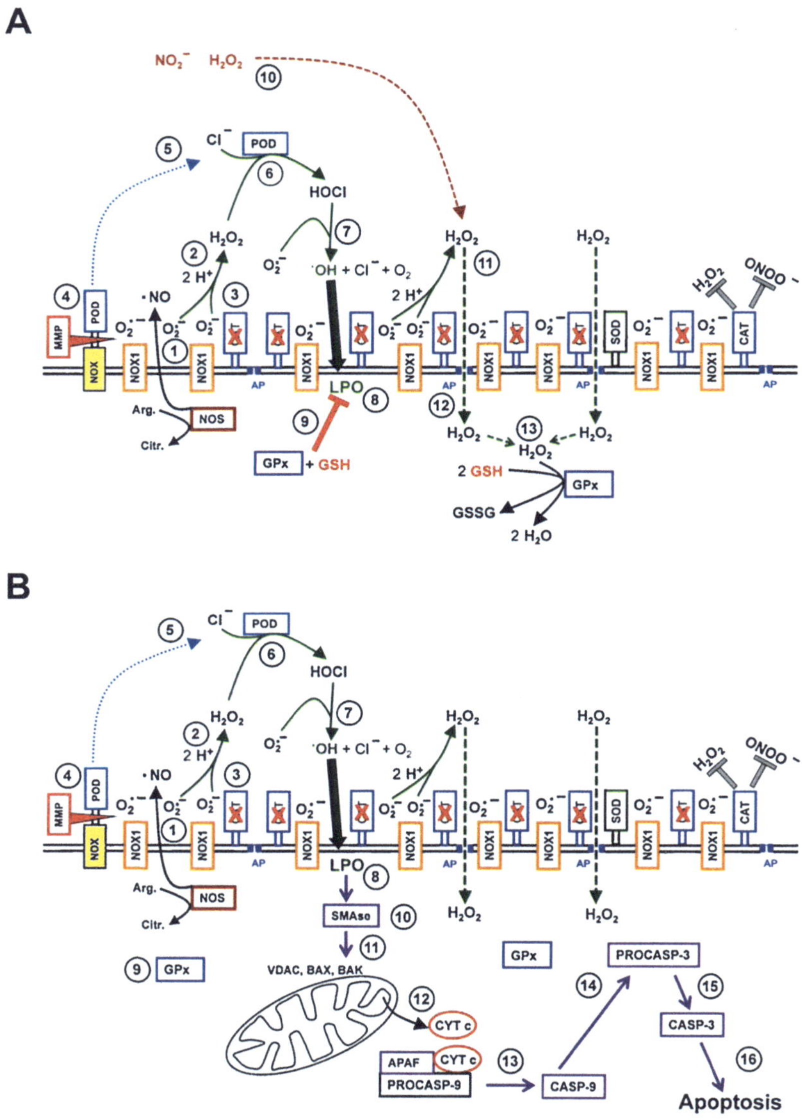


**Fig. 5.12** The dominant role of aquaporins during CAP/PAM-mediated antitumor action. (**A**) After an extensive generation of secondary $^1O_2$ according to Figs. 5.10 and 5.11, a substantial concentration of membrane-associated catalase is inactivated. This has two major consequences: i) intercellular ROS-mediated signaling is activated (#1–8) and (2) extracellular $H_2O_2$ from CAP, PAM and generated by the tumor cells themselves can enter the cells through aquaporins that are no longer gated by catalase (#11, 12). However, lipid peroxidation (LPO) by intercellular HOCl signaling has no apoptosis-inducing effect initially, as lipid peroxides are repaired by glutathione peroxidase 4/glutathione (#9). Only when sufficient extracellular $H_2O_2$ has entered the cell and depleted the intracellular glutathione level through reaction #13, the negative effect of glutathione peroxidase/glutathione on apoptosis induction can be abrogated. This scenario

catalase due to the missing stimulus $H_2O_2$ [69]. Whereas the tumor cell phenotype was characterized by active NOX1 expression and tight control of intercellular ROS/RNS signaling through membrane-associated catalase, the cells with the non-malignant phenotype did not express these two specific markers of tumor cells. Figure 5.13 demonstrates that increasing concentrations of $H_2O_2$ (generated by increasing concentrations of glucose oxidase (GOX)) caused apoptosis in the cells with the non-malignant phenotype at 0.006 mU/ml GOX and higher, whereas a more than 30-fold higher concentration of GOX was required for comparable apoptosis induction in cells with the tumor cell phenotype. The effect on cells with the non-malignant phenotype was independent of NOX1 activity, as it was not inhibited by AEBSF, whereas the effect on tumor cells at high concentrations of GOX required the activity of their NADPH oxidase. These findings show that the sole application of $H_2O_2$ induces apoptosis much more efficiently in non-malignant cells compared to tumor cells. Therefore, apoptosis induction solely by $H_2O_2$ cannot explain the selective effect of CAP and PAM on tumor cells. The resistance of tumor cells towards $H_2O_2$ generated by GOX up to 0.18 mU/ml is due to membrane-associated catalase. The response of the tumor cells at higher concentrations of GOX is mediated by $H_2O_2$-dependent inhibition of membrane-associated SOD, resulting in catalase inhibition by free $O_2^{\bullet-}$ and subsequent auto-amplification of $^1O_2$ generation [71]. The addition of $NO_2^-$ to this experimental system did not affect $H_2O_2$-dependent apoptosis induction of non-malignant cells but caused a strong synergistic effect with $H_2O_2$ on tumor cells. Apoptosis induction was observed at 0.00008 mU/ml GOX in combination with $NO_2^-$. The underlying process was dependent on $O_2^{\bullet-}$ generation by NOX1, as AEBSF inhibited it. Control experiments ensured that this process followed the chemical biology that is described in Figs. 5.11 and 5.12 of this manuscript. Figure 5.13 demonstrates a clearly defined window of selectivity for the action of $H_2O_2$ combined with $NO_2^-$, targeting tumor cells and sparing non-malignant cells, in vitro. These data show that the synergistic effect between $NO_2^-$ and $H_2O_2$ causes apoptosis in tumor cells very efficiently, but that selectivity of this process for the malignant phenotype is only warranted in a window of concentrations of $H_2O_2$ that do not damage non-malignant cells.

Analogous experiments using a portable air plasma "corona pen" plasma source [155] confirmed the "window of selectivity" for CAP as well as PAM treatment in vitro [85]. These results are summarized in Fig. 5.14. Short treatment times with

**Fig. 5.12** (continued) explains how the inhibition of the aquaporin function can have a dominant control function on CAP and PAM-mediated apoptosis induction. (**B**) After $^1O_2$-mediated catalase inactivation and sufficient influx of extracellular $H_2O_2$ through unprotected aquaporins, intracellular glutathione is depleted. Therefore, lipid peroxidation after HOCl signaling (#1–8) leads to the activation of sphingomyelinase (SMAse), generation of ceramides that affect the mitochondrial voltage channel, as well as BAK and BAX (#10, 11) [201]. Cytochrome c is released from the mitochondria and leads to the generation of active caspase-9 (#12, 13). Subsequently, caspase-3 is activated by caspase-9 and finalizes apoptotic cell death (#14-16)

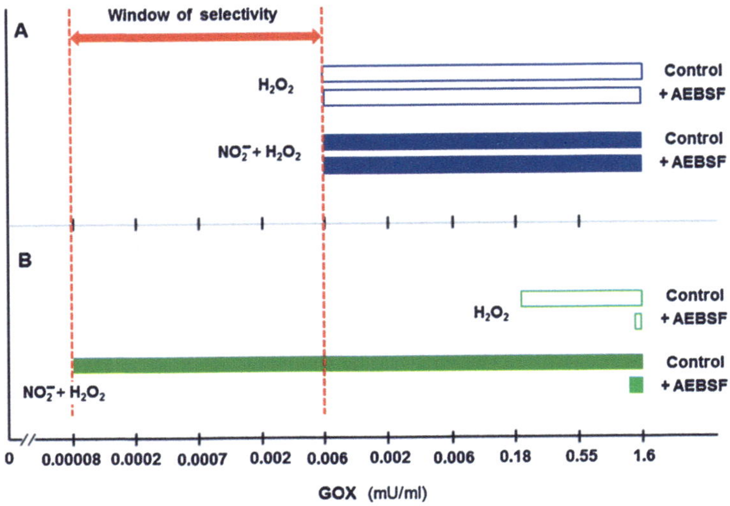


**Fig. 5.13** The window of selective apoptosis induction in tumor cells by $H_2O_2$ and $NO_2^-$. This figure summarizes experimental data on the action of $H_2O_2$ with and without $NO_2^-$ on non-malignant cells (**A**) and tumor cells (**B**) [81]. Glucose oxidase (GOX) was used as a steadily producing $H_2O_2$ source and $NO_2^-$ was added at 1 mM were indicated. The bars define the concentration ranges in which significant apoptosis induction was observed. This figure shows that tumor cells are more resistant to $H_2O_2$ than non-malignant cells when $H_2O_2$ was applied alone. Tumor cells are subject to a remarkable synergistic effect between $H_2O_2$ and $NO_2^-$, which does not affect non-malignant cells. This allows defining a window of selective apoptosis induction in tumor cells for the concentration of $H_2O_2$ in the system. The figure also shows that NOX1 controlled the tumor cell-specific effects, as they were inhibited by the NOX inhibitor AEBSF

CAP and low concentrations of PAM caused selective apoptosis induction in tumor cells, which was dependent on NOX1-derived $O_2^{\bullet-}$, as it was inhibited by the NOX1 inhibitor AEBSF. Parallel control ensured that the underlying process was based on auto-amplification of secondary $^1O_2$, catalase inactivation, and intercellular ROS-dependent apoptosis induction. When the time of CAP treatment was extended or the concentration of PAM was increased, nonselective apoptosis induction in non-malignant cells was gradually increasing. This process seems to be driven by the intracellular Fenton chemistry of the non-malignant cells.

- These findings demonstrate a window for the time of treatment with CAP and the concentrations of PAM, in which selective action towards tumor cells can be achieved in vitro. This window needs to be defined for the in vivo situation of different tumor systems, using CAP sources with a defined composition of ROS/RNS.

These findings also support the conclusion that nonselective apoptosis induction by CAP and PAM, as found by few groups, might be explained by the application of CAP and PAM doses outside the window of the selective application, most likely due to direct effects of excess $H_2O_2$ on non-malignant cells.

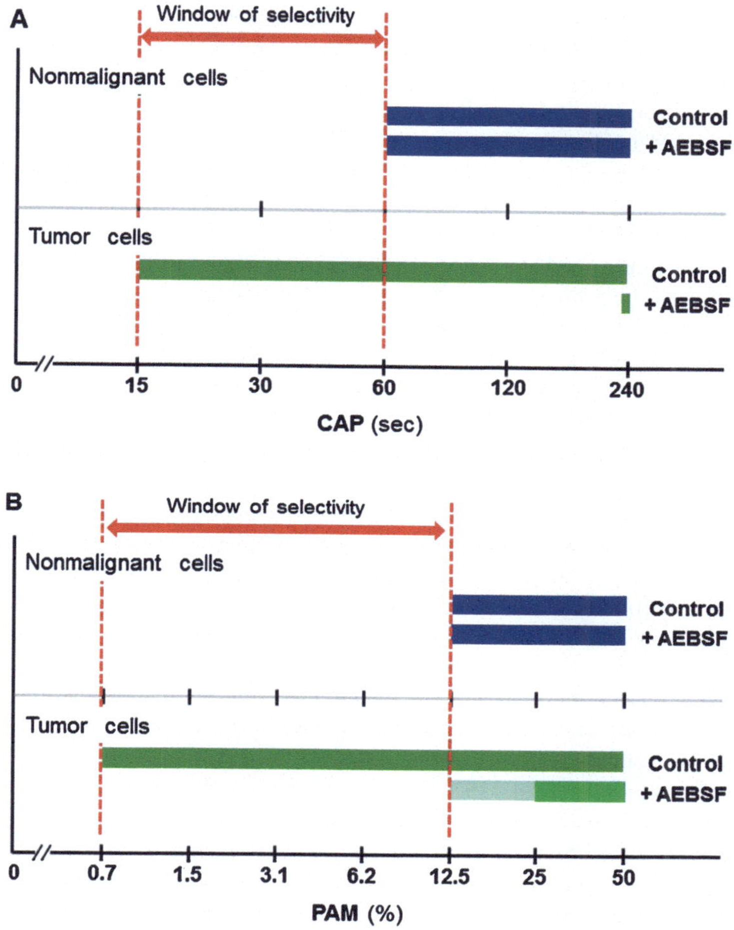


**Fig. 5.14** The window of selective apoptosis induction in tumor cells by CAP and PAM. This figure summarizes experimental findings related to the treatment of non-malignant cells and tumor cells with CAP (**A**) or PAM (**B**) [86]. The bars indicate the time of CAP treatment (**A**) or the concentration of PAM (**B**) that causes apoptosis induction in nonmalignant and tumor cells, both in the absence and presence of the NOX inhibitor AEBSF. These data define a window of selective action of CAP and PAM towards NOX1-expressing tumor cells

## *5.2.8 Divergent and Overlapping Concepts Related to the Mechanism of CAP- and PAM Action on Tumor Cells*

The mechanistic model of CAP and PAM action as presented in Figs. 5.10, 5.11, and 5.12 in this chapter was initiated by findings related to ROS/RNS interactions with non-malignant and malignant cells [16, 23, 25, 34–37, 39, 58, 66–69, 71, 72, 193, 194] and the ROS/RNS composition of the aqueous and liquid phases of CAP and PAM, as summarized in [24, 82, 83, 87–89]. This knowledge was the basis for Bauer and Graves [82] and Bauer [24, 70] to deduce the initial hypothesis on CAP- and PAM-mediated antitumor effects. This hypothesis was transformed to an experimentally founded model through kinetic inhibition studies and dissection of the experimental system into discrete steps, using defined sources of the long-lived species $H_2O_2$ and $NO_2^-$ [81, 82], as well as a plasma source [85, 86]. The important aspect of aquaporin-mediated control effects established by the work of Keidar's group [152, 153] was experimentally confirmed and became an integral part of the model, though there remains a disagreement on the underlying kinetic aspects of aquaporin function. The final model as presented in Figs. 5.10, 5.11, and 5.12 of this chapter comprises primary $^1O_2$ generation from CAP or PAM, auto-amplification of secondary $^1O_2$ by the tumor cells, inactivation of membrane-associated catalase, influx of $H_2O_2$ through aquaporins, and reactivation of intercellular ROS/RNS-dependent apoptosis-inducing signaling. The outstanding feature of this model, differentiating it from all other published models, is its dynamic nature: Treatment of few tumor cells with CAP or PAM allows these cells to transmit secondary $^1O_2$ generation and the subsequent process up to apoptotic cell death to their untreated neighboring cells. In contrast, most other models on the mechanism of CAP and PAM action towards tumor cells are based on the concept that the cell death-inducing compound is already present in CAP and PAM at a concentration sufficient to induce the observed effects and that the targeted cells are responders to CAP and PAM, but not enhancers of the signaling process. In this context, $H_2O_2$ from CAP and PAM was often regarded as a central and direct player.

Keidar's group (Yan et al. [152, 153]) suggested that the increased concentration of aquaporins on tumor cells [151] was the key determinant of selective antitumor action of CAP and PAM, as it should allow for an increased influx of CAP- or PAM-derived $H_2O_2$ into tumor cells, compared to non-malignant cells. This would then result in tumor cell apoptosis through classical intracellular Fenton reaction. Their concept was based on the very efficient inhibition of CAP- or PAM-mediated apoptosis induction in the presence of aquaporin inhibitors or after siRNA-mediated knockdown of aquaporins.

The model by Yan et al. [152, 153] predicts that tumor cells respond to $H_2O_2$ more readily than non-malignant cell. This is not in line with the experimental finding that tumor cells are more resistant to apoptosis induction by exogenous $H_2O_2$ than non-malignant cells [69]. As this resistance is based on membrane-associated catalase of tumor cells [13–17, 68] catalase seems to prevent the influx of $H_2O_2$ through aquaporins via decomposition of $H_2O_2$. However, when membrane-associated catalase has been inactivated after auto-amplification of secondary $^1O_2$ and intercellular ROS/RNS-mediated apoptosis-inducing signaling becomes effective that the aquaporins show their dominant controlling function [81, 85]. This then is in perfect agreement with the experimental findings by Keidar's group, but is explained by a more complex mechanism: an influx of $H_2O_2$ through aquaporins, followed by intracellular glutathione depletion, is necessary to allow intercellular

ROS/RNS signaling to become effective. Tumor cell aquaporins therefore control the step of intracellular sensitization of tumor cells for apoptosis induction, provided catalase activity has been lowered and ROS/RNS signaling is active.

Therefore, the concept formulated by Yan et al. [152, 153] is of crucial importance to understand ROS/RNS-mediated apoptosis induction in tumor cells, though it does not explain the basis for selectivity of CAP and PAM. Nevertheless, central findings by Yan et al. [152, 153] are an integral and central part of the model presented in Figs. 5.11 and 5.12 of this manuscript [81, 84–86].

The consumption of $H_2O_2$ by tumor cells as described by Yan et al. [90, 152, 222] can be explained by the activity of membrane-associated catalase on tumor cells and is consistent with similar findings in reference [81]. The reciprocal correlation between cytotoxicity of CAP on tumor cells and the scavenging rate of the tumor cells for extracellular $H_2O_2$ [222] is also in line with Bauer's model. It can be explained through the control function of membrane-associated catalase on tumor cells.

Yan et al. [148] described that the treatment of tumor cells with CAP, but not with PAM, caused strong cell-based $H_2O_2$ generation. As, due to a low volume of medium above the cells, these were accessible to direct plasma treatment in their experiments, inactivation of membrane-associated catalase directly by $^1O_2$ from the gaseous phase of CAP might have been sufficient to cause a net increase in cell-derived $H_2O_2$ after direct inactivation of membrane-associated catalase by $^1O_2$ from the gaseous phase of CAP. The relatively slow and time-requiring effect of $^1O_2$ generated from the long-lived species $H_2O_2$ and $NO_2^-$ seemed to be suboptimal to give a similar rise in free $H_2O_2$ within a short time and therefore might not have been detected in these experiments.

Finally, the activated state of tumor cells after direct CAP treatment as reported by Yan et al. [223] might be explained by partial inactivation of membrane-associated catalase and the observed desensitization can be explained by $H_2O_2$—mediated dynamic control of catalase expression, according to Böhm et al. [69].

Van der Paal et al. [224] suggested selective CAP and PAM effects directed towards tumor cells were determined by the decreased cholesterol content of tumor cells compared to non-malignant cells, as the ingress of ROS into cells is hampered by cholesterol. This model also depends on the assumption that ROS/RNS in CAP and PAM are *directly* responsible and sufficient for the induction of cell death in the target cells. Also in this model, $H_2O_2$ is discussed as the major effector from CAP and the only effector from PAM. This model (1) is not in line with the experimentally determined resistance of tumor cells to exogenous $H_2O_2$ [13–17] which is based on the activity of membrane-associated catalase [66–69], (2) does not allow to explain the generation and auto-amplification of $^1O_2$, and (3) cannot explain the dynamic nature of CAP and PAM action.

In line with the previous concept of Keidar's group [152, 153], Semmler et al. [225] published a model that is mainly focusing on intracellular ROS/RNS effects after CAP action. Determining steps in this model are the influx of ROS/RNS through aquaporins and through leaks that are generated after lipid peroxidation. In line with the model presented in this chapter, the model by Semmler et al. involves execution of the mitochondrial pathway of apoptosis. However, due to the lack of consideration of membrane-associated catalase, the role of NOX1, $^1O_2$ generation, and extracellular ROS/RNS signaling, the model by Semmler cannot explain the experimental findings presented in [81, 84–86], such as the generation of $^1O_2$ by the cells, the role of extracellular ROS/RNS signaling, and the dynamics of the process.

Attri and Bogaerts [226] have combined selected elements of models described by Adachi et al., Yan et al., and Bauer [70, 140, 146, 153] in their recent model that describes the effects of plasma-treated solutions on tumor cells. The previous model from the same group [224] plays no role in their new model. Attri and Bogaerts [226] follow Bauer's findings and concept [24, 70] that the interaction between long-lived species in PAM leads to the generation of $^1O_2$, which inactivates membrane-associated catalase of tumor cells, but do not present own data to support or extend these findings. Following the findings by Adachi et al. [140], the authors speculate that the subsequent influx of PAM-derived $H_2O_2$ through aquaporins (in line with the findings by Yan et al. [153]) is sufficient to cause tumor cell death through intracellular Fenton chemistry. In the model by Attri and Bogaerts [226], NOX1-derived $O_2^{\bullet -}$ and the generation of secondary $^1O_2$ play no role, whereas they are central and essential elements in the experiments and the model published by Bauer [70], on which they mainly based their modified model [226]. Furthermore, the model by Attri and Bogaerts is apparently discrepant with the experimentally determined consequences of $^1O_2$-mediated inactivation of membrane-associated catalase. Whereas Attri and Bogaerts assume that intracellular Fenton chemistry of PAM-derived $H_2O_2$ causes cell death, it has been experimentally proven that inactivation of membrane-associated catalase by (1) long-lived species contained in PAM, (2) by direct application of a $^1O_2$ donor, and (3) by application of CAP and PAM generated by a plasma source strictly requires NOX-driven HOCl signaling for significant induction of apoptosis [81, 84, 85, 194]. The dominant role of HOCl signaling was proven, as inhibition of HOCl synthesis, scavenging of HOCl, and prevention of HOCl/$O_2^{\bullet -}$ interaction caused complete inhibition of apoptosis induction [81, 85]. The direct influx of $H_2O_2$ through aquaporins under these conditions is required for sensitization of the cells for the signaling effects of lipid peroxide, but is not sufficient to trigger apoptosis induction. If it was sufficient, apoptosis induction should not be completely blocked by inhibitors of extracellular HOCl signaling [81, 85].

Privat-Maldonaldo et al. [227] recently presented a model for CAP action which essentially follows the central concepts published by Bauer [70, 81, 84] and Bauer et al. [85, 86], except for the requirement for FAS receptor-mediated enhancement of NOX1 and NOS activities [81, 84–86] and the dynamic auto-amplification of $^1O_2$ generation. Compared to the model by Attri and Bogaerts [226], intracellular Fenton chemistry has now been replaced by extracellular ROS/RNS signaling, which is in line with the experimental findings [81, 84–86] However, Privat-Maldonaldo et al. state that mathematical modeling would indicate that (1) apoptosis-inducing $^{\bullet}OH$ radicals were only generated through the $^{\bullet}NO/ONOO^-$ pathway, but not through the HOCl pathway, and (2) that 99% of membrane-associated catalase have to be inactivated before apoptosis-inducing signaling through $^{\bullet}NO/ONOO^-$ signaling can become effective. Both theoretically derived conclusions are in direct contradiction to central experimental findings in this system: (1) Apoptosis-inducing $^{\bullet}OH$ in the experiments described in [81, 84] were solely derived from $HOCl/O_2^{\bullet -}$ interaction, as demonstrated by detailed inhibitor studies. (2) $^{\bullet}NO/ONOO^-$ signaling, instead of HOCl signaling, was only achieved after CAP/PAM treatment when excess exogenous $^{\bullet}NO$ was added through application of an $^{\bullet}NO$ donor [81, 85]. (3) Reactivation specifically of $^{\bullet}NO/ONOO^-$ signaling can only be achieved by a relatively low degree of catalase inhibition or inactivation [68, 193, 194, 202, 204, 205], as (4) a high degree of catalase inhibition causes complete interference with $^{\bullet}NO/ONOO^-$ signaling through iron-mediated interfering effects of excess $H_2O_2$ [203]. Therefore it seems that more efforts will be required to establish a mathematical model that is genuinely relevant to the experimental findings before conclusions derived from modeling can be translated into application of CAP or PAM in vivo.

Bekeschus et al. [228] demonstrated that hypochlorite/hypochlorous acid generated through the interaction between CAP-derived atomic oxygen and $Cl^-$ was responsible for apoptosis induction in leukemia cells in vitro. This finding is consistent with the previous finding by Bauer [201] on selective apoptosis induction in tumor cells by exogenous HOCl, dependent on active NOX1 of the tumor cells. The resultant generation of apoptosis-inducing $^{\bullet}OH$ was independent of catalase inactivation, as there was no need for cellular HOCl synthesis in the presence of exogenous HOCl. It was even shown that inhibition of tumor cell catalase had a negative effect on apoptosis induction by exogenous HOCl, as free $H_2O_2$ was consuming HOCl through direct interaction [201]. The findings by Bekeschus et al. [228] in combination with the detailed knowledge on the chemical biology of hypochlorite/hypochlorous acid and dichloride anion radicals [89, 229] should encourage further studies on the effects of chlorine/chloride-related compounds in CAP and PAM. For example, so far it is an open and interesting question whether HOCl-mediated inactivation of catalase through formation of compound I and compound II, followed by HOCl-mediated heme destruction [230] might play a possible role in this context.

### 5.2.9 *The Importance of Immunogenic Cell Death for CAP- and PAM-Mediated Antitumor Effects*

A mechanistic view allows to assume that ROS/RNS-mediated induction of tumor cell apoptosis, as outlined in this chapter, may be essential, but not sufficient for tumor regression after treatment with CAP or PAM. Major arguments for this view are (1) the obvious dependency of ROS/RNS-mediated signaling from cell density, which would not necessarily lead to complete eradication of remaining tumor cells at low cell density, (2) the strong evidence for the more or less general requirement for immunogenic cell death for the finalization of several regimes for cancer treatment [231–236], and (3) the established findings on the strong connection between CAP/PAM tumor treatment and immunogenic cell death [237–245]. These aspects will be presented in detail in chapter of this book.

## 5.3 Conclusions and Outlook

The data and concepts presented in this chapter demonstrate that the long-lived species $H_2O_2$ and $NO_2^-$, which are contained in CAP and PAM, are sufficient to cause the generation of primary $^1O_2$ [81, 85]. $^1O_2$ then interacts specifically with catalase on the surface of tumor cells. This relatively rare and limited effect [84] induced by CAP and PAM compounds is followed by sustained and highly dynamic generation of secondary $^1O_2$ by the tumor cells themselves. This central step requires the activity of NOX1 (which is a hallmark of the malignant phenotype) and the contribution of NOS. This is followed by substantial catalase inactivation within the cell population, the influx of $H_2O_2$ through aquaporins, and the reactivation of intercellular ROS/RNS-driven apoptosis-inducing signaling [81, 85]. The generation of secondary $^1O_2$, the availability of intruding $H_2O_2$ as well as intercellular apoptosis-inducing signaling are determined by active NOX1. This aspect explains the central part of selective CAP and PAM action towards tumor cells. The other part of selective action is determined by the concentration of CAP and PAM-derived $H_2O_2$ that must not exceed a concentration that endangers non-malignant cells, as these are not protected by membrane-associated catalase [68, 69].

The synergistic effect between $H_2O_2$ and $NO_2^-$, based on the generation of primary $^1O_2$, is the clue that warrants selective apoptosis induction in tumor cells. Therefore, these findings explain how the relatively simple ROS/RNS composition of PAM can exert a similar effect on the highly sophisticated ROS/RNS composition of CAP.

However, the data and concepts presented here do not imply that the molecular species that have not been in the focus of the completed experimental work would have no additional function or might bring no additional benefit to CAP and PAM action.

It is conceivable that species like $^\bullet NO$ and $ONOO^-$ that were not sufficient to induce cell death in tumor cells when applied alone may enhance $^1O_2$ generation as well as $^1O_2$-dependent processes when they are applied together with $H_2O_2$ and $NO_2^-$. Such effects have been found in an analogous experimental system [71].

It is also conceivable that an increase in the concentration of $^1O_2$ in the CAP source, combined with a decrease of the layer of cell culture medium or direct accessibility of tumor tissue, will allow direct $^1O_2$ interaction with catalase on the tumor cell surface and thus trigger secondary $^1O_2$ generation. The same consideration may apply to $O_3$ and its potential to generate $^1O_2$ after its reaction with specific amino acids [246, 247].

Though $NO_2^-$ and $H_2O_2$ are the species of central importance for the activity of PAM, the application of PAM rather than the use of the defined chemicals is advantageous [91]. This discrepancy might be explained by reaction products between short-lived molecular species from CAP with medium components (such as lipids or proteins) that are present in PAM but would be missed when pure chemicals were applied. It is not unlikely that such hypothetical reaction products exhibit biological signal functions.

Furthermore, molecular species that played no role in the experimental systems discussed here may play a role through direct positive effects on the immune system, which seems to be required for the finalization of tumor regression.

Further systematic experimentation with defined plasma sources, defined and variable doses of individual CAP components, variable conditions of CAP and PAM application in vivo and in vitro will hopefully contribute to a rational optimization of CAP and PAM application for tumor treatment. Finally, the established knowledge about ROS/RNS reactions and their interactions with tumor cells will allow testing for additional synergistic effects through compounds that affect NO metabolism and control or directly act on catalase or SOD. Such approaches might enhance efficacy, selectivity and minimize side effects of treatment through conceivable synergistic effects.

We like to emphasize that the conditions for initial tumor treatment with CAP and PAM, followed by immunogenic cell death and action of the immune system, are most likely mechanistically different from the approach to destroy residual tumor cells through CAP or PAM after surgical resection of a tumor. Based on the considerations outlined in this chapter, and on the assumption that individual residual tumor cells might not be optimally affected by cell density-dependent processes, the latter approach might require a sufficient dose of $^1O_2$ for selective sensitization of tumor cells through inhibition of catalase, in combination with a sufficient dose of either $^{\bullet}NO$, $ONOO^-$, or $H_2O_2$ (or a combination thereof). Thereby the doses of compounds affecting tumor cells need to be below a concentration that would damage normal tissue.

**Acknowledgments** Georg Bauer gratefully acknowledges intellectual support by many colleagues during IWPCT 2016, IWPCT 2017, IWPCT 2018, ICPM-7, Gordon Research Conference on Plasma Processing Science 2018 and European Corporation in Science & Technoloy action CM0603-CHEMBIORADICAL. The support of laboratory and office space by the University Medical Center Freiburg is acknowledged. I thank J. Brandel (Freiburg) for technical support. This work was possible through constant encouragement from my family.

Sander Bekeschus acknowledges funding by the German Federal Ministry of Education and Research (grant number 03Z22ND11) and the European Social Fund (grant number ESF/14-BM-A55-0001/18).

## References

1. R.A. Weinberg, Oncogenes, anti-oncogenes and the molecular basis of multistep carcinogenesis. Cancer Res. **49**, 3713–3721 (1989)
2. D. Hanahan, R.A. Weinberg, The hallmarks of cancer. Cell **100**, 57–70 (2014)
3. D. Hanahan, R.A. Weinberg, Hallmarks of cancer: the next generation. Cell **144**(5), 646–674 (2011). https://doi.org/10.1016/j.cell.2011.02.013
4. B. Vogelstein, K.W. Kinzler, Cancer genes and the pathways they control. Nat. Med. **10**, 789–799 (2004)
5. H.M. Temin, Control by factors in serum of multiplication of uninfected and cells infected and converted by avian sarcoma viruses, in *Growth Regulatory Substances for Animal Cells in Culture*, The Wistar Symposium Monograph, ed. by V. Defendi, M. Stoker, vol. 7, (Wistar Institute Press, Philadelphia, 1967), pp. 103–116
6. M.B. Sporn, G.J. Todaro, Autocrine secretion and malignant transformation. New Engl. J. Med. **303**, 878–880 (1980)
7. C.H. Heldin, B. Westermark, Growth factors as transforming proteins. Eur. J. Biochem. **184**, 487–496 (1989)
8. M.G.P. Stoker, M. Shearer, C. O'Neill, Growth inhibition of polyoma-transformed cells by contact with static normal fibroblasts. J. Cell Sci. **1**, 297–310 (1966)
9. J.G. Delinassios, Fibroblasts against cancer cells in vitro. Anticancer Res. **7**, 1005–1010 (1987)
10. J.E. Trosko, C.C. Chang, B.V. Madhukar, J.E. Klaunig, Chemical, oncogene and growth factor inhibition of gap junctional intercellular communication: an integrative hypothesis of carcinogenesis. Pathobiology **58**, 265–278 (1990)
11. M.H. Barcellos-Hoff, It takes a tissue to make a tumor. Epigenetics, cancer and microenvironment. J. Mammary Gland Biol. Neoplasia **6**, 213–221 (2001)
12. E. Flaberg, L. Markasz, G. Petranyi, G. Stuber, F. Dicso, N. Alchihabi, E. Olah, I. Csizy, T. Jozsa, O. Andren, J.E. Johansson, S.O. Andersson, G. Klein, L. Szekely, High-throughput live-cell imaging reveals differential inhibition of tumor cell proliferation by human fibroblasts. Int. J. Cancer **128**, 2793–2802 (2011)
13. G.I. Deichman, E.L. Vendrov, Characteristics of in vitro transformed cells essential for their in vivo survival, selection and metastatic activity. Int. J. Cancer. **37**, 401–409 (1986)
14. G.I. Deichman, T.E. Kluchareva, V.A. Matveeva, N.E. Kushlinsky, L.S. Bassalyk, E.L. Vendrov, Clustering of discrete cell properties essential for tumorigenicity and metastasis. I. Studies of Syrian hamster embryo fibroblasts spontaneously transformed in vitro. Int. J. Cancer **44**, 904–907 (1989)
15. G. Deichman, V.A. Matveeva, L.M. Kashkina, N.A. Dyakova, E.N. Uvarova, M.A. Nikiforov, A.V. Udkov, Cell transforming genes and tumor progression: in vivo unified secondary phenotypic cell changes. Int. J. Cancer **75**, 277–283 (1998)
16. G. Deichman, Natural selection and early changes of phenotype of tumor cells in vivo: Acquisition of new defense mechanisms. Biochemist **65**, 78–94 (2000)
17. G. Deichman, Early phenotypic changes of in vitro transformed cells during in vivo progression: possible role of the host innate immunity. Semin. Cancer Biol. **12**, 317–326 (2002)
18. T.G. Graeber, C. Osmanian, T. Jacks, D.E. Housman, C.J. Koch, S.W. Lowe, A.J. Giaccia, Hypoxia-mediated selection of cells with diminished apoptotic potential in solid tumours. Nature **379**, 88–91 (1996)
19. K.W. Kinzler, B. Vogelstein, Life (and death) in a malignant tumour. Nature **379**, 19–20 (1996)
20. J. Folkman, Fundamental concepts of the angiogenic process. Curr. Mol. Med. **3**, 643–651 (2003)
21. R.A. Cairns, I.S. Harris, T.W. Mak, Regulation of cancer cell metabolism. Nat. Rev. Cancer **11**, 85–95 (2011)

22. G. Bauer, Tumor cell protective catalase as a novel target for rational therapeutic approaches based on specific intercellular ROS signaling. Anticancer Res. *32*, 2599–2624 (2012)
23. G. Bauer, Targeting extracellular ROS signaling of tumor cells. Anticancer Res. **34**, 1467–1482 (2014)
24. G. Bauer, Signal amplification by tumor cells: clue to the understanding of the antitumor effects of cold atmospheric plasma and plasma-activated medium. IEEE Trans. Radiat. Plasma Med. Sci. **2**, 87–98 (2018)
25. G. Bauer, HOCl and the control of oncogenesis. J. Inorg. Biochem. **179**, 10–23 (2018). https://doi.org/10.1016/j.jinorgbio.2017.11.005
26. D.I. Feig, T.M. Reid, L.A. Loeb, Reactive oxygen species in tumorigenesis. Cancer Res. **54**, 1890–1894 (1994)
27. P. Storz, Reactive oxygen species in tumor progression. Front. Biosci. **10**, 1881–1896 (2005)
28. P.A. Cerutti, Prooxidant states and tumor promotion. Science **227**, 375–381 (1985)
29. K. Irani, Y. Xia, J.L. Zweier, S.J. Sollott, C.J. Der, E.R. Fearon, M. Sundaresan, T. Finkel, P.J. Goldschmidt-Clermont, Mitogenic signalling by oxidants in Ras-transformed fibroblasts. Science *275*, 1649–1652 (1997)
30. K. Irani, P.J. Goldschmidt-Clermont, Ras, superoxide and signal transduction. Biochem. Pharmacol. **55**, 1339–1346 (1998)
31. Y.-A. Suh, R.S. Arnold, B. Lassegue, J. Shi, X. Xu, D. Sorescu, A.B. Chung, K.K. Griendling, J.D. Lambeth, Cell transformation by the superoxide-generating oxidase Mox1. Nature **401**, 79–82 (1999)
32. F. Bittinger, J.L. Gonzalez-Garcia, C.L. Lein, C. Brochhausen, F. Offner, C.J. Kirkpatrick, Production of superoxide by human malignant melanoma cells. Melanoma Res. **8**, 381–387 (1998)
33. J.Q. Yang, S. Li, F.E. Domann, G. Buettner, L.W. Oberley, Superoxide generation in *v-Ha-ras*-transduced human keratinocyte HaCaT cells. Mol. Carcinog. **26**, 180–188 (1999)
34. M. Herdener, S. Heigold, M. Saran, G. Bauer, Target cell-derived superoxide anions cause efficiency and selectivity of intercellular induction of apoptosis. Free Radic. Biol. Med. *29*, 1260–1271 (2000)
35. I. Engelmann, S. Dormann, M. Saran, G. Bauer, Transformed target cell-derived superoxide anions drive apoptosis induction by myeloperoxidase. Redox Rep. *5*, 207–214 (2000)
36. A. Schwieger, L. Bauer, J. Hanusch, C. Sers, R. Schäfer, G. Bauer, *Ras* oncogene expression determines sensitivity for intercellular induction of apoptosis. Carcinogenesis *22*, 1385–1392 (2001)
37. B. Ivanovas, G. Bauer, Selective and nonselective apoptosis induction in transformed and nontransformed fibroblasts by exogenous reactive oxygen and nitrogen species. Anticancer Res. *22*, 841–856 (2002)
38. S.S. Brar, T.P. Kennedy, A.B. Sturrock, T.P. Huecksteadt, M.T. Quinn, A.R. Whorton, An NADPH oxidase regulates growth and transcription in melanoma cells. Am. J. Phys. **282**, C1212–C1224 (2002)
39. S. Heigold, C. Sers, W. Bechtel, B. Ivanovas, R. Schäfer, G. Bauer, Nitric oxide mediates apoptosis induction selectively in transformed fibroblasts compared to nontransformed fibroblasts. Carcinogenesis *23*, 929–941 (2002)
40. W. Chamulitrat, R. Schmidt, P. Tomakidi, W. Stremmel, W. Chunglok, T. Kawahara, K. Rokutan, Association of GP91PHOX homolog NOX1 with anchorage-independent growth and MAP kinase activation of transformed human keratinocyte. Oncogene *22*, 6045–6053 (2003)
41. J. Mitsushita, J.D. Lambeth, T. Kamata, The superoxide-generating oxidase Nox1 is functionally required for Ras oncogenic transformation. Cancer Res. **64**, 3580–3585 (2004)
42. A.Y. Alexandrova, P.B. Kopnin, J.M. Vasilev, P.B. Kopnin, ROS up-regulation mediates Ras-induced changes of cell morphology and motility. Exp. Cell Res. **312**, 2066–2073 (2006)
43. E. Laurent, J.W. McCoy, R.A. Maccina, W. Liu, G.J. Cheng, S. Robine, J. Papkoff, J.D. Lambeth, Nox1 is overexpressed in human colon cancers and correlates with activating mutations in *K-Ras*. Int. J. Cancer **123**, 100–107 (2008)

44. K. Tominaga, T. Kawahara, T. Sano, K. Toida, Y. Kuwano, H. Sasaki, T. Kawai, Evidence for cancer-associated expression of NADPH oxidase 1 (Nox1)-base oxidase system in the human stomach. Free Radic. Biol. Med. **43**, 1627–1638 (2007)
45. Q. Ma, L.E. Cavallin, B. Yan, S. Zhu, E.M. Duran, H. Wang, L.P. Hala, C. Dong, E. Cesarman, E.A. Mesri, P.J. Goldschmidt-Clermont, Antitumorigenesis of antioxidants in a transgenic Rac1 model of Kaposi's sarcoma. Proc. Natl. Acad. Sci. U. S. A. **106**, 8683–8688 (2009)
46. E.-Y. Kim, J.-M. Seo, C. Kim, J.-E. Lee, K.-M. Lee, J.-H. Kim, BLT2 promotes the invasion and metastasis of aggressive bladder cancer through a reactive oxygen species-linked pathway. Free Radic. Biol. Med. ***49***, 1072–1081 (2010)
47. J. Du, J. Liu, B.J. Smith, M.S. Tsao, J. Cullen, Role of rac-1-dependent NADPH oxidase in the growth of pancreatic cancer. Cancer Gene Ther. **18**, 135–143 (2011)
48. L. Behrend, G. Henderson, R.M. Zwacka, Reactive oxygen species in oncogenic transformation. Biochem. Soc. Trans. **31**, 1441–1444 (2003)
49. M. Lopez-Lazaro, Excessive superoxide anion generation plays a key role in carcinogenesis. Int. J. Cancer **120**, 1378–1380 (2007)
50. T. Kamata, Roles of Nox1 and other Nox isoforms in cancer development. Cancer Sci. **100**, 1382–1388 (2009)
51. F. Weinberg, N.S. Chandel, Reactive oxygen species-dependent signaling regulates cancer. Cell. Mol. Life Sci. ***66***, 3663–3673 (2009)
52. J. Jürgensmeier, C.P. Schmitt, E. Viesel, P. Höfler, G. Bauer, TGF-ß-treated normal fibroblasts eliminate transformed fibroblasts by induction of apoptosis. Cancer Res. **54**, 393–398 (1994)
53. J. Jürgensmeier, P. Höfler, G. Bauer, TGF-ß-induced elimination of transformed fibroblasts by normal cells: independence of cell-to-cell contact and dependence on reactive oxygen species. Int. J. Oncol. **5**, 525–531 (1994)
54. C. Langer, J.M. Jürgensmeier, G. Bauer, Reactive oxygen species act both at TGF-ß-dependent and -independent steps during induction of apoptosis of transformed cells by normal cells. Exp. Cell Res. ***222***, 117–124 (1996)
55. J. Panse, M.-L. Hipp, G. Bauer, Fibroblasts transformed by chemical carcinogens are sensitive for intercellular induction of apoptosis: implications for the control of oncogenesis. Carcinogenesis ***18***, 259–264 (1997)
56. E. Beck, R. Schäfer, G. Bauer, Sensitivity of transformed fibroblasts for intercellular induction of apoptosis is determined by their transformed phenotype. Exp. Cell Res. ***234***, 47–56 (1997)
57. P. Kundrát, G. Bauer, P. Jacob, W. Friedland, Mechanistic modelling suggests that the size of preneoplastic lesions is limited by intercellular induction of apoptosis in oncogenically transformed cells. Carcinogenesis **33**, 253–259 (2012)
58. S. Pottgiesser, S. Heinzelmann, G. Bauer, Intercellular HOCl-mediated apoptosis induction in malignant cells: interplay between NOX1-dependent superoxide anion generation and DUOX-related HOCl-generating peroxidase activity. Anticancer Res. **35**, 5927–5943 (2015)
59. M. Jaganjac, A. Cipak, R.J. Schaur, N. Zarkovic, Pathophysiology of neutrophil-mediated extracellular redox reactions. Front. Biosci. **21**, 839–855 (2016)
60. C.L.L. Chiang, J.A. Ledermann, A.N. Rad, D.R. Katz, B.M. Chain, Hypochlorous acid enhances immunogenicity and uptake of allogeneic ovarian tumor cells by dendritic cells to cross-prime tumor-specific T cells. Cancer Immunol. Immunother. **55**, 1384–1395 (2006)
61. C.L.L. Chiang, J.A. Ledermann, E. Aitkens, E. Benjamin, D.R. Katz, B.M. Chain, Oxidation of ovarian epithelial cancer cells by hypochlorous acid enhances immunogenicity and stimulates T cells that recognize autologous primary tumor. Clin. Cancer Res. **14**, 4898–4907 (2008)
62. C.L.L. Chiang, G. Coukos, L.E. Kandalaft, Whole tumor antigen vaccines: where are we? Vaccine **3**, 344–372 (2015)
63. Z.M. Prokopowicz, F. Arce, R. Biedron, C.L.L. Chiang, M. Ciszek, D.R. Katz, M. Nowakowska, S. Zapotoczny, J. Marcinkiewicz, B.M. Chain, Hypochlorous acid: a natural adjuvant that facilitates antigen processing, cross-priming, and the induction of adaptive immunity. J. Immunol. **184**, 824–835 (2010)

64. R. Biedron, M.K. Konopinski, J. Marcinkiewicz, S. Jozefowski, Oxidation by neutrophil-derived HOCl increases immunogenicity of proteins by converting them into ligands of several endocytic receptors involved in antigen uptake by dendritic cells and macrophages. PLoS ONE **10**, e01123293 (2015)
65. R. Zhou, W.-L. Huang, C. Ma, Y. Zhou, Y.-Q. Yao, Y.-X. Wang, L.-T. Gou, Y. Chen, J.L. Yang, HOCl oxidation-modified CT26 cell vaccine inhibits colon tumor growth in a mouse model. Asian Pac. J. Cancer Prev. **13**, 4037–4043 (2012)
66. W. Bechtel, G. Bauer, Catalase protects tumor cells against apoptosis induction by intercellular ROS signaling. Anticancer Res. **29**, 4541–4557 (2009)
67. W. Bechtel, G. Bauer, Modulation of intercellular ROS signaling of human tumor cells. Anticancer Res. **29**, 4559–4570 (2008)
68. S. Heinzelmann, G. Bauer, Multiple protective functions of catalase against intercellular apoptosis-inducing ROS signaling of human tumor cells. Biol. Chem. **391**, 675–693 (2010)
69. B. Böhm, S. Heinzelmann, M. Motz, G. Bauer, Extracellular localization of catalase is associated with the transformed state of malignant cells. Biol. Chem. **396**, 1339–1356 (2015)
70. G. Bauer, Targeting the protective catalase of tumor cells with cold atmospheric plasma-treated medium (PAM). Anticancer Agents Med. Chem. **18**, 784–804 (2018). https://doi.org/10.2174/1871520617666170801103708
71. G. Bauer, Autoamplificatory singlet oxygen generation sensitizes tumor cells for intercellular apoptosis-inducing signaling. Mech. Ageing Dev. **172**, 59–77 (2018)
72. G. Bauer, Increasing the endogenous NO level causes catalase inactivation and reactivation of intercellular apoptosis signaling specifically in tumor cells. Redox Biol. **6**, 353–371 (2015)
73. K. Sato, H. Ito, H. Kohara, Y. Yamaguchi, K. Adachi, H. Endo, Negative regulation of catalase gene expression in hepatoma cells. Mol. Cell. Biol. **12**, 2525–2533 (1992)
74. D.B. Coursin, H.P. Cihla, J. Sempf, T.D. Oberley, L.W. Oberley, An immunohistochemical analysis of antioxidant and glutathione-S-transferase enzyme levels in normal and neoplastic human lung. Histol. Histopathol. **11**, 851–860 (1996)
75. A.M. Baker, L.W. Oberley, M.B. Cohen, Expression of antioxidant enzymes in human prostatic carcinoma. Prostate **32**, 229–233 (1997)
76. D.G. Bostwick, E.E. Alexander, R. Singh, A. Shan, J. Qian, R.M. Santella, L.W. Oberley, T. Yan, W. Zhong, T.D. Jiang, Antioxidant enzyme expression and reactive oxygen species damage in prostatic intraepithelial neoplasia and cancer. Cancer **89**, 123–134 (2000)
77. H.J. Chung-Man, S. Zheng, S.A. Comhair, C. Farver, S.C. Erzurum, Differential expression of manganese superoxide dismutase and catalase in lung cancer. Cancer Res. **61**, 8578–8585 (2001)
78. J.J. Cullen, F.A. Mitros, L.W. Oberley, Expression of antioxidant enzymes in diseases of the human pancreas: another link between chronic pancreatitis and pancreatic cancer. Pancreas **26**, 23–27 (2003)
79. K.A. Kwei, J.S. Finch, E.J. Thompson, G.T. Bowden, Transcriptional repression of catalase in mouse skin tumor progression. Neoplasia **6**, 440–448 (2004)
80. R.S. Arnold, J. Shi, E. Murad, A.M. Whalen, C.Q. Sun, R. Palavarapu, S. Parthasarathy, J.A. Petros, J.D. Lambeth, Hydrogen peroxide mediates the cell growth and transformation caused by the mitogenic oxidase Nox1. Proc. Natl. Acad. Sci. U. S. A. **98**, 5550–5555 (2001)
81. G. Bauer, The synergistic effect between hydrogen peroxide and nitrite, two long-lived molecular species from cold atmospheric plasma, triggers tumor cells to induce their own cell death. Redox Biol. **26**, 101291 (2019)
82. G. Bauer, D.B. Graves, Mechanisms of selective antitumor action of cold atmospheric plasma-derived reactive oxygen and nitrogen species. Plasma Process. Polym. **13**, 1157–1178 (2016)
83. G. Bauer, Cold atmospheric plasma and plasma-activated medium: antitumor cell effects with inherent synergistic potential. Plasma Med. **9**, 57–88 (2019). https://doi.org/10.1615/PlasmaMed.1019029462
84. G. Bauer, Intercellular singlet oxygen-mediated bystander signaling triggered by long-lived species of cold atmospheric plasma and plasma-activated medium. Redox Biol. **26**, 101301 (2019)

85. G. Bauer, D. Sersenova, D.B. Graves, Z. Machala, Cold atmospheric plasma and plasma-activated medium trigger RONS- based tumor cell apoptosis. Sci. Rep. **9**(1), 14210 (2019). https://doi.org/10.1038/s41598-019-50291-0
86. G. Bauer, D. Sersenova, D.B. Graves, Z. Machala, Dynamics of singlet oxygen-triggered, RONS-based apoptosis induction after treatment of tumor cells with cold atmospheric plasma or plasma-activated medium. Sci. Rep. **9**(1), 13931 (2019). https://doi.org/10.1038/s41598-019-50329-3
87. D.B. Graves, The emerging role of reactive oxygen and nitrogen species in redox biology and some implications for plasma applications to medicine and biology. J. Phys. D. Appl. Phys. **45**, 263001 (2012). https://doi.org/10.1088/0022-3727/45/26/263001
88. D.B. Graves, Mechanisms of plasma medicine: coupling plasma physics, biochemistry, and biology. IEEE Trans. Radiat. Plasma Med. Sci. **1**, 281–292 (2017). https://doi.org/10.1109/TRPMS.2017.2710880
89. K. Wende, P. Williams, J. Dalluge, W. Van Gaens, H. Akoubakr, J. Bischof, T. von Woedtke, S.M. Goyal, K.-D. Weltmann, A. Bogaerts, K. Masur, P.J. Bruggeman, Identification of biologically active liquid chemistry induced by nonthermal atmospheric pressure plasma jet. Biointerphases **10**, 029518 (2015). https://doi.org/10.1116/1.4919710
90. A. Yan, N. Talbot, X. Nourmokammadi, J. Cheng, J. Canady, J.H. Sherman, M. Keidar, Principles of using cold atmospheric plasma stimulated media for cancer treatment. Sci. Rep. **5**, 1833901–18339017 (2015). https://doi.org/10.1038/srep18339
91. N. Kurake, H. Tanaka, K. Ishikawa, T. Kondo, M. Sekine, K. Nakamura, H. Kajiyama, F. Kikkawa, M. Mizuno, M. Hori, Cell survival of glioblastoma grown in medium containing hydrogen peroxide and/or nitrite, or in plasma-activated medium. Arch. Biochem. Biophys. **605**, 102–108 (2016)
92. P.-M. Girard, A. Arbabian, M. Fleury, G. Bauville, V. Puech, M. Dutreix, J.S. Sousa, Synergistic effect of $H_2O_2$ and $NO_2^-$ in cell death induced by cold atmospheric He plasma. Sci. Rep. **6**, 29098 (2016). https://doi.org/10.1038/srep29098.
93. G. Uchida, A. Nakajima, K. Takenaka, T. Kawasaki, K. Koja, M. Shiratani, Y. Setsuhara, Effects of nonthermal plasma jet irradiation on the selective production of $H_2O_2$ and $NO_2^-$ in liquid water. J. Appl. Phys. **120**, 201102 (2016)
94. G. Fridman, G. Friedman, A. Gutsol, A.B. Shekhter, V.N. Vasilets, A. Fridman, Applied plasma medicine. Plasma Process. Polym. **5**, 503–533 (2008)
95. E. Stoffels, Y. Sakiyama, D.B. Graves, Cold atmospheric plasma: charged species and their interactions with cells and tissues. IEEE Trans. Plasma Sci. **36**, 1441–1457 (2008)
96. D. Dobrynin, G. Fridman, G. Friedman, A. Fridman, Physical and biological mechanisms of direct plasma interaction with living tissue. New J. Phys. **11**, 115020 (2009)
97. M. Vandamme, S. Pesnel, E. Barbosa, S. Dozias, J. Sobilo, S. Lerondel, A. Le Pape, J.-M. Pouvesle, Antitumor effect of plasma treatment on U87 glioma xenografts: preliminary results. Plasma Process. Polym. **7**, 264–273 (2010)
98. T. Von Woedtke, H.-R. Metelmann, K.-D. Weltmann, Clinical plasma medicine: state and perspectives of in vivo application of cold atmospheric plasma. Contrib. Plasma Physics **54**, 104–117 (2014)
99. M. Laroussi, From killing bacteria to destroying cancer cells: 20 years of plasma medicine. Plasma Process. Polym. **11**, 1138–1141 (2014)
100. M. Laroussi, Low-temperature plasma jet for biomedical applications: a review. IEEE Trans. Plasma Sci. **43**, 703–712 (2015)
101. M. Keidar, Plasma for cancer treatment. Plasma Sources Sci. Technol. **24**, 033001 (2015)
102. D.Y. Yan, J.H. Sherman, M. Keidar, Cold atmospheric plasma, a novel promising anti-cancer treatment modality. Oncotarget **8**, 15977–15995 (2017)
103. H.-R. Metelmann, D.S. Nedrelow, C. Seebauer, M. Schuster, T. von Woedtke, K.-D. Weltmann, S.P. Kindler, P.H. Metelmann, S.E. Finkelstein, D.D. Von Hoff, Head and neck cancer treatment and physical plasma. Clin. Plasma Med. **3**, 17–23 (2015)
104. H.-R. Metelmann, C. Seebauer, V. Miller, A. Fridman, G. Bauer, D.B. Graves, J.-M. Pouvesle, R. Rutkowski, M. Schuster, S. Bekeschus, K. Wende, K. Masur, S. Hasse, T. Gerling, M. Hori,

H. Tanaka, E.H. Choi, K.-D. Weltmann, P.H. Metelmann, D.D. von Hoff, T. von Woedtke, Clinical experience with cold plasma in the treatment of locally advanced head and neck cancer. Clin. Plasma Med. **9**, 6–13 (2018). https://doi.org/10.1016/j.cpme.2017.09.001
105. H.R. Metelmann, C. Seebauer, R. Rutkowski, M. Schuster, S. Bekeschus, P. Metelmann, Treating cancer with cold physical plasma: on the way to evidence-based medicine. Contrib. Plasma Physics **58**, 415–419 (2018)
106. M. Schuster, C. Seebauer, R. Rutkowski, A. Hauschild, F. Podmelle, C. Metelmann, B. Metelmann, T. Von Woedtke, S. Hasse, K.-D. Weltmann, H.-R. Metelmann, Visible tumor surface response to physical plasma and apoptotic cell kill in head and neck cancer. J. Cranio-Maxillofac. Surg. **44**, 1445–1452 (2016)
107. K.D. Weltmann, T. von Woedtke, Plasma medicine-current state of research and medical application. Plasma Phys. Control. Fusion **59**, 014031 (2017). https://doi.org/10.1088/0741-3335/59/1/014031
108. M. Keidar, R. Walk, A. Shashurin, P. Srinivasan, P. Sandler, A. Sandler, S. Dasgupta, R. Ravi, R. Guerrero-Preston, B. Trink, Cold plasma selectivity and the possibility of a paradigm shift in cancer therapy. Br. J. Cancer **105**, 1295–1301 (2011)
109. M. Keidar, A. Shashurin, O. Volotskova, M.A. Stepp, P. Srinivasan, A. Sandler, B. Trink, Cold atmospheric plasma in cancer therapy. Phys. Plasma **20**, 057101 (2013). https://doi.org/10.1063/1.4801516
110. M. Ishaq, M. Evans, K. Ostrikov, Effect of atmospheric gas plasmas on cancer cell signaling. Int. J. Cancer **134**, 1517–1528 (2013). https://doi.org/10.1002/ijc.28323
111. N. Barekzi, M. Laroussi, Effects of low temperature plasmas on cancer cells. Plasma Process. Polym. **10**, 1039–1050 (2013)
112. J. Schlegel, J. Köritzer, V. Boxhammer, Plasma in cancer treatment. Clin. Plasma Med. **1**, 2–7 (2013)
113. H. Tanaka, M. Mizuno, K. Ishikawa, K. Takeda, K. Nakamura, F. Utsumi, H. Kajiyama, H. Kano, Y. Okazaki, S. Toyokuni, S. Maruyama, F. Kikkawa, M. Hori, Plasma medical science for cancer therapy: toward cancer therapy using nonthermal atmospheric pressure plasma. IEEE Trans. Plasma Sci. **42**, 3760–3764 (2013). https://doi.org/10.1109/TPS.014.2353659
114. D.B. Graves, Reactive species from cold atmospheric plasma: implications for cancer therapy. Plasma Process. Polym. **11**, 1120–1127 (2014)
115. E.A. Ratovitski, X. Heng, D. Yan, J.H. Sherman, J. Canady, B. Trink, M. Keidar, Anti-cancer therapies of 21st century: novel approach to treat human cancers using cold atmospheric plasma. Plasma Process. Polym. **11**, 1128–1137 (2014)
116. P. Babington, K. Rajjoub, J. Canady, A. Siu, M. Keidar, J.H. Sherman, Use of cold atmospheric plasma in the treatment of cancer. Biointerphases **10**, 029403 (2015)
117. J. Gay-Mimbrera, M.C. Garcia, B.I. Tejera, A. Rodero-Serrano, A.V. Garcia-Nieto, J. Ruano, Clinical and biological principles of cold atmospheric plasma application in skin cancer. Adv. Ther. **33**, 894–909 (2016)
118. S. Kalghati, C. Kelly, E. Cerchar, J. Azizkhan-Clifford, Selectivity of non-thermal atmospheric-pressure microsecond-pulsed dielectric barrier discharge plasma induced apoptosis in tumor cells over healthy cells. Plasma Med. **1**, 249–263 (2011)
119. H. Tanaka, K. Ishikawa, K. Nakamura, H. Kajiyama, H. Komo, T. Kikkawa, M. Hori, Plasma-activated medium selectively kills glioblastoma brain tumor cells by down-regulating a survival signaling molecule, AKT kinase. Plasma Med. **1**, 265–277 (2011). https://doi.org/10.1615/PlasmaMed.2012006275
120. S.N. Zucker, J. Zirnheld, A. Bagati, T.M. DiSanto, B. Des Soye, J.A. Jwawrzyniak, K. Eternadi, M. Nikiforov, R. Berezney, Preferential induction of apoptotic cell death in melanoma cells as compared with normal keratinocytes using a non-thermal plasma torch. Cancer Biol. Ther. **13**, 1299–1306 (2012). https://doi.org/10.4161/cbt.21787
121. M. Wang, B. Holmes, X. Cheng, W. Zhu, M. Keidar, L.G. Zhang, Cold atmospheric plasma for selectively ablating metastatic breast cancer cells. PLoS One **8**, e73741 (2013)
122. R. Guerrero-Preston, T. Ogawa, M. Uemura, G. Shumulinsky, B.L. Valle, F. Pirini, R. Ravi, D. Sidransky, M. Keidar, B. Trink, Cold atmospheric plasma treatment selectively targets head and neck squamous cell carcinoma cells. Int. J. Mol. Med. **34**, 941–946 (2014)

123. N. Kaushik, N. Kumar, C.H. Kim, N. Kaushik, E.H. Choi, Dielectric barrier discharge plasma efficiently delivers an apoptotic response in human monocytic lymphoma. Plasma Process. Polym. **11**, 1175–1187 (2014). https://doi.org/10.1002/ppap.201400102
124. F. Utsumi, H. Kajiyama, K. Nakamura, H. Tanaka, M. Hori, F. Kikkawa, Selective cytotoxicity of indirect nonequilibrium atmospheric pressure plasma against ovarian clear-cell carcinoma. Springerplus **3**, 398–316 (2014). https://doi.org/10.1186/2193-1801-3-398
125. A. Siu, O. Volotskova, X. Cheng, S.S. Khaisa, K. Bian, F. Murad, M. Keidar, J.H. Sherman, Differential effects of cold atmospheric plasma in the treatment of malignant glioma. PLoS ONE **10**, e0126313 (2015). https://doi.org/10.1371/journal.pone.0126313
126. S.J. Kim, T.H. Chung, Cold atmospheric plasma jet-generated RONS and their selective effects on normal and carcinoma cells. Sci. Rep. **6**, 20332 (2016). https://doi.org/10.1038/srep.20332
127. J. Duan, X. Lu, G. He, The selective effect of plasma-activated medium in an in vitro co-culture of liver cancer and normal cells. J. Appl. Phys. **121**, 013302 (2017)
128. C. Canal, R. Fontelo, I. Hamouda, J. Guillem-Marti, U. Cvelbar, M.-P. Ginebra, Plasma-induced selectivity in bone cancer cell death. Free Radic. Biol. Med. **110**, 72–80 (2017)
129. K.R. Liedtke, S. Bekeschus, A. Kaeding, C. Hackbarth, J.P. Kuehn, C.D. Heidecke, W. von Bernstorff, T. von Woedtke, L.I. Partecke, Non-thermal plasma-treated solution demonstrates antitumor activity against pancreatic cancer cells in vitro and in vivo. Sci. Rep. **7**, 8319 (2017). https://doi.org/10.1038/sa1598-017-08560-3
130. K. Wende, S. Straßenburg, B. Haertel, M. Harms, S. Holtz, A. Barton, K. Masur, T. von Woedtke, U. Lindequist, Atmospheric pressure plasma jet treatment evokes transient oxidative stress in HaCat keratinocytes and influences cell physiology. Cell Biol. Int. **38**, 412–425 (2014)
131. A.M. Hirst, M.S. Simms, V.M. Mann, N.J. Maitland, D. O'Connell, F.M. Frame, Low temperature plasma treatment induces DNA damage leading to necrotic cell death in primary prostate epithelial cells. Br. J. Cancer **112**, 1536–1545 (2015). https://doi.org/10.1038/bjc.2025.113
132. S. Bekeschus, K. Masur, J. Kolata, K. Wende, A. Schmidt, L. Bundscherer, A. Barton, A. Kramer, B. Bröker, K.-D. Weltmann, Human mononuclear cell survival and proliferation is modulated by cold atmospheric plasma jet. Plasma Process. Polym. **10**, 706–713 (2013)
133. S. Bekeschus, S. Iseni, S. Reuter, K. Masur, K.-D. Weltmann, Nitrogen shielding of an argon plasma jet and its effects on human immune cells. IEEE Trans. Plasma Sci. **43**, 776–781 (2015). https://doi.org/10.1109/TPS.205.2393379
134. L. Bundscherer, S. Bekeschus, H. Tresp, S. Hasse, S. Reuter, K.-D. Weltmann, U. Lindequist, K. Masur, Viability of human blood leucocytes compared with their respective cell lines after plasma treatment. Plasma Med. **3**, 71–80 (2013)
135. F. Girard-Sahun, V. Badets, P. Lefrançois, N. Sojic, F. Clement, S. Arbault, Reactive oxygen species generated by cold atmospheric plasmas in aqueous solution: successful electrochemical monitoring in situ under a high voltage system. Anal. Chem. **91**, 8002–8007 (2019). https://doi.org/10.1021/acs.analchem.9b01912
136. M. Vandamme, E. Robert, S. Lerondel, V. Sarron, D. Ries, S. Dozias, J. Sobilo, D. Gosset, C. Kieda, B. Legrain, J.-M. Pouvesle, A. Le Pape, ROS implication in a new antitumor strategy based on nonthermal plasma. Int. J. Cancer **130**, 2194–2185 (2012)
137. F. Utsumi, H. Kajiyama, K. Nakamura, H. Tanaka, M. Mizuno, K. Ishikawa, H. Kondo, H. Kano, M. Hori, F. Kikkawa, Effect of indirect nonequilibrium atmospheric pressure plasma on anti-proliferative activity against chronic chemo-resistant ovarian cancer cells *in vitro* and *in vivo*. PLoS ONE **8**, e81576 (2013). https://doi.org/10.1371/journal.pone.0081576
138. D. Yan, J.H. Sherman, X. Cheng, E. Ratovitski, J. Canady, M. Keidar, Controlling plasma stimulated media in cancer treatment application. Appl. Phys. Lett. **105**, 22410101–22410104 (2014). https://doi.org/10.1063/1.4902875
139. S. Mohades, M. Laroussi, J. Sears, N. Barekzi, H. Razavi, Evaluation of the effects of a plasma-activated medium on cancer cells. Phys. Plasmas **22**, 122001 (2015)

140. T. Adachi, H. Tanaka, S. Nonomura, H. Hara, S.-I. Kondo, M. Hori, Plasma-activated medium induces A459 cell injury via a spiral apoptotic cascade involving the mitochondrial-nuclear network. Free Radic. Biol. Med. **79**, 28–44 (2015)
141. N. Kumar, J.H. Park, S.N. Jeon, B.S. Park, E.H. Chori, P. Attri, The action of microsecond -pulsed plasma-activated media on the inactivation of human lung cancer cells. J. Phys. D. Appl. Phys. **49**, 11540101–11540109 (2016). https://doi.org/10.1088/0022-3727/49/11/115401
142. D. Yan, N. Nourmohammadi, K. Bian, F. Murad, J.H. Sherman, M. Keidar, Stabilizing the cold plasma-stimulated medium by regulating medium‘s composition. Sci. Rep. **6**, 2602 (2016). https://doi.org/10.1038/srep26016
143. Y. Li, M.H. Kang, H.S. Uhm, G.J. Lee, E.H. Choi, I. Han, Effects of atmospheric-pressure non-thermal bio-compatible plasma and plasma activated nitric oxide water on cervical cancer cells. Sci. Rep. **7**, 45781 (2017). https://doi.org/10.1038/srep45781
144. D. Koensgen, I. Besic, D. Gumbel, A. Kaul, M. Weiss, K. Diesing, A. Kramer, S. Bekeschus, A. Mustea, M.B. Stope, Cold atmospheric plasma (CAP) and CAP-stimulated cell culture media suppress ovarian cancer cell growth – a putative treatment option in ovarian cancer therapy. Anticancer Res. **37**, 6739–6744 (2017)
145. J.-I. Ikeda, H. Tanaka, K. Ishikawa, H. Sakakita, Y. Ikehara, M. Hori, Plasma-activated medium (PAM) kills human cancer-initiating cells. Pathol. Int. **68**, 23–30 (2018)
146. D. Yan, J.H. Sherman, M. Keidar, The application of cold-atmospheric plasma-activated solutions in cancer treatment. Anti Cancer Agents Med. Chem. **18**, 769–775 (2018)
147. F. Saadati, H. Mahdikia, H.-A. Abbaszadeh, M.A. Abdollahifar, M.S. Khoramgah, B. Shokri, Comparison of direct and indirect cold atmospheric-pressure plasma methods in the b16f10 melanoma cancer cells treatment. Sci. Rep. **8**, 7689 (2018). https://doi.org/10.1038/s41598-018-25990-9
148. D. Yan, H. Cui, W. Zhu, A. Talbot, L.G. Zhang, J.H. Sherman, M. Keidar, The strong cell-based hydrogen peroxide generation triggered by cold atmospheric plasma. Sci. Rep. **7**, 10831 (2017). https://doi.org/10.1038/s41598-017-11480-x
149. F. Judée, C. Fongia, B. Ducommun, M. Yousfi, V. Lobjois, N. Merbahi, Short and long time effects of low temperature plasma activated media on 3d multicellular tumor spheroids. Sci. Rep. **6**, 21421 (2016)
150. E. Freund, K.R. Liedtke, J. van der Linde, H.-R. Metelmann, C.-D. Heidecke, L.-I. Partecke, S. Bekeschus, Physical plasma-treated saline promotes an immunogenic phenotype in Ct26 colon cancer cells in vitro and in vivo. Sci. Rep. **9**, 634 (2019). https://doi.org/10.1038/s441598-018-37169-3
151. A.S. Verkman, M. Hara-Chikuma, M.C. Papadopoulos, Aquaporins – new players in cancer biology. J. Mol. Med. **86**, 523–529 (2008)
152. D.Y. Yan, A. Talbot, N. Nourmohammadi, J.H. Sherman, X.Q. Cheng, M. Keidar, Toward understanding the selective anticancer capacity of cold atmospheric plasma- a model based on aquaporins. Biointerphases **10**, 040801 (2015)
153. D. Yan, H. Xiao, W. Zhu, N. Nourmohammadi, L.G. Zhang, K. Bian, M. Keidar, The role of aquaporins in the anti-glioblastoma capacity of the cold plasma-stimulated medium. J. Phys. D. Appl. Phys. **50**, 055401 (2017)
154. A. Lin, Y. Gorbanev, J. De Backer, W. Van Boxem, F. Lemière, P. Cos, S. Dewilde, E. Smits, A. Bogaerts, Non-thermal plasma as a unique delivery system of short-lived reactive oxygen and nitrogen species for immunogenic cell death in melanoma cells. Adv. Sci. **6**, 1802062 (2019). https://doi.org/10.1002/advs.201802062
155. Z. Machala, D.B. Graves, Frugal biotech applications of low-temperature plasma. Trends Biotechnol. **36**, 579–581 (2018)
156. B. Halliwell, Reactive species and antioxidants. Redox biology is a fundamental theme of aerobic life. Plant Physiol. **141**, 312 (2006)
157. B. Halliwell, Free radicals and antioxidants: updating a personal review. Nutr. Rev. **70**, 257–265 (2012)
158. J. Herrmann, T. Dick, Redox biology on the rise. Biol. Chem. **393**, 999–1004 (2012)

159. K. Holmstrom, T. Finkel, Cellular mechanisms and physiological consequences of redox-dependent signalling, nature reviews. Mol. Cel. Biol. **15**, 411–421 (2014)
160. D. Jones, H. Sies, The redox code. Antioxid. Redox Signal. **23**, 734–746 (2015)
161. B. Kalyanaraman, Teaching the basics of redox biology to medical and graduate students: oxidants, antioxidants and disease mechanisms. Redox Biol. **1**, 244–257 (2013)
162. R.B. Mikkelsen, P. Wardman, Biological chemistry of reactive oxygen and nitrogen and radiation-induced signal transduction mechanisms. Oncogene **22**, 5734–5754 (2003)
163. C.C. Winterbourn, A.J. Kettle, Redox reactions and microbial killing in the neutrophil phagosome. Antioxid. Redox Signal. **18**, 642–660 (2013)
164. H. Sies, Role of metabolic H2O2. Generation, redox signaling and oxidative stress. J. Biol. Chem. **289**, 8735–8741 (2014)
165. H. Sies, Hydrogen peroxide as a central redox signaling molecule in physiological oxidative stress: oxidative eustress. Redox Biol. **11**, 613–618 (2017)
166. L.K. Folkes, L.P. Candeias, P. Wardman, Kinetics and mechanisms of hypochlorous acid reactions. Arch. Biochem. Biophys. **323**, 120–126 (1995)
167. L.P. Candeias, K.B. Patel, M.R.L. Stratford, P. Wardmann, Free hydroxyl radicals are formed on reaction between the neutrophil-derived species superoxide anion and hypochlorous acid. FEBS **333**, 151–153 (1993)
168. C. Steinebach, G. Bauer, An alternative signalling pathway based on nitryl chloride during intercellular induction of apoptosis. In Vitro Appl. Mol. Toxicol. **14**, 107–120 (2001)
169. A.M. Held, D.J. Halko, J.K. Hurst, Mechanisms of chlorine oxidation of hydrogen peroxide. J. Am. Chem. Soc. **100**, 5732–5740 (1978)
170. M. Saran, C. Michel, W. Bors, Reaction of NO with $O_2^{-}$. Implication for the action of endothelium-derived relaxing factor (EDRF). Free Radic. Res. Commun. **10**, 221–226 (1990)
171. J.S. Beckman, T.W. Beckman, J. Chen, P.A. Marshall, B.A. Freeman, Apparent hydroxyl radical production by peroxynitrite: implications for endothelial injury form nitric oxide and superoxide. Proc. Natl. Acad. Sci. U. S. A. **87**, 1620–1624 (1990)
172. W.H. Koppenol, J.J. Moreno, W.A. Pryor, H. Ischiropoulos, J.S. Beckman, Peroxynitrite, a cloaked oxidant formed by nitric oxide and superoxide. Chem. Res. Toxicol. **5**, 834–842 (1992)
173. R.E. Huie, S. Padmaja, The reaction of NO with superoxide. Free Radic. Res. Commun. **18**, 195–199 (1993)
174. S. Goldstein, G. Czapski, The reaction of NO with $O_2^{\cdot-}$ and $HO_2^{\cdot-}$: a pulse radiolysis study. Free Radic. Biol. Med. **19**, 505–510 (1995)
175. S. Goldstein, D. Meyerstein, R. van Eldik, G. Czapski, Peroxynitrous acid decomposes via homolysis: evidence from high-pressure pulse radiolysis. J. Phys. Chem. A **103**, 6587–6590 (1999)
176. L. Brunelli, V. Yermilov, J.S. Beckman, Modulation of catalase peroxidatic and catalytic activity by nitric oxide. Free Radic.Biol. Med. **30**, 709–714 (2001)
177. D.A. Wink, J.B. Mitchell, Chemical biology of nitric oxide: insights into regulatory, cytotoxic, and cytoprotective mechanisms of nitric oxide. Free Radic. Biol. Med. **25**, 434–456 (1998)
178. Y. Kono, T. Yamasaki, A. Ueda, H. Shibata, Catalase catalyzes of peroxynitrite-mediated phenolic nitration. Biosci. Biotechnol. Biochem. **62**, 448–452 (1998)
179. L. Gebicka, J. Didik, Catalytic scavenging of peroxynitrite by catalase. Int. J. Inorg. Biochem. **103**, 1375–1379 (2009)
180. G.C. Brown, Reversible binding and inhibition of catalase by nitric oxide. Eur. J. Biochem. **232**, 188–191 (1995)
181. S. Moncada, E.A. Higgs, The L-arginine-nitric oxide pathway. New Engl. J. Med. **329**, 2002–2012 (1993)
182. P.R. Gardner, L.A. Martin, D. Hall, A.M. Gardner, Dioxygen-dependent metabolism of nitric oxide in mammalian cells. Free Radic. Biol. Med. **31**, 191–204 (2001)
183. P.R. Gardner, Assay and characterization of the NO dioxygenase activity of flavohemoglobins. Globins and other nitric oxide-reactive proteins. Methods Enzymol. **436**, 217–237 (2008)

184. C.K. Hallstrom, A.M. Gardner, P.R. Gardner, Nitric oxide metabolism in mammalian cells: substrate and inhibitor profiles of a NADPH-cytochrome P450 oxidoreductase-coupled microsomal nitric oxide dioxygenase. Free Radic. Biol. Med. **37**, 216–228 (2004)
185. J.O. Lundberg, F. Weitzberg, M.T. Gladwin, The nitrate-nitrite-nitric oxide pathway in physiology and therapeutics. Nat. Rev. Drug Discov. **7**, 156–167 (2008)
186. S. Goldstein, G. Czapski, Kinetics of nitric oxide autooxidation in aqueous solution in the absence and presence of various reductants. The nature of the oxidizing intermediates. J. Am. Chem. Soc. **117**, 12078–12084 (1995)
187. A. Denicola, B.A. Freeman, M. Trujillo, R. Radi, Peroxynitrite reaction with carbon dioxide/bicarbonate: kinetics and influence on peroxynitrite-mediated reactions. Arch. Biochem. Biophys. **333**, 49–58 (1996)
188. S. Goldstein, G. Czapski, Formation of peroxynitrate from the reaction of peroxynitrite with $CO_2$: evidence for carbonate radical production. J. Am. Chem. Soc. **120**, 3458–3463 (1998)
189. G.L. Squadrito, W.A. Pryor, Oxidative chemistry of nitric oxide: the roles of superoxide, peroxynitrite, and carbon dioxide. Free Radic. Biol. Med. **25**, 392–403 (1998)
190. S. Goldstein, J. Lind, G. Merenyi, Chemistry of peroxynitrites as compared to peroxynitrates. Chem. Rev. **105**, 2457–2470 (2005)
191. J.A. Escobar, A. Rubio, E.A. Lissi, SOD and catalase inactivation by singlet oxygen and peroxyl radicals. Free Radic. Biol. Med. **20**, 285–290 (1996)
192. Y.K. Kim, O.J. Kwon, J.W. Park, Inactivation of catalase and superoxide dismutase by singlet oxygen derived from photoactivated dye. Biochimie **83**, 437–444 (2001)
193. K. Scheit, G. Bauer, Direct and indirect inactivation of tumor cell protective catalase by salicylic acid and anthocyanidins reactivates intercellular ROS signaling and allows for synergistic effects. Carcinogenesis **36**, 400–411 (2015)
194. M. Riethmüller, N. Burger, G. Bauer, Singlet oxygen treatment of tumor cells triggers extracellular singlet oxygen generation, catalase inactivation and reactivation of intercellular apoptosis-inducing signaling. Redox Biol. **6**, 157–168 (2015)
195. G. Bauer, N. Zarkovic, Revealing mechanisms of selective, concentration-dependent potentials of 4-hydroxy-2-nonenal to induce apoptosis in cancer cells through inactivation of membrane-associated catalase. Free Radic. Biol. Med. **81**, 128–144 (2015)
196. G.P. Bienert, J.K. Schjoerring, T.P. Jahn, Membrane transport of hydrogen peroxide. Biochem. Biophys. Acta **1758**, 994–1003 (2006)
197. G.P. Bienert, A.L. MØller, K.A. Kristiansen, A. Schulz, I.M. MØller, J.K. Schjoerring, T.P. Jahn, Specific aquaporins facilitate the diffusion of hydrogen peroxide across membranes. J. Biol. Chem. **282**, 1183–1192 (2007)
198. A.B. Abdelrazzak, S.J. Pottgießer, M.A. Hill, P. O'Neill, G. Bauer, Enhancement of peroxidase release from non-malignant and malignant cells through low-dose irradiation with different radiation quality. Radiat. Res. **185**, 199–213 (2016)
199. S. Heinzelmann, G. Bauer, Site-specific effects of DUOX1-related peroxidase on intercellular apoptosis signaling. Anticancer Res. **35**, 5955–5971 (2015)
200. H. Krüger, G. Bauer, Lactobacilli enhance reactive oxygen species-dependent apoptosis-inducing signalling. Redox Biol. **11**, 715–724 (2017)
201. G. Bauer, HOCl-dependent singlet oxygen and hydroxyl radical generation modulate and induce apoptosis of malignant cells. Anticancer Res. **33**, 3589–3602 (2013)
202. G. Bauer, Central signaling elements of intercellular reactive oxygen/nitrogen species-dependent induction of apoptosis in malignant cells. Anticancer Res. **37**, 499–514 (2017)
203. G. Bauer, Nitric oxide contributes to selective apoptosis induction in malignant cells through multiple reaction steps. Crit. Rev. Oncog. **21**, 365–398 (2016)
204. G. Bauer, SiRNA-based analysis of the abrogation of the protective function of membrane-associated catalase of tumor cells. Anticancer Res. **37**, 567–582 (2017)
205. G. Bauer, M. Motz, The antitumor effect of single-domain antibodies directed towards membrane-associated catalase and superoxide dismutase. Anticancer Res. **36**, 5945–5956 (2016)

206. G. Merenyi, J. Lind, S. Goldstein, G. Czapski, Peroxynitrite homolyzes into .OH and .NO2 radicals. Chem. Res. Toxicol. **11**, 712–713 (1998)
207. H. Christensen, K. Sehested, H. Corfitzen, Reactions of hydroxyl radicals with hydrogen peroxide at ambient and elevated temperature. J. Phys. Chem. **86**, 1588–1590 (1982)
208. S. Miyamoto, G.E. Ronsein, T.C. Correa, G.R. Martinez, M.H.G. Medeiros, P. Di Mascio, Direct evidence of singlet molecular oxygen generation form peroxynitrate, a decomposition product of peroxynitrite. Dalton Trans. **7**, 5720–5729 (2009). https://doi.org/10.1039/b905560f
209. E.J. Szili, J.W. Bradley, R.D. Short, A "tissue model" to study the plasma delivery of reactive oxygen species. J. Phys. D Appl. Phys. **47**, 152002 (2014)
210. E.J. Szili, J.S. Oh, S.H. Hong, A. Hatta, R.D. Short, Probing the transport of plasma-generated RONS in an agarose target as surrogate for real tissue: dependency on time, distance and material composition. J. Phys. D. Appl. Phys. **48**, 202001 (2015). https://doi.org/10.1088/0022-3727/48/20/202001
211. E.J. Szili, S.-H. Hong, J.-S. Oh, N. Gaur, R.D. Short, Tracking the penetration of plasma reactive species in tissue models. Trends Biotechnol. **36**, 594–602 (2018)
212. E.J. Szili, J.S. Oh, H. Fukuhara, R. Bhatia, N. Gaur, C.K. Nguyen, S.H. Hong, S. Ito, K. Ogawa, C. Kawada, T. Shuin, M. Tsuda, M. Furikata, A. Kurabayashi, H. Furuta, M. Ito, K. Inoue, A. Hatta, R.D. Short, Modelling the helium plasma jet delivery of reactive species into a 3D cancer tumor. Plasma Sour. Technol. **27**, 014001 (2018)
213. L. Nie, Y. Yang, J. Duan, F. Sun, X. Lu, G. He, Effect of tissue thickness and liquid composition on the penetration of long-lifetime reactive oxygen and nitrogen species (RONS) generated by plasma jet. J. Phys. D. Appl. Phys. **51**, 345204 (2018)
214. H. Sies, V.S. Sharov, L.-O. Klotz, K. Briviba, Glutathione peroxidase protects against peroxynitrite-mediated oxidation. J. Biol. Chem. **272**, 27812–27817 (1997)
215. M. Steinmann, N. Moosmann, M. Schimmel, C. Gerhardus, G. Bauer, Differential role of extra- and intracellular superoxide anions for nitric oxide-mediated apoptosis induction. In Vivo **18**, 293–310 (2004)
216. C. Whiteside, H.M. Hassan, Role of oxyradicals in the inactivation of catalase by ozone. Free Radic. Biol. Med. **5**, 305–312 (1988)
217. Y.-K. Lee, S.M.K. Kim, S. Hand, Ozone-induced inactivation of antioxidant enzymes. Biochimie **85**, 947–952 (2003)
218. W. Van Boxem, J. Van der Paal, Y. Gorbanev, S. Vanuytsel, E. Smits, S. Dewilde, A. Bogaerts, Anti-cancer capacity of plasma-treated PBS: effect of chemical composition on cancer cell cytotoxicity. Sci. Rep. **7**, 16478 (2017). https://doi.org/10.1038/s41598-017-16758-8
219. P. Lukes, E. Dolezalova, I. Sisrova, M. Clupek, Aqueous-phase chemistry and bactericidal effects from an air discharge plasma in contact with water: evidence for the formation of peroxynitrite through a pseudo-second-order post-discharge reaction of $H_2O_2$ and $HNO_2$. Plasma Sources Sci. Technol. **23**, 015019 (2014)
220. P. Di Mascio, E.J.H. Bechara, M.H.G. Medeiros, K. Briviba, H. Sies, Singlet molecular oxygen production in the reaction of peroxynitrite with hydrogen peroxide. FEBS Lett. **355**, 287–289 (1994)
221. H. Imai, Y. Nakagawa, Biological significance of phospholipid hydroperoxide glutathione peroxidase (PHGPx, GPx4) in mammalian cells. Free Radic. Biol. Med. **34**, 145–169 (2003)
222. D. Yan, L. Lin, J.H. Sherman, J. Candady, B. Trink, M. Keidar, The correlation between the cytotoxicity of cold atmospheric plasma and the extracellular $H_2O_2$-scavenging rate. IEEE Trans. Radiat. Plasma Med. Sci. **2**, 618–623 (2018)
223. D. Yan, W. Xu, X. Yao, L. Lin, J.H. Sherman, M. Keidar, The cell activation phenomena in the cold atmospheric plasma cancer treatment. Sci. Rep. **8**, 15418 (2018). https://doi.org/10.1038/s41598-018-33914-w
224. J. Van der Paal, C. Verheyen, E.C. Neyts, A. Bogaerts, Hampering effect of cholesterol on the permeation of reactive oxygen species through phospholipid bilayer: possible explanation for plasma cancer selectivity. Sci. Rep. **7**, 39526 (2017)

225. M.L. Semmler, S. Bekeschus, M. Schäfer, T. Bernhardt, T. Fischer, K. Witzke, C. Seebauer, H. Rebl, E. Grambow, B. Vollmer, J.B. Nebe, H.-R. Metelmann, T. von Woedtke, S. Emmert, L. Boeckmann, Molecular mechanisms of the efficacy of cold atmospheric pressure plasma (CAP) in cancer treatment. Cancer **12**, 269 (2020)
226. P. Attri, A. Bogaerts, Perspectives of plasma-treated solutions as anticancer drugs. Anti Cancer Agents Med. Chem. **19**, 436–418 (2019)
227. A. Privat-Maldonado, C. Bengtson, J. Razzokov, E. Smits, A. Bogaerts, Modifying the tumor microenvironment: challenges and future perspectives for anticancer plasma treatment. Cancer **2019**, 11 (1920)
228. S. Bekeschus, K. Wende, M.M. Hefny, K. Rödder, H. Jablonowski, A. Schmidt, T. von Woedtke, K.-D. Weltmann, J. Benedikt, Oxygen atoms are critical in rendering THP-1 leukaemia cells susceptible to cold atmospheric plasma-induced apoptosis. Sci. Rep. **7**, 2791 (2017)
229. V. Jirásek, P. Lukeš, Formation of reactive chlorine species in saline solution treated by non-equilibrium atmospheric pressure He/O2 plasma jet. Plasma Sources Sci. Technol. **28**, 035015 (2019)
230. I. Ali, S.N. Khan, C. Chatzicharalampous, D. Bai, H.M. Abu-Soud, Catalase protects MPO self-destruction in response to oxidative stress. J. Inorg. Biochem. **197**, 110706 (2019)
231. L. Apetoh, A. Tesnier, F. Ghiringhelli, G. Kroemer, L. Zitvogel, Interactions between dying tumor cells and the innate immune system determine the efficiency of conventional antitumor therapy. Cancer Res. **68**, 4026–4030 (2008)
232. D.R. Green, T. Ferguson, L. Zitvogel, G. Kroemer, Immunogenic and tolerogenic cell death. Nat. Rev. Immunol. **9**, 353–363 (2009)
233. D.V. Krysko, A.D. Garg, A. Kaczmarek, O. Krysko, P. Agostinis, P. Vandenabeele, Immunogenic cell death and DAMPs in cancer therapy. Nat. Rev. Cancer **12**, 860–875 (2012)
234. G. Kroemer, L. Galluzzi, O. Kepp, L. Zitvogel, Immunogenic cell death in cancer therapy. Annu. Rev. Immunol. **31**, 51–72 (2013)
235. A.D. Garg, P. Agostinos, ER stress, autophagy and immunogenic cell death in photodynamic therapy-induced anti-cancer immune responses. Photochem. Photobiol. Sci. **13**, 474–487 (2014)
236. S.M. Candeias, U.S. Gaipl, The immune system in cancer prevention, development and therapy. Anticancer Agents Med. Chem. **16**, 101–107 (2016)
237. A. Lin, B. Truong, A. Pappas, L. Kirifides, A. Oubarri, S. Chen, S. Lin, D. Dobrynin, G. Fridman, A. Fridman, N. Sang, V. Miller, Uniform nanosecond pulsed dielectric barrier discharge plasma enhances anti-tumor effects by induction of immunogenic cell death in tumors and stimulation of macrophages. Plasma Process. Polym. **12**, 1392–1399 (2015)
238. A. Lin, B. Truong, S. Patel, N. Kaushik, E.H. Choi, G. Fridman, A. Fridman, V. Miller, Nanosecond-pulsed DBD plasma-generated reactive oxygen species trigger immunogenic cell death in A549 lung carcinoma cells through intracellular oxidative stress. Int. J. Mol. Sci. **18**, 966 (2017)
239. A. Lin, B. Truong, G. Fridman, A. Fridman, V. Miller, Immune cells enhance selectivity of nanosecond-pulsed DBD plasma against tumor cells. Plasma Med. **7**, 85–96 (2017)
240. A.G. Lin, B. Xiang, D.J. Merlino, T.R. Baybutt, J. Sahu, A. Fridman, A.E. Snook, V. Miller, Non-thermal plasma induces immunogenic cell death in vivo in murine CT26 colorectal tumors. Oncoimmunology. **7**, e148978 (2018)
241. V. Miller, A. Lin, A. Fridman, Why target immune cells for plasma treatment of cancer. Plasma Chem. Plasma Process. **36**, 259–268 (2016)
242. K. Mizuno, Y. Yonetamari, Y. Shirakawa, T. Akiyama, R. Ono, Anti-tumor immune response induced by nanosecond pulsed streamer discharge in mice. J. Phys. D Appl. Phys. **50**, 12LT01 (2017)
243. S. Bekeschus, A. Mueller, V. Miller, U. Gaipl, K.-D. Weltmann, Physical plasma elicits immunogenic cancer cell death and mitochondrial singlet oxygen. IEE Trans. Radiat. Plasma Med. Sci. **2**, 138–147 (2018). https://doi.org/10.1109/TRPMS.2017.2766027

244. S. Bekeschus, R. Clemen, H.-R. Metelmann, Potentiating anti-tumor immunity with physical plasma. Clin. Plasma Med. **12**, 17–22 (2018)
245. N.K. Kaushik, N. Kaushik, B. Min, K. Choi, H.Y.J. KH, V. Miller, A. Fridman, E.H. Choi, Cytotoxic macrophage-released tumour necrosis factor-alpha (TNF-alpha) as a killing mechanism for cancer cell death after cold plasma activation. J. Phys. D. Appl. Phys. **49**, 084001 (2018)
246. J.R. Kanofsky, P. Sima, Singlet oxygen production from the reactions of ozone with biological molecules. J. Biol. Chem. **266**, 9039–9042 (1991)
247. W. Adam, D.V. Kazakov, V. Kazakov, Singlet-oxygen chemiluminescence in peroxide reactions. Chem. Rev. **105**, 3371–3387 (2005)

# Chapter 6
# Plasma-Activated Solution in Cancer Treatment

**Hiromasa Tanaka, Mounir Laroussi, Sander Bekeschus, Dayun Yan, Masaru Hori, and Michael Keidar**

## Contents



**Abstract** Cold plasma-activated solution (PAS), particularly the cold plasma-activated medium (PAM), is a type of chemotherapy used in the cold plasma-based cancer treatment. Compared with the direct cold plasma treatment, PAM can be stored for a long time and can then be used without dependence on cold plasma sources or devices. Many in vitro and in vivo experiments have demonstrated anti-tumor effects of PAM against a variety of cancer cells. PAM contains a variety of reactive oxygen species (ROS), reactive nitrogen species (RNS), other cold plasma-activated species, and the generated compounds. These species either interact with

H. Tanaka · M. Hori
Center for Low-Temperature Plasma Sciences, Nagoya University, Nagoya, Japan

M. Laroussi (✉)
Plasma Engineering and Medicine Institute, Old Dominion University, Norfolk, VA, USA
e-mail: mlarouss@odu.edu

S. Bekeschus
INP Greifswald, Leibniz Institute for Plasma Science and Technology, Greifswald, Germany

D. Yan
School of Engineering and Applied Science, The George Washington University, Washington, DC, USA

M. Keidar
Mechanical and Aerospace Engineering, School of Engineering and Applied Science, The George Washington University, Washington, DC, USA

 
M. Keidar (ed.), *Plasma Cancer Therapy*, Springer Series on Atomic, Optical, and Plasma Physics 115, https://doi.org/10.1007/978-3-030-49966-2_6

the cellular membrane or the intracellular function to activate signal pathways and the expression of many genes. Several important signaling pathways are affected by PAM. The effectiveness of PAM has been demonstrated in animal studies using peritoneal metastasis model mice.

## 6.1 Cold Plasma-Activated Medium in Current Plasma Cancer Treatment

Cold plasma has been studied as a next-generation cancer therapy [1–3]. Although cancer cells in vitro or tumorous tissues in vivo can be treated directly with cold plasma, the indirect cold plasma treatments have also been developed over the past years [4–6]. The indirect cold plasma treatments include the cold plasma-assisted immunotherapy [7] and the cold plasma-activated solutions (PAS) [8]. PAS includes medium, phosphate-buffered saline (PBS), or Ringer's solutions, which are used to inhibit the growth of tumorous tissue by subcutaneous or intraperitoneal injection in vivo. To date, the CAP-activated medium (PAM) is the most extensively investigated PAS. PAM can be made by treating the medium using the CAP jet or DBD just above its surface (Fig. 6.1) [8–14]. Moreover, PAM can also be made by the direct discharge in the medium [15]. A number of studies have reported that PAM is suitable for the use in treating various cancers, not only causing the apoptosis but also suppressing the migration and invasion abilities of cancer cells [3, 11, 16–27]. The anti-tumor capacity of PAM over the colon adenocarcinoma multicellular tumor spheroids (MCTS) has also been demonstrated in vitro [28]. Noticeable DBS occurred at the peripheral part of the PAM-treated MCTS [28].

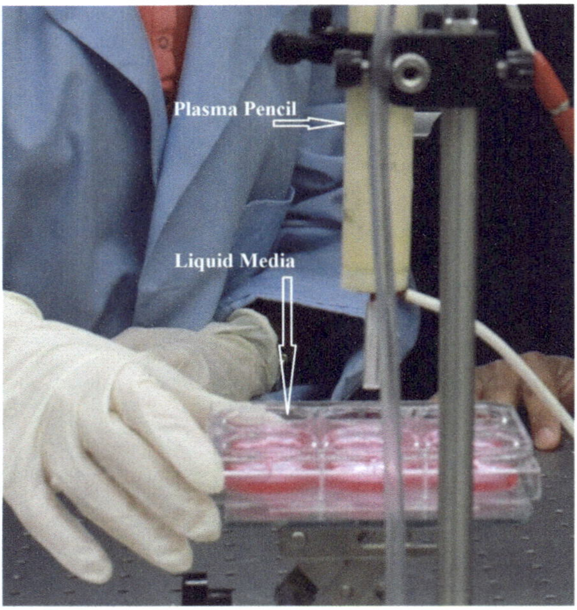


**Fig. 6.1** Biological liquid medium being exposed to the plume of a plasma jet to make PAM

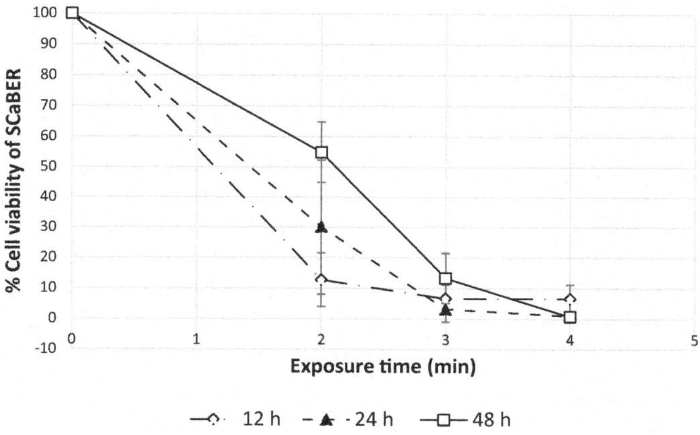


**Fig. 6.2** The anti-tumor effect of PAM on SCaBER cells. PAM was produced by exposing 1 mL of MEM growth media (containing 10% calve bovine serum, 1% penicillin-streptomycin-glutamine) to the plasma plume of the plasma pencil, a pulsed plasma jet. The produced PAM was transferred to wells with SCaBER cells, and the suspension was incubated at 37 °C and 5% $CO_2$. Exposure times were 2, 3.5, and 5 min. Cell viability was assessed at three PAM application times, 12, 24, and 48 h. Experiments were conducted in triplicate and cell viability was normalized to the control and expressed in percent. (Reproduced from Mohades et al. (2016) [29])

As an illustrative example of the effects of PAM on cancer cells, we present in the following section an investigation on the killing of bladder squamous cell carcinoma cells (SCaBER) by PAM (Fig. 6.2). As can be seen in the figure, the PAM created by longer exposure times to cold plasma shows higher kill rates. Cell viability evaluated at all the PAM application times have similar killing trends with the highest killing efficiency (more than 90%) achieved by PAM created by a 5 min of plasma exposure.

Hiromasa Tanaka et al. first demonstrated the selective anti-tumor capacity of PAM over glioblastoma (U251SP) cells [8]. PAM can selectively kill glioblastoma cells rather than normal astrocytes (ACBRI-371) [8]. After that, many studies have validated a similar trend (Fig. 6.3). PAM selectively kills colon cancer cells, lung cancer cells, cervical cancer cells, melanoma cells, breast cancer cells, bladder cancer cells, liver cancer cells, and osteosarcoma cells [12, 13, 15, 29–34]. In addition, the selective anti-tumor effect of PAM has also been observed in a co-culture system with the normal cell line L02 and liver cancer cell line HepG2 [34]. A similar selective anti-tumor effect is also found in a co-culture system with ovarian cancer cell line ES2 and human peritoneal mesothelial cells (HPMCs) [27].

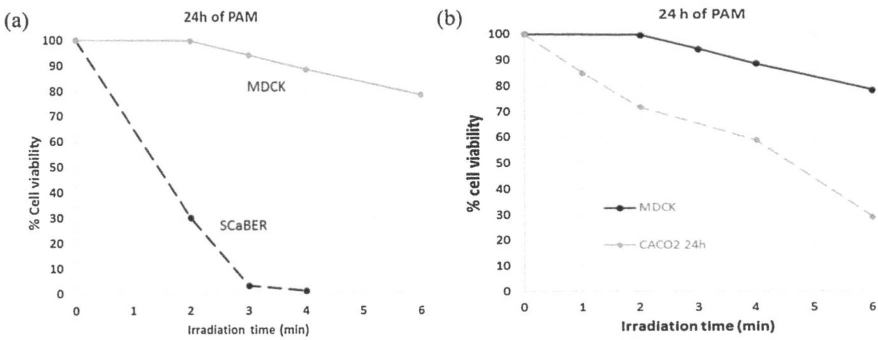


**Fig. 6.3** The selective anti-tumor effect of PAM 24 h after the treatment. (**a**) SCaBER (human colorectal adenocarcinoma) cells and non-cancerous MDCK (Madin–Darby Canine Kidney) cells. (**b**) Caco-2 (human colorectal adenocarcinoma) cells and non-cancerous MDCK (Madin–Darby Canine Kidney) cells. Irradiation time refers to the time the media was exposed to plasma to make PAM. It was found that there is an optimum dose of PAM to suppress the viability of cancer cells significantly while inducing minimum damage to non-malignant cells. (Reproduced from Laroussi et al. (2019) [30])

## 6.2 Reactive Species in PAM

The exposure of liquid media to plasma results in diffusion and solvation of ROS and RNS into the treated liquid. The concentration of reactive species produced in the liquid medium is a function of the duration of plasma exposure, gas type and flow rate, and the chemical composition of the liquid [19]. However, many of the reactive species generated by cold plasma in the gas phase (nitric oxide (NO), ozone ($O_3$), hydroxyl radicals (OH.), singlet oxygen ($^1O_2$), and superoxide anion ($O_2^-$), etc.) may be short-lived and do not diffuse into the bulk of the liquid [35, 36]. Fortunately, some of the solvated species produce more stable molecules such as hydrogen peroxide, $H_2O_2$, and peroxynitrite, $ONOO^-$. $H_2O_2$ is stable and can diffuse into liquid media relatively efficiently (Fig. 6.4).

The anti-tumor effect of PAM is mainly due to the chemical factors, which includes the change of pH, the cold plasma-generated species, and the components generated during the interaction between cold plasma and other components formed in solution. Due to the buffering components in PAM, the CAP treatment only results in a negligible change in pH [37, 38]. ROS and RNS play key roles in the cellular function of PAM. For example, NO in PAM may contribute to its cytotoxicity. The dissolution of NO in PAM significantly increases the cytotoxicity of PAS over HeLa cells [39]. $NO_2^-$ alone will not cause observable cytotoxicity to cancer cells [14].

Among diverse cold plasma-generated species, $H_2O_2$ plays an important role in PAM (Fig. 6.5) [6, 11, 19, 28, 29, 31, 40–42]. $H_2O_2$ alone can cause noticeable inhibition and cell death on cancer cells, though it cannot cause the same cytotoxicity as PAM [6]. Even synergistically using $H_2O_2/NO_2^-$ or $H_2O_2/NO_2^-/NO_3^-$

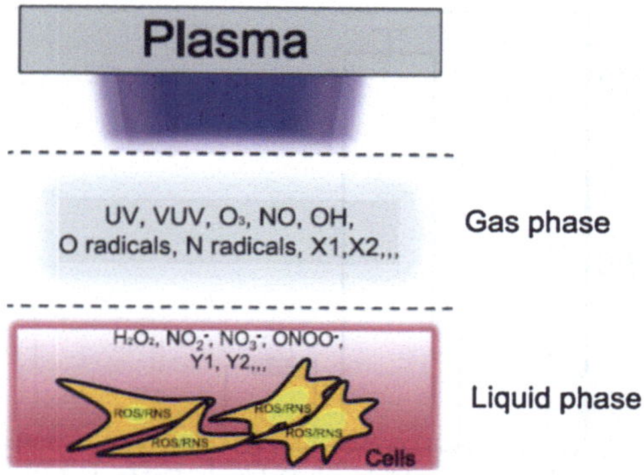


**Fig. 6.4** Cold plasma-generated ROS and RNS in the gas phase, liquid phase, and inside cells. The cold plasma interacts with oxygen, nitrogen, and humidity in the air to produce oxygen, nitrogen, and hydroxyl radicals. Various long-lived reactive species are further formed due to the secondary reaction in the liquid phase, such as hydrogen peroxide, nitrites, and nitrates. These species, in turn, will either directly or indirectly trigger the generation of the intracellular ROS or RNS. (Reproduced from Tanaka et al. (2014) [20])

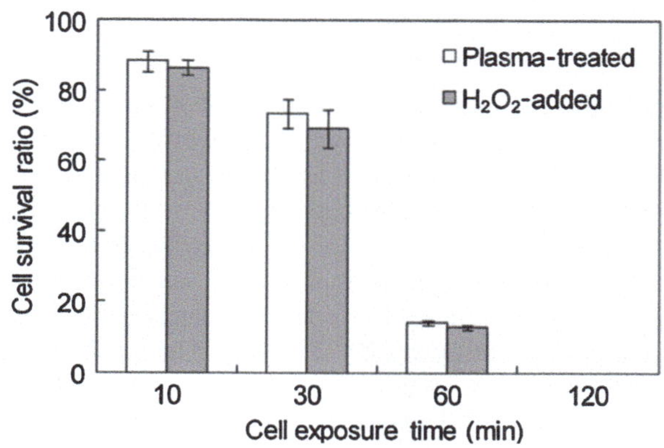


**Fig. 6.5** The earliest evidence that the $H_2O_2$ plays important role in the anti-tumor effect of PAM. Cell survival ratios against the exposure time to cold plasma-treated and $H_2O_2$-added culture media. The survival ratios are normalized by the absorbance intensity of a control which was exposed to the regular culture medium. (Reproduced from Sato et al. (2011) [11])

in the medium also cannot achieve similar selective cytotoxicity as PAM does [12, 43]. Although $H_2O_2$ is cytotoxic, the cytotoxicity depends on its concentration. The concentration of $H_2O_2$ in PAM is typically below the threshold required for synergy between $H_2O_2$ and $NO_2^-$ that produces a cytotoxic effect (Fig. 6.6).

In addition to reactive species, the cold plasma-activated organic compounds exhibit important anti-tumor effects. As an example, we introduced the study on the cold plasma-activated ringer's lactate solution (PAL) here. PAL exhibits anti-tumor effects on glioblastoma cells [44]. Ringer's lactate solution contains NaCl, KCl, CaCl2, and L-sodium lactate, which functions as an anti-tumor factor upon cold plasma treatment (Fig. 6.7).

Similarly, the cold plasma-treated fetal bovine serum (FBS) has also shown cytotoxicity over cervical cancer HeLa cells and hamster cells CHO-K1 [13]. The

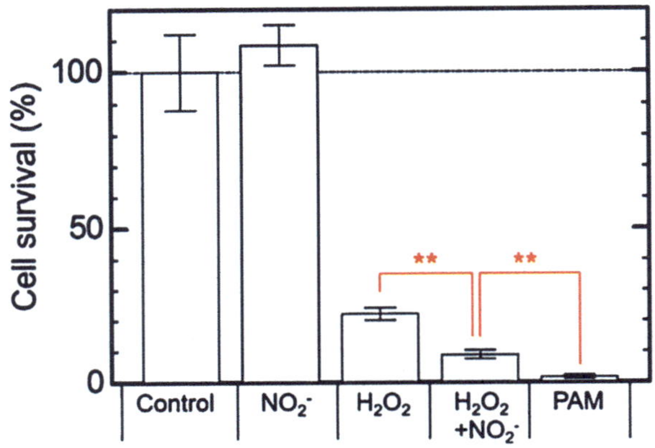


**Fig. 6.6** The synergistic anti-tumor effects of $H_2O_2$ and $NO_2^-$ on glioblastoma cells. For the cold plasma-treated medium (180 s), the PAM contained 63 μM $H_2O_2$ and 1890 μM $NO_2^-$. At a concentration of 63 μM, $H_2O_2$ alone exhibited only weak anti-tumor effect, whereas $NO_2^-$ alone (1890 μM) did not exhibit anti-tumor effects. However, the combination of 63 μM $H_2O_2$ and 1890 μM $NO_2^-$ exhibited an anti-tumor effect almost equivalent to that caused by PAM (Reproduced from Kurake et al. (2016) [43])

potential chemical modification on the contents of FBS is thought to be the main reason for such cytotoxicity [13]. However, due to the extremely long cold plasma treatment time length, such as 10 min in that study, the thermal deactivation of the proteins in FBS may also happen.

In addition to chemical factors, the carrier of the reactive species, such as medium or PBS, is another factor to affect the anti-tumor efficacy of PAM. Specific cancer cells may be vulnerable to one PAS such as PAM than another PAS such as the cold plasma-activated PBS (Fig. 6.8). The glioblastoma cells U87MG and PA-TU-8988T cells were specifically vulnerable to the cold plasma-activated PBS and PSM, respectively [14]. The specific rise of intracellular ROS in U87MG cells and PA-TU-8988T cells may cause the different vulnerabilities of cancer cells to PAM and the cold plasma-activated PBS [14].

The specific cellular response in the nutrient-starved environment and the nutrient-rich environment may cause these cancer cells to show quite different responses to the extracellular reactive species such as $H_2O_2$ [14]. The cytotoxicity of reactive species on cancer cells disappears when the PAM was made of simple buffered solutions such as PBS [44, 45]. The dilution of PBS in DMEM drastically weakens the anti-tumor effect of PAM on pancreatic adenocarcinoma PA-TU-8988T cells and breast cancer MDA-MB-231 cells [14]. Similarly, compared with PAM, PAL induced the generation of a lower amount of intracellular ROS (Fig. 6.9). Compared with medium, the ringer's lactate solution contains much less nutrient for cell growth. These results suggest that both PAM and PAL exert intracellular ROS-based anti-tumor effects against glioblastoma cells, albeit by different molecular mechanisms.

(a)

| | Plasma-treated 2X solutions | Complementary 2X solutions | Treat cells with mixed 1X solutions for 1h |
|---|---|---|---|
| ① | N/A | NaCl, KCl, $CaCl_2$, L-sodium lactate | Untreated Lactec |
| ② | NaCl | KCl, $CaCl_2$, L-sodium lactate | NaCl-GOF |
| ③ | KCl | NaCl, $CaCl_2$, L-sodium lactate | KCl-GOF |
| ④ | $CaCl_2$ | NaCl, KCl, L-sodium lactate | CaCl2-GOF |
| ⑤ | L-sodium lactate | NaCl, KCl, $CaCl_2$ | L-sodium lactate-GOF |
| ⑥ | KCl, $CaCl_2$, L-sodium lactate | NaCl | NaCl-LOF |
| ⑦ | NaCl, $CaCl_2$, L-sodium lactate | KCl | KCl-LOF |
| ⑧ | NaCl, KCl, L-sodium lactate | $CaCl_2$ | CaCl2-LOF |
| ⑨ | NaCl, KCl, $CaCl_2$ | L-sodium lactate | L-sodium lactate-LOF |
| ⑩ | NaCl, KCl, $CaCl_2$, L-sodium lactate | Milli-Q | Plasma-treated Lactec |
| ⑪ | Milli-Q | NaCl, KCl, $CaCl_2$, L-sodium lactate | Treated water |

(b)

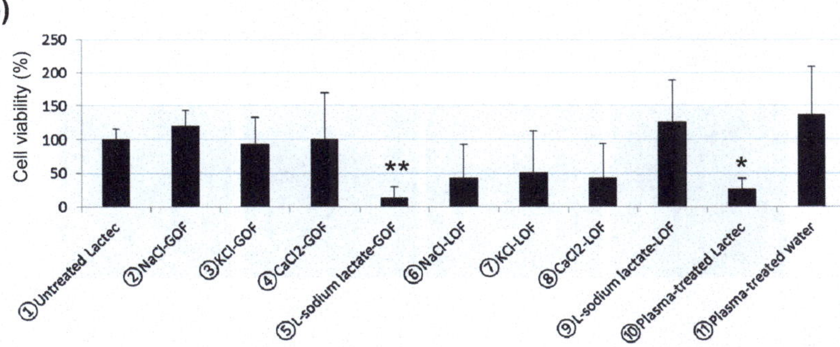


**Fig. 6.7** Identification of PAL components exhibiting anti-tumor effects. (**a**) Schematic of experiments for identifying the anti-tumor components in Ringer's lactate solution by cold plasma irradiation. Each doubly concentrated NaCl, KCl, $CaCl_2$, and L-sodium lactate solution were treated with cold plasma for 2 min and then mixed with the complementary doubly concentrated solutions. These solutions are referred to as NaCl-LOF, KCl-LOF, CaCl2-LOF, and L-sodium lactate-LOF (6–9). Doubly concentrated Ringer's lactate solution was treated with plasma and mixed with the same volume of Milli-Q water (10), and vice versa (11). (**b**) 10,000 U251SP cells were seeded in 200 μL medium in a 96-well plate. On the following day, the medium of the cells in the 96-well plate was replaced with 200 μL of the solutions described in (**a**). (Reproduced from Tanaka et al. (2016) [44])

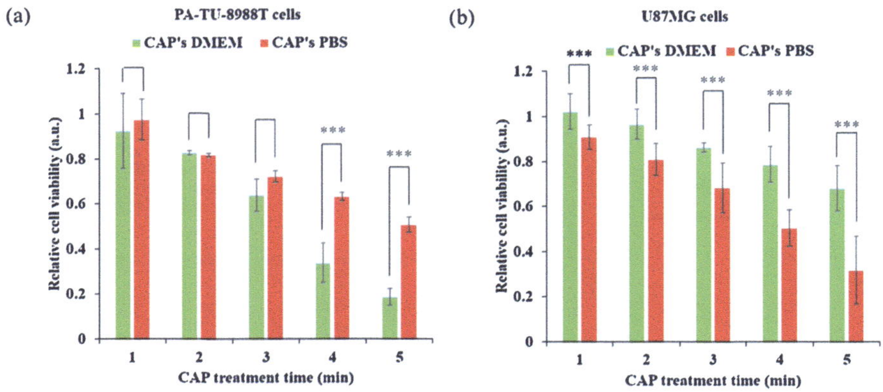


**Fig. 6.8** The different anti-tumor effect of the cold plasma-activated DMEM (CAP's DMEM) and the cold plasma-activated PBS (CAP's PBS) on two cancer cell lines. (**a**) PA-TU-8988T cells. (**b**) U87MG cells. Results are presented as the mean ± s.d. of two independently repeated experiments performed in sextuplicate. Student's *t*-test was performed and the significance is indicated as $^{***}p < 0.005$, $^{**}p < 0.01$, and $^{*}p < 0.05$. (Reproduced from Yan et al. (2017) [14])

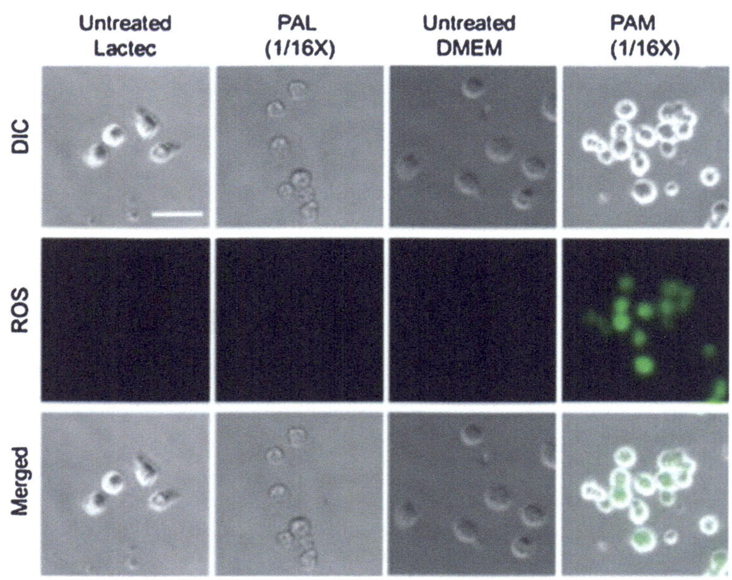


**Fig. 6.9** Identification of components in PAL exhibiting anti-tumor effects. Ten thousand U251SP cells were seeded in an 8-well chamber slide. On the following day, the medium of the cells was replaced with 200 μL of CM-H2DCFDA (10 μM) in PBS. Lactate (8 mL) in a 60 mm dish was treated with plasma (L = 3 mm, 2.0 slm), and the PAL was diluted 16 times with lactate. After 1 h, the 200 μL CMH2DCFDA (10 μM) in PBS was replaced with the PAL and untreated lactate. After 2 h, the PALs or untreated lactate were replaced with 200 μL culture medium. After 2 h, the cells were observed using the fluorescent microscope. (Reproduced from Tanaka et al. (2016) [44])

## 6.3 Intracellular Molecular Mechanism of Cancer Cell Death Induced by PAM Treatment

The long-lived reactive species, such as $H_2O_2$ and $NO_2^-$, are thought to play an important role as sources for singlet oxygen formation on the cell membrane [46, 47]. The reaction between $H_2O_2$ and $NO_2^-$ will generate peroxynitrite ($ONOO^-$), followed by the primary singlet oxygen. Primary singlet oxygen locally inactivates membrane-associated catalase, which leads to the generation of secondary singlet oxygen, thus triggering anto-amplification of singlet oxygen generation. Secondary singlet oxygen is generated by $H_2O_2$ and $ONOO^-$ supplied by NADPH oxidase-1 (NOX1) and nitric oxide synthase (NOS). $NOX_1$ generates superoxide anions ($O_2^-$) extracellularly, whereas NOS generates NO, which is capable of passing through the cell membrane.

The biological effects of $H_2O_2$ include the peroxidation of lipids, induction of DNA damage, and playing a role in mitogenic stimulation and cell cycle regulation. Exposure of cells to $ONOO^-$ can trigger caspase activation followed by apoptosis, or can cause lipid peroxidation, protein nitration or oxidation, leading to necrosis.

Intracellular levels of ROS are generally elevated in tumor cells, and the treatment with PAM can induce further intracellular ROS generation in these cells. It appears that cancer cells, which are under high oxidative stress, cannot withstand further increase in ROS and RNS such as OH·, O, $H_2O_2$, and $ONOO^-$. A substantial increase in the concentrations of these species can induce cell death, including apoptosis and/or necrosis [30]. The threshold for the apoptosis is lower in the tumor cells than the corresponding non-malignant cell counterparts [48]. This "boost" hypothesis explains the oxidative stress-mediated selective killing of cancer cells.

Intracellular RONS generally damage proteins and DNA and act as signaling molecules. Many oxidative stress responses signal pathways have been studied in the cold plasma-treated and the PAM-treated cells. Interactions between cell membrane components and the reactive species might be important, as the cell membrane serves as an interface between the extracellular environment and the intracellular biochemical processes.

Aquaporins (AQPs) might play an important role in determining the anti-tumor effects of PAM against cancer cells [49]. Indeed, knockdown of the gene encoding AQP8, which functions as an efficient $H_2O_2$ channel in glioblastoma cells, significantly reduced the anti-tumor effects of PAM. Hydroxyl radical, which is generated from $H_2O_2$ via the Fenton reaction, is the most harmful ROS, and it was reported that iron mediates PAM-induced cell injury [31].

The development of most cancers likely involves multiple mutations. Mutations in genes associated with cell survival and proliferation signaling networks are especially important in terms of cancer development and therapeutic targeting. Two particularly important signaling pathways associated with survival and proliferation signaling networks are the phosphoinositide 3-kinase (PI3K)/AKT and rat sarcoma (RAS)/mitogen-activated protein kinase (MAPK) signaling pathways. As many as 50% of glioblastoma cells harbor somatic mutations in the PI3K/AKT pathway [50, 51]. An essential regulator of the PI3K/AKT pathway is the phosphatase

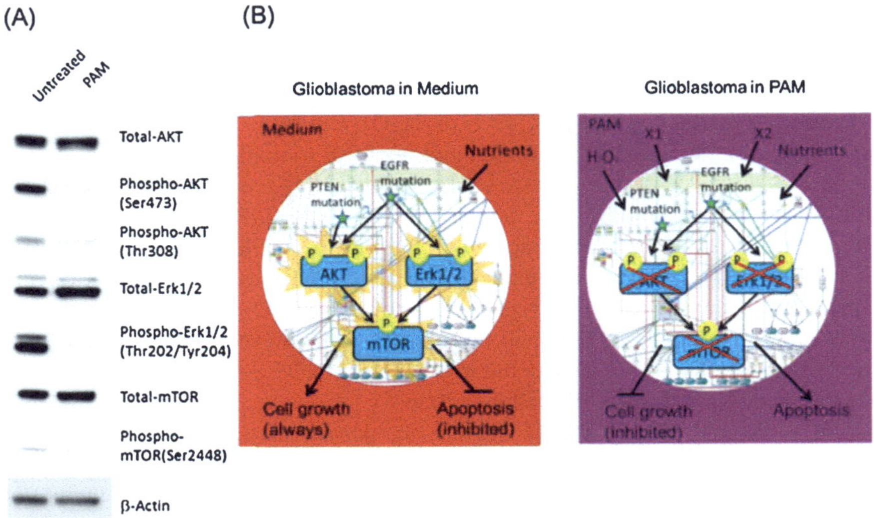


**Fig. 6.10** PAM downregulates both the PI3K/AKT and RAS/MAPK signaling pathways in U251SP glioblastoma cells. Western blotting analysis (**a**) and models illustrating the intracellular molecular mechanisms of apoptosis induced by PAM (**b**). (Reproduced from Tanaka et al. (2014) [20])

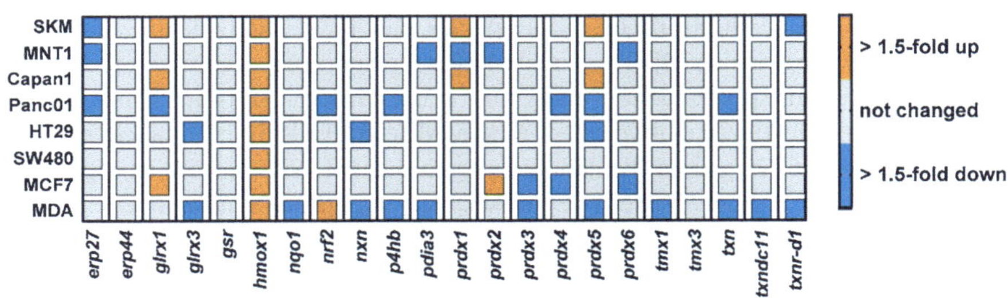


**Fig. 6.11** HMOX1 is a biomarker for the treatment of cancer cells exposed to the plasma-treated medium. (Reproduced from Bekeschus et al. (2018) [54])

and tensin homolog (PTEN) tumor suppressor gene [52]. In U251SP glioblastoma cells harboring mutations in survival and proliferation signaling network genes such as PTEN, both the PI3K/AKT and RAS/MAPK signaling pathways are constitutively activated. Treatment with PAM downregulated both the PI3K/AKT and the RAS/MAPK signaling pathways in U251SP cells (Fig. 6.10) [8, 20, 53]. These results suggest that PAM induces the apoptosis of glioblastoma cells by downregulating survival and proliferation signaling networks.

Bekeschus and co-workers conducted a comprehensive screening of eight human cancer cell lines and 27 genes associated with redox regulation. To that end they identified HMOX1 as a biomarker to expose interaction of plasma-treated medium with cell lines (Fig. 6.11) [54]. Interestingly, HMOX1 expression intensity showed

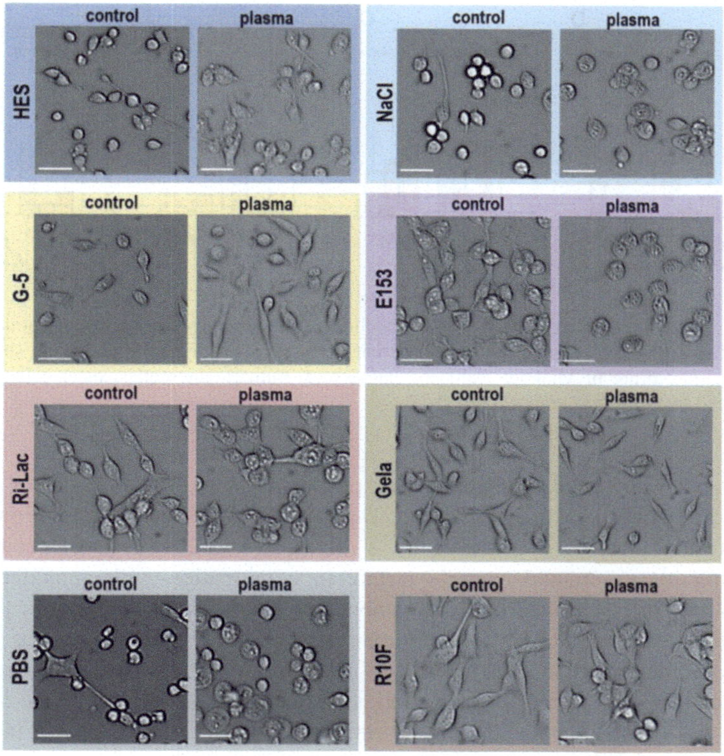


**Fig. 6.12** Morphological changes in colon cancer cells exposed to six clinically approved plasma-treated liquids (as well as PBS and RPMI cell culture medium) compared to the untreated counterparts. (Reproduced from Freund et al. (2019) [55])

a significant positive correlation with the tumor cells not responding to plasma-treated medium in cytotoxic fashion. This means that tumors low in expression or upregulation of HMOX1 might be more prone to plasma-treated medium-induced toxicity.

In another mechanistic study from that group, a comprehensive analysis of the "ideal" plasma-treated liquid was performed. This is because the hurdles of receiving the accreditation of complex liquid such as cell culture medium for the clinical practice are high, while at the same time, a number of clinically accredited liquids exist. The study performed storage, redox chemical, cell morphological, and cytotoxic analysis of six clinically accepted liquids, and compared them to PBS and RPMI cell culture medium (Fig. 6.12). The results revealed that sodium chloride and ringer's lactate are the most promising candidates to pursue this promising novel treatment avenue further [55].

Since cancer patients often receive drugs, Bekeschus and colleagues next asked the question whether there might be an additive or synergistic activity with oxidizing liquids and chemotherapy. After screening a drug library of 80 kinase-inhibitors,

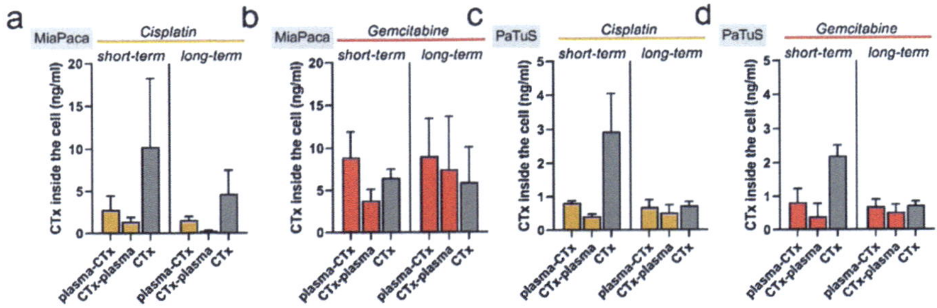


**Fig. 6.13** Mass spectrometry analysis of intracellular chemotherapy drug concentrations (Gemcitabine, Cisplatin) upon exposure to plasma-treated Ringer's Lactate solution during short-term and long-term culture in MiaPaca (a,b) and PatuS (c,d) human pancreatic cancer cells *in vitro*. (Reproduced from Liedtke et al. (2020) [57])

they found a number of targets showing promising synergistic activity in several human cancer cell lines assayed under 2D and 3D tumor spheroid conditions [56]. To extend this idea closer to the clinic, a subsequent investigation was concerned with using plasma-treated ringer's lactate in combination with gemcitabine or cisplatin, two drugs applied in the treatment schemes of pancreatic cancer patients in the clinical setting [57]. The main findings were that additive toxicity was present in the combination treatment, leading to a decline in metabolic activity and cell growth, and an increase of apoptosis and cell cycle arrest. Importantly, the additive toxicity was found in both 2D cancer cell lines and 3D tumors grown in the in ovo model. Mechanistically, this study is important because it analyzed intracellular drug concentrations by mass spectrometry. Interestingly, drug concentrations were in the tendency higher in cells receiving drugs alone, while combination treatment with plasma-treated ringer's lactate decreased the concentration (Fig. 6.13). This might be an indicator of accelerated cell death with the combination treatment.

## 6.4 Some Guidelines to Make PAM

Extending the cold plasma treatment time is the easiest strategy to enhance the anti-tumor capacity of PAM [6, 8, 12, 13, 44]. Three general principles to improve the anti-tumor effect of PAM have been demonstrated by regulating three factors: the gap between the cold plasma source and the surface of the solution, the contacting area between cold plasma and surface of the solution, as well as the volume of solution (Fig. 6.14). Recently, these principles have been tested in other cold plasma-activated solutions such as PBS by other groups [58]. A larger containing area between the cold plasma and the surface of the solution will obtain higher reactive species concentration in PAM [6]. For instance, the PAM made in a well on a 6-well plate will be much more toxic to cancer cells than the PAM made in a well on a 96-well plate. In addition, a shorter gap between the plasma source and medium will

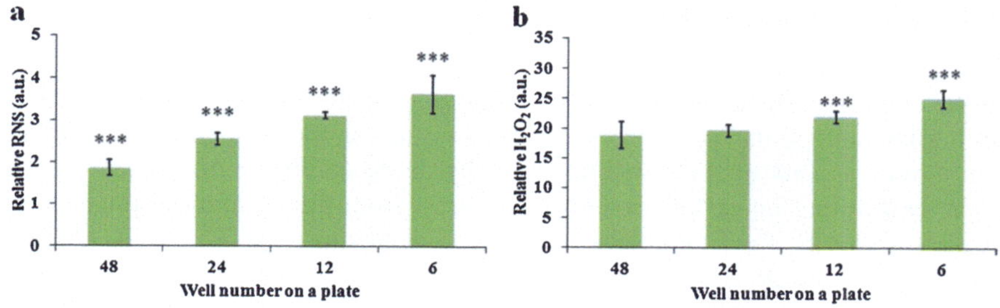


**Fig. 6.14** The well size-dependent ROS/RNS accumulation in the PAM. (**a**) Relative RNS concentration in 1 mL of PAM. (**b**) Relative H2O2concentration in 1 mL of PAM. Results are presented as the mean ± s.d. of three repeated experiments performed in triplicate. Student's *t*-test was performed, and the significance compared with the first bar is indicated as $*p < 0.05$, $**p < 0.01$, $***p < 0.005$. (Reproduced from Yan et al. (2015) [6])

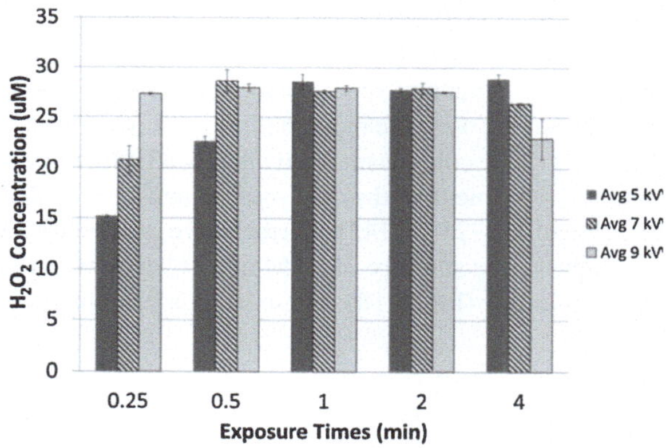


**Fig. 6.15** The $H_2O_2$ concentration in micromoles versus cold plasma exposure time for three different applied voltages. (Reproduced from Mohades et al. (2015) [19])

also make a more toxic PAM with a higher reactive species concentration [6, 58]. The stronger anti-tumor capacity of PAM can also be achieved by using the medium with a smaller volume during the CAP treatment [6, 58].

In addition to these guidelines, the basic operational parameters of cold plasma can also determine the anti-tumor of PAM. For example, the applied voltage to cause the discharge of cold plasma can affect the generation of $H_2O_2$ in the PAM, which directly affects the anti-tumor effect (Fig. 6.15) [19].

Diverse, complicated reactions exist at the interface between cold plasma and aqueous solutions. Several chemicals are highly reactive with the reactive species in PAM. For example, FBS a standard component of a typical complete cell culture medium, is highly reactive to ROS in PAM. FBS eliminates the anti-tumor effect of PAS on glioblastoma cells [10]. In addition, pyruvate is also highly reactive with the $H_2O_2$ in PAM [41]. Cysteine is a common component in nearly all standard mediums. Among all 20 amino acids, however, cysteine is most reactive with the $H_2O_2$ in PAM [10]. In short, an ideal PAM should avoid containing FBS, pyruvate, and cysteine.

## 6.5 The Storage of PAM

In the earliest study about the anti-tumor effect of PAM, the degradation of PAM during the storage has been observed. The anti-tumor species and corresponding anti-tumor capacity of PAM will be gradually degraded during the storage over a wide temperature range for less than just 1 day, from the room temperature, few degrees above the freezing point of water, and as low as −20 °C ~ −30 °C (Fig. 6.16) [6, 10, 28, 29, 41]. Such rapid degradation can be inhibited by storage at a low-temperature environment (−80 °C in a freezer or −150 °C in the liquid nitrogen) [28, 41, 44]. The ideal storage temperature range of most media on the market is between 2 °C and 8 °C. An ideal PAM should be stably stored under such a temperature range.

The degradation mechanism of PAM has been investigated recently. By comparing the $H_2O_2$ concentration in the cold plasma-activated PBS and the cold plasma-activated DMEM after the storage at 8 °C, 22 °C, and at −25 °C, it is found that the degradation of PAM was mainly due to the reaction between the cold plasma-generated species and some components in the medium [42]. Cysteine and methionine were the main medium components causing the degradation of PAM (Fig. 6.17). The cold plasma-activated DMEM without cysteine and methionine was much more stable than the cold plasma-activated DMEM during the storage at 8 °C, 22 °C, and at −25 °C [42]. Furthermore, a tyrosine derivative, 3-nitro-L-tyrosine, can improve the stability of PAM at 8 °C even when the medium contains cysteine and methionine (Fig. 6.18). These studies demonstrate that a noticeable degradation

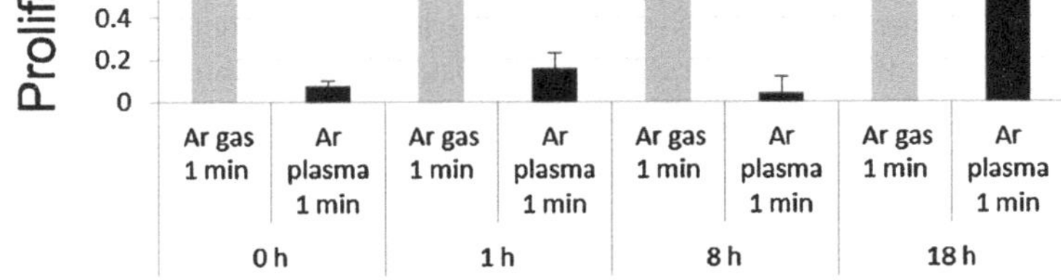


**Fig. 6.16** The duration of the effectiveness of PAM. U251SP glioblastoma cells were plated at a density of $1 \times 10^3$ cells in 200 μL medium in a 96-well plate. On the following day, 3 mL fresh medium (DMEM with 10% FBS, 1% P/S) in a 6-well plate was treated with cold plasma or Ar gas for 1 min (L = 13 mm, 2.0 slm), and 1, 8, or 18 h after plasma treatment of the medium, 200 μL of this (PAM) was used to replace the medium on the cells in the 96-well plate. On the following day, cell viability was assayed using the MTS assay. (Reproduced from Tanaka et al. (2011) [8])

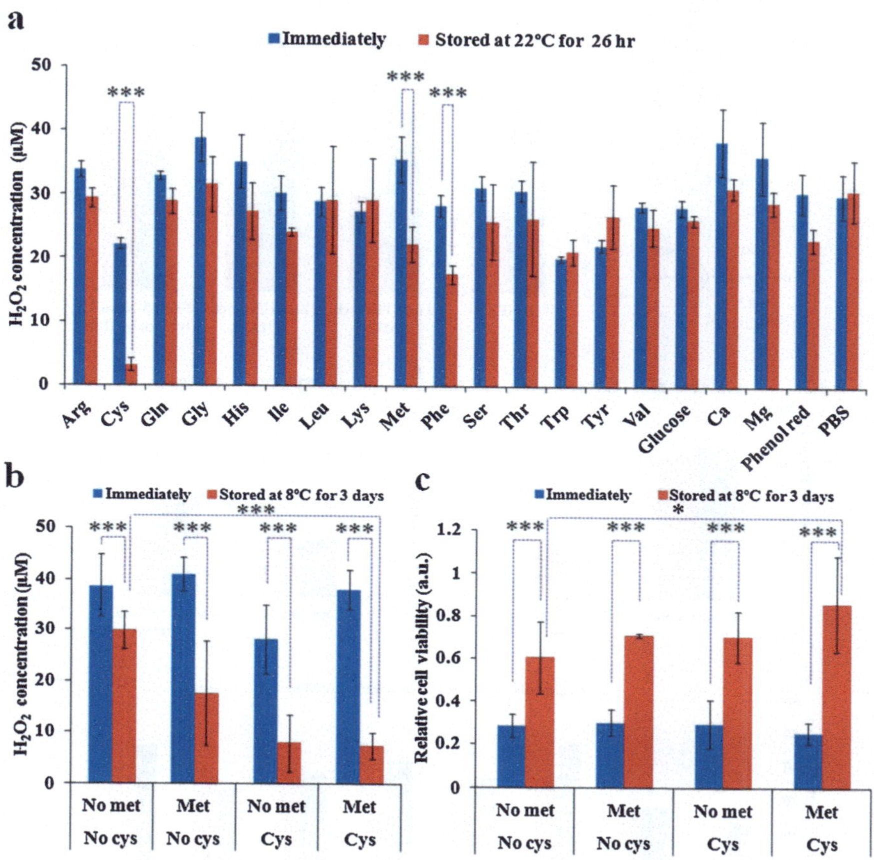


**Fig. 6.17** Cysteine and methionine mainly cause the degradation of PAM at 22 °C and 8 °C. (**a**) The change of $H_2O_2$ concentration in the cold plasma-activated PBS containing a specific component during the storage at 22 °C for 26 h. (**b**) The change of $H_2O_2$ concentration in the cold plasma-activated cys/met-free DMEM, cys-free DMEM, met-free DMEM, and standard DMEM during the storage at 8 °C for 3 days. (**c**) The change of anti-tumor effect of the cold plasma-activated cys/met-free DMEM, cys-free DMEM, met-free DMEM, and standard DMEM during the storage at 8 °C for 3 days. (Reproduced from Yan et al. (2016) [42])

of PAM at around 8 °C will not occur when the corresponding medium does not contain these reactive components such as cysteine and methionine [42].

## 6.6 Animal Studies of PAM

The ultimate aim of the study on PAM is the clinical application. The future clinical application of PAM will require its mechanism of action to be fully elucidated and

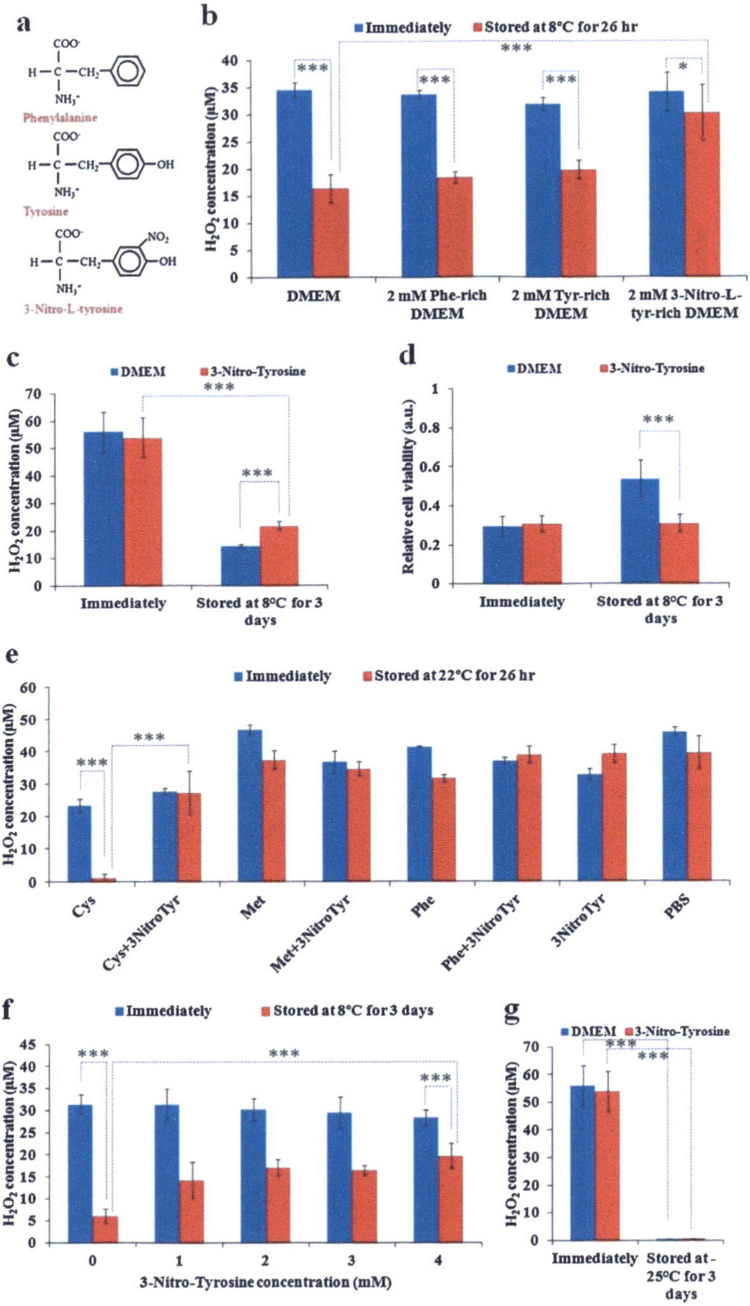


**Fig. 6.18** 3-Nitro-L-Tyrosine inhibits the degradation of PAM. (**a**) Chemical formulas of phenylalanine, tyrosine, and 3-Nitro-L-Tyrosine. (**b**) The change of $H_2O_2$ concentration in the cold

its effectiveness and safety to be sufficiently demonstrated in animal studies. The studies on the anti-tumor effect of PAM in vivo are a key step to build such a foundation. The anti-tumor effects of PAM have been demonstrated in various in vivo experiments. Fumi Utsumi et al. first demonstrated that injecting PAM into the subcutaneous tumors grown from the xenografted chemical-resistant ovarian cancer cells could drastically inhibit the tumor growth (Fig. 6.19) [16]. Similar anti-tumor effects of PAM were reported in a xenograft mouse model of pancreatic cancer [59].

Hiromasa Tanaka et al. demonstrated the strong anti-tumor effects of the CAP-activated Ringer's lactate solution via injecting it into the subcutaneously xenografted human cervical cancer cell line (SiHa)-originated tumors in mice. Such a solution effectively reduced tumor volumes and showed no apparent adverse effects on the mice [44]. Recently, the anti-tumor effect of PAM has been tested using intraperitoneal injections in mouse models of intraperitoneal metastasis of gastric cancer [58] and ovarian cancer (Fig. 6.20) [23]. A similar study was shown in Fig. 6.24. Shigeomi Takeda et al. demonstrated that PAM effectively decreased the formation of peritoneal metastatic nodules by 60% in the mouse model without causing observable adverse events [58]. Importantly, PAM treatment did not affect body weight in these studies, suggesting that treatment using PAM is both efficacious and safe. Another study involving a mouse model of intraperitoneal pancreatic cancer demonstrated that PAL also exhibits anti-tumor effects [61].

The biological safety of PAM in clinical trials is paramount. In one recent study, Liedtke, Bekeschus, and colleagues tested the therapeutic efficacy and safety of plasma-treated RPMI cell culture medium in a syngeneic (same genetic background of tumor cells and mice), orthotopic (tumor located at the anatomic location where it would occur in patients) model of peritoneal pancreatic carcinomatosis [62]. Retarded tumor growth was observed by means of magnetic resonance imaging and weighing of the total tumor mass (Fig. 6.21). The tumor lesions showed extensive apoptosis and a significant decline in the number of proliferating cells, with a mean

**Fig. 6.18** (continued) plasma-activated DMEM, phe-containing DMEM, tyr-containing DMEM, and 3-Nitro-L-tyr-containing DMEM during the storage at 8 °C for 26 h. (**c**) The change of $H_2O_2$ concentration in the cold plasma-activated DMEM and 3-Nitro-L-tyr-containing DMEM during the storage at 8 °C for 3 days. (**d**) The change of anti-tumor effect of the cold plasma-activated DMEM and 3-Nitro-L-tyr-containing DMEM during the storage at 8 °C for 3 days. (**e**) The change of $H_2O_2$ concentration in the cold plasma-activated single amino acids-containing PBS and double amino acids-containing PBS during the storage at 22 °C for 26 h. Corresponding concentration of cys, met, phe, and 3-Nitro-L-tyr in PBS were 0.2 mM, 0.2 mM, 0.4 mM, and 2 mM, respectively. (**f**) The change of $H_2O_2$ concentration in the cold plasma-activated 3-Nitro-L-tyr-containing DMEM with different concentrations during the storage at 8 °C for 3 days. (**g**) The change of $H_2O_2$ concentration in the cold plasma-activated standard DMEM and 3-Nitro-L-tyr-containing DMEM during the storage at −25 °C for 3 days. For all experiments, the volume of solution in each well was 1 mL. The treatment time was 1 min for (**b**, **e**, **f**) and was 2 min for (**c**, **d**, **g**), respectively. Results are presented as the mean ± s.d. of three independently repeated experiments performed in triplicate (**b**, **c**, **e**–**g**) or in sextuplicate (**d**). For the cell viability, the data have been normalized to the corresponding control group. Student's *t*-test was performed and the significance is indicated as $*p < 0.05$, $**p < 0.01$, $***p < 0.005$. (Reproduced from Yan et al. (2016) [42])

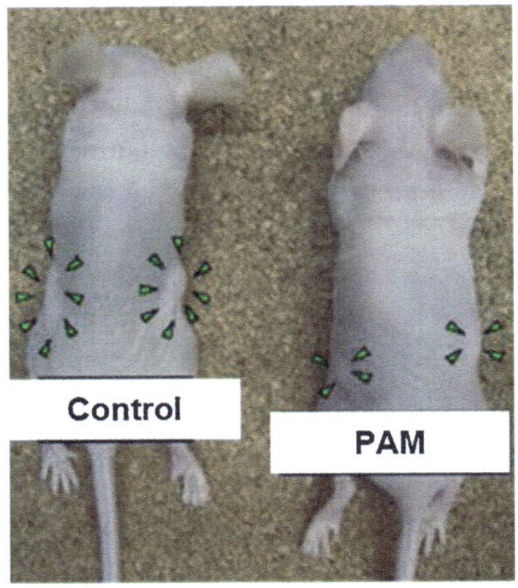


**Fig. 6.19** Anti-tumor effect of PAM in a xenograft mouse model of ovarian cancer. (Reproduced from Utsumi et al. (2013) [16])

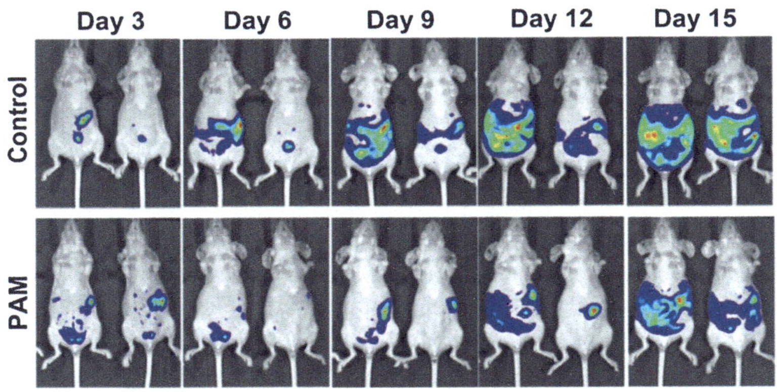


**Fig. 6.20** Anti-tumor effects demonstrated in a mouse model of intraperitoneal metastasis of ovarian cancer. (Reproduced from Utsumi et al. (2017) [23])

penetration depth of 250 μm. In the same study in vitro, a selective toxic effect of the plasma-treated medium was confirmed in pancreatic cancer cells against murine primary non-malignant fibroblasts in terms of metabolic activity and abrogation of cellular growth. Simultaneously, the study confirmed the safety of the approach. None of the animals receiving the plasma-treated medium showed off-target side effects. Extensive analysis of all major murine blood leukocyte populations (e.g., neutrophils, monocytes, cytotoxic T-cells, B-cells, NK-cells) did not indicate a negative influence of plasma-treated cell culture medium compared to animals not receiving such treatment. Moreover, over a dozen of hematological parameters were assessed (e.g., red blood cell count, hematocrit, and mean platelet value) and, similar to the analysis of a range of blood cytokines (e.g., IL-6, IL-10, IL-12, and IFNγ), no significant changes were observed in the group of mice receiving repeated injections of plasma-treated medium. In a follow-up study of the Bekeschus lab

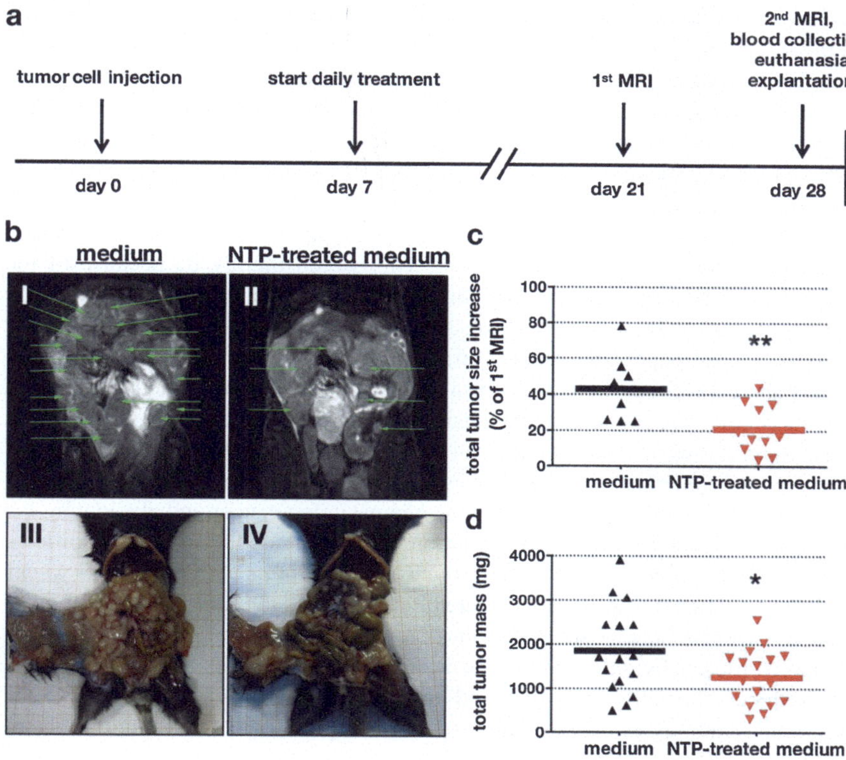


**Fig. 6.21** PAM decreases the number, growth, and size of pancreatic lesions in vivo. The experimental timeline (**a**) is given. The number of tumor nodes in the peritoneum was followed by MRI and was elevated in the control animals (I) compared to the NTP-treatment group (II) at the end of treatment (d28). Also, corresponding macroscopic findings for control (III) and NTP-treated (IV) animals are shown (**b**). MRI-based calculation of tumor volume revealed a significantly decreased total tumor growth in the treatment group (**c**). On day 28, animals were sacrificed and tumor nodes were excised and weighed, showing a significantly decreased total tumor mass (**d**). Representative images of 13 animals are given; green arrows indicate tumor nodes (one green arrow per lesion) (**b**). Data are presented as mean of 8–17 animals (**c**, **d**). (Reproduced from Liedtke et al. (2017) [62])

in Greifswald using tumor lesions from this experiment, the Immunobiology of the therapy with the plasma-treated medium was explored [63]. It was discovered that the treatment group tumor lesions contained significantly more macrophages compared to the animals not receiving medium alone (without plasma treatment). Macrophage infiltration in pancreatic cancer, however, is known to correlate with a poor outcome in patients. This is because of the macrophages acquiring a so-called tumor-associated macrophage (TAM) M2-phenotype that supports tumor growth [64]. By contrast, our study revealed a significant decrease of the M2-macrophage marker CD206 in the tumor lesions exposed to plasma-treated medium, indicating their tumor-toxic nature that might explain the growth retardation seen with the treatment. Multicolor flow cytometry of digested tumor nodules moreover revealed a significant increase of T-cells in mice receiving plasma-treated medium. These cells are known to have the ability to promote anti-tumor effects. This corresponds to the findings of elevated calreticulin exposure, a protein known to enhance pro-immunogenic effects [65], in pancreatic tumor nodules exposed to the plasma-treated medium.

While these results were promising, the applicability of the plasma-treated medium in a clinical setting may be low. This would mainly be because of the organic compounds cell culture medium contains, making accreditation of such liquid problematic. To this end, Bekeschus et al. tested plasma-treated PBS for its activity against pancreatic cancer [66]. The selective apoptosis-inducing effects towards pancreatic cancer cells vs. non-malignant fibroblasts, as found with plasma-treated medium, were replicated when using plasma-treated PBS. This was not only the case for cells cultured in 2D but also in a three-dimensional in ovo tumor model, showing growth retardation and apoptosis (Fig. 6.22).

This study was further extended, testing against colon cancer and elucidating the cellular and molecular mechanisms of action. To investigate the storage conditions, the solution was stored at −20 °C and found to maintain its activity up to 3 weeks post plasma treatment [67]. Plasma-treated PBS oxidized the tumor cells, decreased growth and metabolic activity, led to morphological changes not associated with epithelial-to-mesenchymal (EMT) transition, mediated apoptosis and cell cycle arrest, reduced the cells' motility, and induced hallmarks of the immunogenic cancer cell death as shown by an increased expression of calreticulin, heat-shock protein 70, and HMGB1. The results were replicated in three cancer cell lines and selective when compared against non-malignant HaCaT keratinocytes. Interestingly, incubation with concentration-matched $H_2O_2$ was able to replicate many of the effects observed. This, however, was not true for cytotoxic effects in 3D tumor spheroids, where plasma-treated PBS outperformed $H_2O_2$ by far. Importantly, plasma-treated saline also effectively reduced the tumor burden in vivo in a syngeneic, orthotopic model of peritoneal colon carcinomatosis (Fig. 6.23).

In another recent study, the safety of PAW in an immunocompromised animal model has been investigated [60]. Xu et al. examined the safety of immune-deficient nude mice by oral lavage treatment of PAW. The growth status, main organs, and biochemical blood indexes were investigated [60]. The acute toxicity test results showed that the maximum dose of plasma treatment for 15 min had no lethal effect and other acute toxicity. There were no significant changes in body weight and survival status of mice after 2 min and 4 min of plasma-activated water (PAW) treatment for 2 weeks. After treatment, the major organs, including heart, liver, spleen, lung, and kidney, were not significantly changed in organ coefficient and tissue structure (Fig. 6.24). Biochemical blood markers showed that blood neutrophils and mononuclear cells were slightly increased, and the others remained unchanged. Liver function, renal function, electrolytes, glucose metabolism, and lipid metabolism were not affected by different doses of PAW treatment. The above

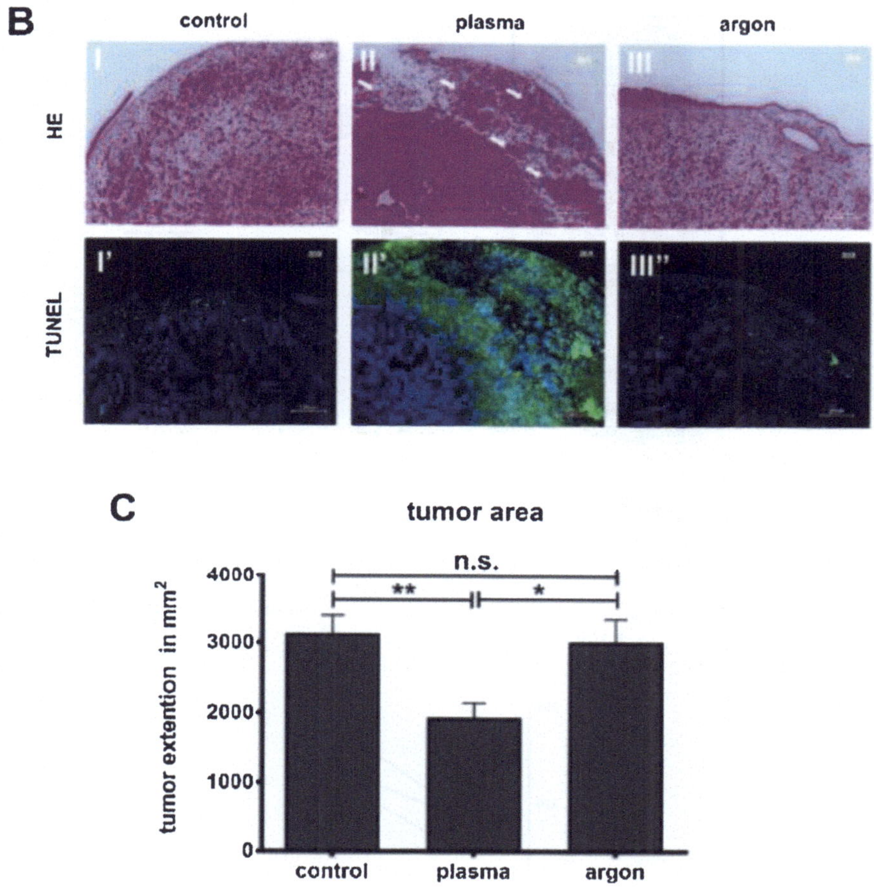


**Fig. 6.22** Reduced pancreatic cancer tumor growth with plasma-treated PBS in an in ovo model. (Reproduced from Bekeschu et al. (2018) [66])

results indicate that PAW treatment can be used to treat immuno-deficient nude mice without significant safety problems [60].

## 6.7 Future Directions of PAM Studies

Cancer therapy using PAM could be combined with personalized cancer treatments in the future. Certain treatments exhibit better effects for some patients than for others, depending on the genetic characteristics of both the patient and the tumors. The development of new types of PAM could enable effective treatments that involve different molecular mechanisms. For example, PAM and PAL exhibit anti-

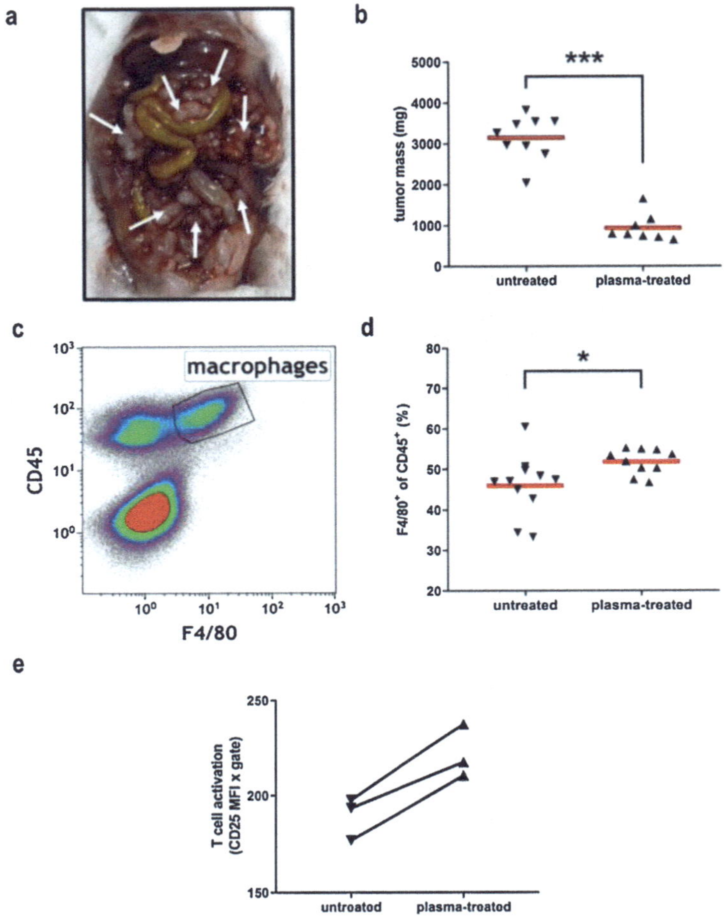


**Fig. 6.23** Plasma-treated saline solution effectively controlled the growth of peritoneal metastasis in a syngeneic, orthotopic tumor model of colon carcinomatosis in mice. The treatment also had an immunological dimensions as seen with an increase in immunogenic cancer cell death, intratumoral macrophages, and enhanced activity of intratumoral T-cells. (Reproduced from Freund et al. (2013) [67])

tumor effects on glioblastoma cells via different mechanisms. Obtaining a more-complete understanding of the mechanism of PAM-induced death of cancer cells could ultimately help make it possible to treat cancer patients using personalized approaches.

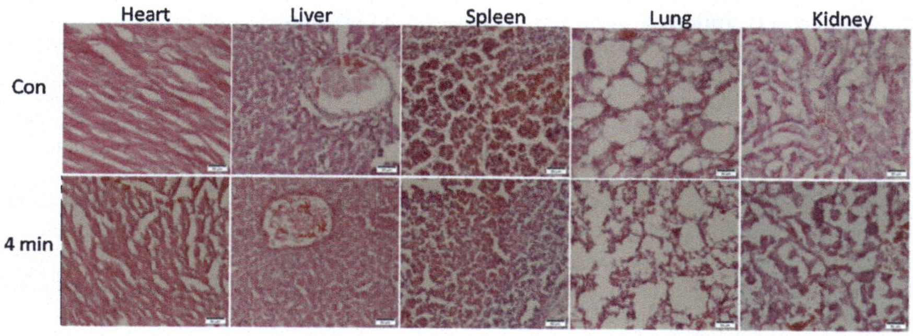


**Fig. 6.24** Effect of cold plasma-activated water (PAW) on the vital organs. (Reproduced from Xu et al. (2018) [60])

# References

1. M. Keidar, A. Shashurin, O. Volotskova, M. Ann Stepp, P. Srinivasan, A. Sandler, B. Trink, Cold atmospheric plasma in cancer therapy. Phys. Plasmas **20**(5), 057101 (2013)
2. N. Barekzi, M. Laroussi, Effects of low temperature plasmas on cancer cells. Plasma Process. Polym. **10**, 1039–1050 (2013)
3. H. Tanaka, K. Ishikawa, State of the art in medical applications using non-thermal atmospheric pressure plasma. Rev. Mod. Plasma Phys. **1**(1), 3 (2017)
4. H.-R. Metelmann, D.S. Nedrelow, C. Seebauer, M. Schuster, T. von Woedtke, K.-D. Weltmann, S. Kindler, P.H. Metelmann, S.E. Finkelstein, D.D. Von Hoff, F. Podmelle, Head and neck cancer treatment and physical plasma. Clin. Plasma Med. **3**(1), 17–23 (2015)
5. A. Shashurin, D. Scott, T. Zhuang, J. Canady, I.I. Beilis, M. Keidar, Electric discharge during electrosurgery. Sci. Rep. **5**, 9946 (2015)
6. D. Yan, A. Talbot, N. Nourmohammadi, X. Cheng, J. Canady, J. Sherman, M. Keidar, Principles of using cold atmospheric plasma stimulated media for cancer treatment. Sci. Rep. **5**, 18339 (2015)
7. V. Miller, A. Lin, A. Fridman, Why target immune cells for plasma treatment of cancer. Plasma Chem. Plasma Process. **36**(1), 259–268 (2016)
8. H. Tanaka, M. Mizuno, K. Ishikawa, K. Nakamura, H. Kajiyama, H. Kano, F. Kikkawa, M. Hori, Plasma-activated medium selectively kills glioblastoma brain tumor cells by down-regulating a survival signaling molecule, AKT kinase. Plasma Med. **1**(3–4), 265–277 (2011)
9. M. Vandamme, E. Robert, S. Lerondel, V. Sarron, D. Ries, S. Dozias, J. Sobilo, D. Gosset, C. Kieda, B. Legrain, J.M. Pouvesle, A. Le Pape, ROS implication in a new antitumor strategy based on non-thermal plasma. Int. J. Cancer **130**(9), 2185–2194 (2012)
10. D. Yan, J.H. Sherman, X. Cheng, E. Ratovitski, J. Canady, M. Keidar, Controlling plasma stimulated media in cancer treatment application. Appl. Phys. Lett. **105**(22), 224101 (2014)
11. E. Boix-garriga, P. Acedo, A. Casadó, T. Sato, M. Yokoyama, K. Johkura, A key inactivation factor of HeLa cell. J. Phys. D. Appl. Phys. **44**(37), 372001 (2011)
12. P.-M. Girard, A. Arbabian, M. Fleury, G. Bauville, V. Puech, M. Dutreix, J.S. Sousa, Synergistic effect of H2O2 and NO2 in cell death induced by cold atmospheric He plasma. Sci. Rep. **6**, 29098 (2016)
13. D. Boehm, C. Heslin, P.J. Cullen, P. Bourke, Cytotoxic and mutagenic potential of solutions exposed to cold atmospheric plasma. Sci. Rep. **6**, 21464 (2016)
14. D. Yan, H. Cui, W. Zhu, N. Nourmohammadi, J. Milberg, L.G. Zhang, J.H. Sherman, M. Keidar, The specific vulnerabilities of cancer cells to the cold atmospheric plasma-stimulated solutions. Sci. Rep. **7**(1), 4479 (2017)

15. N. Kumar, J.H. Park, S.N. Jeon, B.S. Park, E.H. Choi, P. Attri, The action of microsecond-pulsed plasma-activated media on the inactivation of human lung cancer cells. J. Phys. D. Appl. Phys. **49**, 115401 (2016)
16. H. Tanaka, M. Mizuno, K. Ishikawa, K. Takeda, K. Nakamura, F. Utsumi, H. Kajiyama, H. Kano, Y. Okazaki, S. Toyokuni, S. Maruyama, F. Kikkawa, M. Hori, Plasma medical science for cancer therapy: toward cancer therapy using nonthermal atmospheric pressure plasma. IEEE Trans. Plasma Sci. **42**, 3760–3764 (2014)
17. H. Tanaka, M. Mizuno, K. Ishikawa, H. Kondo, K. Takeda, H. Hashizume, K. Nakamura, F. Utsumi, H. Kajiyama, H. Kano, Y. Okazaki, S. Toyokuni, S. Akiyama, S. Maruyama, S. Yamada, Y. Kodera, H. Kaneko, H. Terasaki, H. Hara, T. Adachi, M. Iida, I. Yajima, M. Kato, F. Kikkawa, M. Hori, Plasma with high electron density and plasma-activated medium for cancer treatment. Clin. Plasma Med. **3**, 72–76 (2015)
18. H. Tanaka, M. Mizuno, S. Toyokuni, S. Maruyama, Y. Kodera, H. Terasaki, T. Adachi, M. Kato, F. Kikkawa, M. Hori, Cancer therapy using non-thermal atmospheric pressure plasma with ultra-high electron density. Phys. Plasmas **22**(12), 122004 (2015)
19. H. Tanaka, M. Mizuno, F. Kikkawa, M. Hori, Interactions between a plasma-activated medium and cancer cells. Plasma Med. **6**(1), 101–106 (2016)
20. H. Tanaka, M. Mizuno, K. Ishikawa, S. Toyokuni, H. Kajiyama, F. Kikkawa, M. Hori, New hopes for plasma-based cancer treatment. Plasma **1**, 150–155 (2018)
21. H. Tanaka, M. Mizuno, K. Ishikawa, S. Toyokuni, H. Kajiyama, F. Kikkawa, M. Hori, Molecular mechanisms of non-thermal plasmainduced effects in cancer cells. Biol. Chem. **400**(1), 87–91 (2018)
22. D. Yan, J.H. Sherman, M. Keidar, The application of the cold atmospheric plasma-activated solutions in cancer treatment. Anti-Cancer. Agents Med. Chem. **18**(16), 769–775 (2018)
23. K. Nakamura, Y. Peng, F. Utsumi, H. Tanaka, M. Mizuno, S. Toyokuni, M. Hori, F. Kikkawa, H. Kajiyama, Novel intraperitoneal treatment with non-thermal plasma-activated medium inhibits metastatic potential of ovarian cancer cells. Sci. Rep. **7**(1), 6085 (2017)
24. F. Utsumi, H. Kajiyama, K. Nakamura, H. Tanaka, M. Mizuno, K. Ishikawa, H. Kondo, H. Kano, M. Hori, F. Kikkawa, Effect of indirect nonequilibrium atmospheric pressure plasma on anti-proliferative activity against chronic chemo-resistant ovarian cancer cells in vitro and in vivo. PLoS One **8**(12), e81576 (2013)
25. S. Mohades, N. Barekzi, M. Laroussi, Efficacy of low temperature plasma against SCaBER cancer cells. Plasma Process Polym. **11**(12), 1150–1155 (2014)
26. M. Laroussi, From killing Bacteria to destroying cancer cells: 20 years of plasma medicine. Plasma Process Polym. **11**(12), 1138–1141 (2014)
27. S. Mohades, M. Laroussi, J. Sears, N. Barekzi, H. Razavi, Evaluation of the effects of a plasma activated medium on cancer cells. Phys. Plasmas. **22**(12), 122001 (2015)
28. F. Judée, C. Fongia, B. Ducommun, M. Yousfi, V. Lobjois, N. Merbahi, Short and long time effects of low temperature plasma activated media on 3D multicellular tumor spheroids. Sci. Rep. **6**, 21421 (2016)
29. S. Mohades, N. Barekzi, H. Razavi, V. Maruthamuthu, M. Laroussi, Temporal evaluation of the anti-tumor efficiency of plasma-activated media. Plasma Process Polym. **13**(12), 1206–1211 (2016)
30. M. Laroussi, Effects of PAM on select normal and cancerous epithelial cells. Plasma Res. Express **1**(2), 025010 (2019)
31. T. Adachi, S. Nonomura, M. Horiba, T. Hirayama, T. Kamiya, H. Nagasawa, H. Hara, Iron stimulates plasma-activated medium-induced A549 cell injury. Sci. Rep. **6**, 20928 (2016)
32. N.H. Nguyen, H.J. Park, S.S. Yang, K.S. Choi, J.-S. Lee, Anti-cancer efficacy of nonthermal plasma dissolved in a liquid, liquid plasma in heterogeneous cancer cells. Sci. Rep. **6**, 29020 (2016)
33. K. Saito, T. Asai, K. Fujiwara, J. Sahara, H. Koguchi, N. Fukuda, M. Suzuki-Karasaki, M. Soma, Y. Suzuki-Karasaki, Tumor-selective mitochondrial network collapse induced by atmospheric gas plasma-activated medium. Oncotarget **7**(15), 19910 (2016)

34. J. Duan, X. Lu, G. He, The selective effect of plasma activated medium in an *in vitro* co-culture of liver cancer and normal cells. J. Appl. Phys. **121**(1), 013302 (2017)
35. D.B. Graves, Oxy-nitroso shielding burst model of cold atmospheric plasma therapeutics. Clin. Plasma Med. **2**(2), 38–49 (2014)
36. Y.F. Yue, S. Mohades, M. Laroussi, X. Lu, Measurements of plasma-generated hydroxyl and hydrogen peroxide concentrations for plasma medicine applications. IEEE Trans. Plasma Sci. **44**(11), 2754–2758 (2016)
37. K. Panngom, K.Y. Baik, M.K. Nam, J.H. Han, H. Rhim, E.H. Choi, Preferential killing of human lung cancer cell lines with mitochondrial dysfunction by nonthermal dielectric barrier discharge plasma. Cell Death Dis. **4**(5), e642 (2013)
38. N. Kaushik, N. Kumar, C.H. Kim, N.K. Kaushik, E.H. Choi, Dielectric barrier discharge plasma efficiently delivers an apoptotic response in human monocytic lymphoma. Plasma Process. Polym. **11**(12), 1175–1187 (2014)
39. Y. Li, M. Ho Kang, H. Sup Uhm, G. Joon Lee, E. Ha Choi, I. Han, Effects of atmospheric-pressure non-thermal bio-compatible plasma and plasma activated nitric oxide water on cervical cancer cells. Sci. Rep. **7**, 45781 (2017)
40. M. Yokoyama, K. Johkura, T. Sato, Gene expression responses of HeLa cells to chemical species generated by an atmospheric plasma flow. Biochem. Biophys. Res. Commun. **450**(4), 1266–1271 (2014)
41. T. Adachi, H. Tanaka, S. Nonomura, H. Hara, S.I. Kondo, M. Hori, Plasma-activated medium induces A549 cell injury via a spiral apoptotic cascade involving the mitochondrial-nuclear network. Free Radic. Biol. Med. **79**, 28–44 (2015)
42. D. Yan, N. Nourmohammadi, K. Bian, F. Murad, J.H. Sherman, M. Keidar, Stabilizing the cold plasma-stimulated medium by regulating medium's composition. Sci. Rep. **6**, 26016 (2016)
43. N. Kurake, H. Tanaka, K. Ishikawa, T. Kondo, M. Sekine, K. Nakamura, H. Kajiyama, F. Kikkawa, M. Mizuno, M. Hori, Cell survival of glioblastoma grown in medium containing hydrogen peroxide and/or nitrite, or in plasma-activated medium. Arch. Biochem. Biophys. **605**, 102–108 (2016)
44. H. Tanaka, K. Nakamura, M. Mizuno, K. Ishikawa, K. Takeda, Non-thermal atmospheric pressure plasma activates lactate in Ringer's solution for anti-tumor effects. Sci. Rep. **6**, 36282 (2016)
45. S. Kalghatgi, C.M. Kelly, E. Cerchar, B. Torabi, O. Alekseev, A. Fridman, G. Friedman, J. Azizkhan-Clifford, Effects of non-thermal plasma on mammalian cells. PLoS One **6**(1), e16270 (2011)
46. G. Bauer, D.B. Graves, Mechanisms of selective antitumor action of cold atmospheric plasma-derived reactive oxygen and nitrogen species. Plasma Process. Polym. **13**, 1157–1178 (2016)
47. M. Riethmüller, N. Burger, G. Bauer, Singlet oxygen treatment of tumor cells triggers extracellular singlet oxygen generation, catalase inactivation and reactivation of intercellular apoptosis-inducing signaling. Redox Biol. **6**, 157–168 (2015)
48. S. Toyokuni, The origin and future of oxidative stress pathology: from the recognition of carcinogenesis as an iron addiction with ferroptosis-resistance to non-thermal plasma therapy. Pathol. Int. **66**, 245–259 (2016)
49. D. Yan, H. Xiao, W. Zhu, N. Nourmohammadi, L.G. Zhang, K. Bian, M. Keidar, The role of aquaporins in the anti-glioblastoma capacity of the cold plasma-stimulated medium. J. Phys. D. Appl. Phys. **50**, 055401 (2017)
50. T.L. Yuan, L.C. Cantley, PI3K pathway alterations in cancer: variations on a theme. Oncogene **27**, 5497–5510 (2008)
51. A.S. Dhillon, S. Hagan, O. Rath, W. Kolch, MAP kinase signalling pathways in cancer. Oncogene **26**, 3279–3290 (2007)
52. D. Koul, PTEN signaling pathways in glioblastoma. Cancer Biol. Ther. **7**, 1321–1325 (2008)
53. H. Tanaka, M. Mizuno, K. Ishikawa, K. Nakamura, F. Utsumi, H. Kajiyama, H. Kano, S. Maruyama, F. Kikkawa, M. Hori, Cell survival and proliferation signaling pathways are downregulated by plasma-activated medium in glioblastoma brain tumor cells. Plasma Med. **2**(4), 207–220 (2012)

54. S. Bekeschus, E. Freund, K. Wende, R. Gandhirajan, A. Schmidt, Hmox1 upregulation is a mutual marker in human tumor cells exposed to physical plasma-derived oxidants. Antioxidants **7**(11), 151 (2018)
55. E. Freund, K.R. Liedtke, R. Gebbe, A.K. Heidecke, L.I. Partecke, S. Bekeschus, In vitro anticancer efficacy of six different clinically approved types of liquids exposed to physical plasma. IEEE Trans. Radiat. Plasma Med. Sci. **3**(5), 588 (2019)
56. E. Freund, K.R. Liedtke, L. Miebach, K. Wende, A. Heidecke, N.K. Kaushik, E.H. Choi, L.I. Partecke, S. Bekeschus, Identification of two kinase inhibitors with synergistic toxicity with low-dose hydrogen peroxide in colorectal cancer cells in vitro. Cancers **12**(1), 122 (2020)
57. K.R. Liedtke, E. Freund, M. Hermes, S. Oswald, C.D. Heidecke, L.I. Partecke, S. Bekeschus, Gas plasma-conditioned Ringer's lactate enhances the cytotoxic activity of cisplatin and gemcitabine in pancreatic cancer in vitro and in vivo. Cancers **12**(1), 123 (2020)
58. S. Takeda, S. Yamada, N. Hattori, K. Nakamura, H. Tanaka, H. Kajiyama, M. Kanda, D. Kobayashi, C. Tanaka, T. Fujii, M. Fujiwara, M. Mizuno, M. Hori, Y. Kodera, Intraperitoneal administration of plasma-activated medium: proposal of a novel treatment option for peritoneal metastasis from gastric cancer. Ann. Surg. Oncol. **24**(5), 1188–1194 (2017)
59. N. Hattori, S. Yamada, K. Tori, S. Takeda, K. Nakamura, H. Tanaka, H. Kajiyama, M. Kanda, T. Fuji, G. Nakayama, H. Sugimoto, M. Koike, S. Nomoto, M. Fujiwara, M. Mizuno, M. Hori, Y. Kodera, Effectiveness of plasma treatment on pancreatic cancer cells. Int. J. Oncol. **47**, 1655–1662 (2015)
60. D. Xu, Q. Cui, Y. Xu, B. Wang, M. Tian, Q. Li, Z. Liu, D. Liu, H. Chen, M.G. Kong, Systemic study on the safety of immuno-deficient nude mice treated by atmospheric plasma-activated water. Plasma Sci. Technol. **20**, 44003 (2018)
61. Y. Sato, S. Yamada, S. Takeda, N. Hattori, Effect of plasma-activated lactated ringer's solution on pancreatic cancer cells in vitro and in vivo. Ann Surg Oncol. **25**(1), 299–307 (2018)
62. K.R. Liedtke, S. Bekeschus, A. Kaeding, C. Hackbarth, J.-P. Kuehn, C.-D. Heidecke, W. von Bernstorff, T. von Woedtke, L.I. Partecke, Non-thermal plasma-treated solution demonstrates antitumor activity against pancreatic cancer cells in vitro and in vivo. Sci. Rep. **7**, 8319 (2017)
63. K.R. Liedtke, E. Freund, C. Hackbarth, C.D. Heidecke, L.I. Partecke, S. Bekeschus, A myeloid and lymphoid infiltrate in murine pancreatic tumors exposed to plasma-treated medium. Clin. Plasma Med. **11**, 10–17 (2018)
64. A. Mielgo, M.C. Schmid, Impact of tumour associated macrophages in pancreatic cancer. BMB Rep. **46**(3), 131 (2013)
65. M. Obeid, A. Tesniere, F. Ghiringhelli, G.M. Fimia, L. Apetoh, J.L. Perfettini, M. Castedo, G. Mignot, T. Panaretakis, N. Casares, D. Métivier, Calreticulin exposure dictates the immunogenicity of cancer cell death. Nat. Med. **13**(1), 54 (2007)
66. S. Bekeschus, A. Käding, T. Schröder, K. Wende, C. Hackbarth, K.R. Liedtke, J. van der Linde, T. von Woedtke, C.D. Heidecke, L.I. Partecke, Cold physical plasma-treated buffered saline solution as effective agent against pancreatic cancer cells. Anti-Cancer Agents Med. Chem. **18**(6), 824–831 (2018)
67. E. Freund, K.R. Liedtke, J. van der Linde, H.R. Metelmann, C.D. Heidecke, L.I. Partecke, S. Bekeschus, Physical plasma-treated saline promotes an immunogenic phenotype in CT26 colon cancer cells in vitro and in vivo. Sci. Rep. **9**(1), 634 (2019)

# Chapter 7
# Plasma and Plasma–Cell Interaction Simulations

**Annemie Bogaerts, Jonas Van der Paal, Pepijn Heirman, Jamoliddin Razzokov, and Maksudbek Yusupov**

**Contents**




**Abstract** In this chapter, we first give an overview of modeling approaches in the literature that are relevant for studying plasma for cancer treatment, including both macro-scale and atomic/molecular scale models. Subsequently, we will present typical results obtained by these models to illustrate the type of information that can be obtained. Finally, we will conclude and discuss some future challenges for modeling plasma for cancer treatment.


## 7.1 Introduction

Although plasma is very promising for cancer treatment, as shown in other chapters of this book, the underlying mechanisms are not yet fully understood. Hence, there is a clear need for a better understanding of these mechanisms. This can be obtained


A. Bogaerts (✉) · J. Van der Paal · P. Heirman · J. Razzokov · M. Yusupov
Research group PLASMANT, Department of Chemistry, University of Antwerp, Antwerp, Belgium
e-mail: annemie.bogaerts@uantwerpen.be





M. Keidar (ed.), *Plasma Cancer Therapy*, Springer Series on Atomic, Optical, and Plasma Physics 115, https://doi.org/10.1007/978-3-030-49966-2_7

from experiments, but measurements cannot always provide detailed information, e.g., on the species concentrations in the plasma, or on the interaction of these species with biomolecules present in or around (cancer) cells. Such information can be obtained from modeling. Different types of models can be applied, depending on the system to be studied and the type of information required. In this chapter, we first give an overview of modeling approaches in literature, relevant for studying plasma for cancer treatment, including both macro-scale and atomic/molecular scale models. Subsequently, we will present typical results obtained by these models, to illustrate the type of information that can be obtained. Finally, we will conclude and discuss some future challenges for modeling plasma for cancer treatment.

## 7.2 Overview of Modeling Approaches in Literature

### *7.2.1 Macro-Scale Models for Plasma Sources and Plasma–Liquid Interaction*

Several different plasma sources are in use for cancer treatment, as outlined in Chap. XX of this book, but most common are plasma jets (in various designs) and dielectric barrier discharges (DBDs). In addition, besides direct plasma treatment, where the plasma is directly applied on the tissue or cells to be treated, plasma can also be applied to a liquid, so that reactive oxygen and nitrogen species (RONS) produced by the plasma can be captured in the liquid, after which the liquid can be injected into the tissue. These so-called plasma-treated liquids (PTLs), or plasma-activated liquids (PALs) or plasma-activated media (PAM), are also gaining increasing interest, as explained in Chap. 6 of this book.

Various biomedical plasma sources, as well as plasma–liquid interaction, have been modeled by either zero-dimensional (0D) chemical kinetics models or 1D and 2D fluid dynamics models to reveal the behavior of the important RONS in the plasma, the effluent, and/or the liquid.

A 0D chemical kinetics model is based on solving balance equations for the various species (i.e., electrons, various ions, radicals, molecules, and excited species) as a function of time, with production and loss rates determined by the chemical reactions between all species. It can describe a large chemistry without too much computation time, which is its main advantage. On the other hand, it is based on some major approximations. For instance, it needs certain input data (e.g., electric field), and it assumes a uniform plasma (or liquid), and thus, it cannot capture spatial variations, e.g., concentration gradients from the gas–liquid interface towards the bulk liquid, although there exist solutions to overcome this limitation, by applying different 0D models for different subsequent layers in the liquid [1]. In addition, for a plasma jet, a 0D model can provide information on the species density evolution as a function of position in the effluent, by converting the calculated temporal evolution into a spatial evolution, based on the velocity profile of the feed gas, yielding in fact a pseudo-1D model [2].

Fluid dynamics models (in 1D, 2D, or in principle also 3D) are based on solving conservation equations for mass, momentum, and energy. Often, the momentum equation is replaced by a drift–diffusion approximation (transport of species by diffusion due to concentration gradients and drift or migration in the electric field), and the energy equation is often only solved for the electrons. Fluid dynamics models for plasma jets in contact with liquid can be quite complicated, as they need to couple both gas and liquid phase, describing the flow dynamics (often turbulent flow), heat transfer, transport of species, and chemistry, in both gas and liquid phase [3].

Several modeling studies were performed to study plasma bullet propagation in plasma jets, used in plasma medicine. Naidis et al. numerically demonstrated that these wavefronts propagate at very high speeds through the helium channel that develops in the ambient air, by including a simple photo-ionization function [4]. Nevertheless, later simulations (all performed for a pre-developed helium channel in ambient air) suggested that photo-ionization might not be crucial for the propagation of these ionization fronts, as the existence of a pre-ionization channel (created by previous bullets) was sufficient for their propagation [5, 6]. Nevertheless, including the photo-ionization phenomena did affect the velocity of the ionization wave [7], although a higher velocity could also be obtained by using a higher degree of pre-ionization. Anyway, the development of any ionization wave required both an electric field high enough in its front (to be able to ionize the gas) and the presence of seed electrons just ahead of the front of the ionization wave. The latter could be caused both "instantaneously" by photo-ionization or by a pre-ionized channel created by preceding bullets. It was shown by Yousfi et al. that the ionization rate rapidly drops when helium is significantly diluted by air (above 1%) [8]. Therefore, this explained why the mixing of helium with ambient air limits the bullet propagation both in the radial and axial direction. Breden et al. and Liu et al. [9] performed simulations in ambient helium and observed that the plasma (formed within the dielectric tube during the rise time of the voltage pulse) propagated along the surface of the dielectric during the rest of the pulse, without transitioning into a streamer. Finally, Kushner and coworkers simulated the propagation of ionization waves through flexible capillary channels, impinging upon a target, through branched tubes and across different channels, thereby mimicking experiments performed with the plasma gun of GREMI [10–12]. In addition, they also studied ionization waves launched from single and multiple plasma jets (pin-electrode setup) after a steady state was obtained for the neutral fluid flow of helium containing 2% oxygen into ambient humid air, taking into account photo-ionization processes. They found that the jets interact through electrostatic, hydrodynamic, and photolytic means [13].

More recently, the effects of various admixtures to a helium feed gas on the evolution of a plasma jet or plasma gun have been investigated. These admixtures include, e.g., nitrogen, oxygen, or air, which are often used in plasma medicine. Naidis et al. [14] focused on the addition of air to a helium-driven plasma jet, whereas Bourdon et al. [15] investigated nitrogen admixtures in a helium-driven plasma gun. Finally, Lazarou et al. [16] employed 2D simulations to study the effect of oxygen admixtures in a helium-driven plasma jet. These studies showed that the dynamics of the plasma devices strongly depend on the composition of the feed gas. Indeed, adding oxygen (1) promoted plasma bullet propagation on the axis of symmetry of the tube while (2) lowering the propagation speed of the plasma bullets. At the same time, much higher electric fields were induced on a dielectric surface treated (in comparison to the pure helium plasma jet) [16]. In the case of nitrogen addition, Bourdon et al. [15] showed that two- and three-body Penning reactions are crucial for the discharge dynamics. Moreover, it was shown that the ionization front velocity increased when increasing the amplitude of the applied voltage. Overall, it was concluded that the ionization front velocity at different levels of nitrogen admixture in the helium gas depends on a complex coupling between the kinetics of the discharge, the photo-ionization, and the 2D structure of the discharge in the tube. In a follow-up study, Viergas et al. [17] showed that the peak electric field is able to propagate outside the plasma jet device. After impact of the ionization front on a metallic target, a rebound front is shown to propagate in the opposite direction, with a velocity 2–3 times higher than the first ionization front.

Recently, 2D models developed by Lietz et al. [18] showed that molecular admixtures ($O_2$ and $H_2O$) can significantly impact the ionization wave propagation speed, the electron density, the electron temperature, and the RONS production in a helium-driven plasma jet. Moreover, they also showed that, in the absence of intentional admixtures, impurities in the feed gas can account for 79–98% of the production of most RONS [18]. Chang et al. [19] investigated the effect of seed electrons on the repeatability of plasma plume propagation in a plasma jet, using helium as feed gas with different concentrations of air added. They found that the minimum seed electron density required for repeatable propagation mode is independent of the air concentration in the feed gas. The main source of the seed electrons, however, depends on the feed gas composition. Indeed, at low air concentrations (0.25%), the main source of seed electrons is due to leftover electrons from previous discharges. At an air concentration of 0.50%, on the other hand, the detachment of $O_2^-$ is the main source of the seed electrons [19]. The rate of this detachment was calculated based on Monte Carlo simulations.

A large number of models have also been developed with specific focus on the plasma chemistry in plasma jets (and their effluent) and DBD plasmas, used in plasma medicine applications. Sakiyama et al. [20] developed two coupled 0D models to describe a surface micro-discharge with 50 species and 600 reactions. One of the 0D models described the plasma zone near the electrodes, with a high electric field, thus solving balance equations for both neutral and charged species, whereas the other model only described the neutrals and highlighted the chemistry of the long-lived species. Babaeva et al. [21] developed a combined 0D–2D fluid model to provide more insight into the effect of filaments on the chemical formation of species and the particle fluxes towards wounded skin. The 2D results were used as input for a 0D model calculation and vice versa. The 2D fluid was able to describe streamer development in air between the electrode and the tissue, which acts as a floating electrode, but only for short time scales due to the computational load [22, 23]. The 0D model was used in a complementary way to evaluate which chemical species were formed on the long term from the short-lived chemical species created in the streamers simulated in 2D. Also interesting is that the authors included a cellular structure in the model, mimicking actual skin tissue, and investigated the induced electric fields in the context of electroporation. Later, this work was extended to wounds covered by liquid, possibly containing blood platelets [24]. Similarly, Naidis coupled his 2D fluid model, describing the streamer or "bullet" behavior of a helium plasma jet, with a small 0D chemical kinetics model (He/$H_2O$) for describing the OH radical production mechanism inside the device, thus not in the effluent [25].

An interesting 0D numerical study was reported by Liu et al. [26] with a quite large He/$H_2O$ chemistry set, containing 46 species and 577 reactions. In this publication, the authors also suggested several reduced chemistries that can be used to run complex simulations. An example of such a model was the work of Waskoenig et al. [27] for a He/$O_2$ mixture with 16 species and 116 reactions. This code was developed to comprise chemical and physical phenomena across the gap between two parallel plate electrodes (hence 1D) of a micro-scaled radio-frequency driven atmospheric pressure plasma jet device (μAPPJ), designed for plasma medicine applications. Results of a similar model were reported by McKay et al. [28] but with the addition of water to the He–$O_2$ mixture. In this study, a very large reaction chemistry set of 61 species and 878 reactions was implemented in a 1D fluid model. The chemistry of the μAPPJ was further studied by Murakami et al. [29–31], by means of a global model. The lower complexity of the latter enabled these authors to do a faster kinetic analysis of a He/$O_2$/$N_2$/$H_2O$/$CO_2$ mixture with 59 species and 1048 reactions.

Another plasma medicine tool that has been studied numerically is the plasma needle, initially by Brok et al. and later by Sakiyama and Graves. Brok was the first one to model in 2D (in the field of plasma medicine at least) and his simulations ran for a fixed He/$N_2$ gas ratio over the entire simulation domain, neglecting the gas flow and using only a limited plasma chemistry [32]. The approach of Sakiyama and Graves was similar, yet already more advanced, as they used an unstructured mesh (and not a Cartesian mesh like in Brok's work), which enabled a high resolution close to the needle electrode tip. Therefore, these authors were able to accurately model the sheath and they found that the corona discharge at low power exhibits a transition to a glow discharge at a critical power, spreading back along the needle surface [33, 34]. Subsequently, they studied this discharge type with an asymmetric electrode in a more fundamental way in 1D spherical coordinates, e.g., looking at the validity of the local field approximation, the influence of nitrogen impurities, and secondary electron emission [35, 36]. Finally, they published the results of a 2D fluid calculation that considered the neutral gas flow (i.e., ambient air diffusion, here represented by $N_2$) [37].

While the above models focused mainly on helium plasma jets, Van Gaens and Bogaerts developed a 0D model for an argon plasma jet flowing into humid air, considering 84 different species and 1880 reactions [2]. It was first applied to the experimental conditions of a plasma jet geometry developed at Eindhoven University of Technology. In a follow-up study [38], the gas phase production and loss pathways of the most important biomedical species were examined, showing that the pathways of the biomedical species drastically change as a function of position in the effluent, i.e., inside the discharge device, the active plasma jet effluent and the afterglow region far from the nozzle, and are also strongly dependent on the conditions, such as ambient air humidity, total deposited power into the plasma, and gas temperature along the jet. In [39] complementary numerical work and experiments were presented on the ozone generation in the same plasma jet from Eindhoven University of Technology. Another combined experimental-computational study was applied to two cold atmospheric pressure plasma jets [40], operating in Ar + 2% air, with a different electrode geometry but the same power dissipated in the plasma. This combined work allowed to derive the most probable spatial power distribution in an RF-driven plasma jet, by comparing measured and calculated species density profiles, because of the strong effect of the power distribution on the NO and O density profiles. In addition, the dominant reaction pathways for both NO and O were identified. In [41], the kINPen plasma jet was studied, operating in argon with different admixture fractions up to 1% pure $N_2$, $O_2$, and $N_2 + O_2$, and a shielding gas of dry air, focusing on the production of $O_3$ and $NO_2$, in comparison with experiments performed in the far effluent by infrared absorption spectroscopy.

Schmidt-Bleker et al. [42] also developed a 0D model for the kINPen plasma jet, with a reduced reaction scheme, to model the chemistry of the effluent without including the electron dynamics. By varying the shield gas humidity and composition (i.e., $O_2/N_2$ ratio) they obtained valuable insights regarding the chemistry of the plasma jet effluent. An extension to this work [43] took different time- and length-scales into account by using an approach involving two separate kinetic models. A first model accounts for the electron impact reactions occurring in the vicinity of the streamer head, whereas the second model only accounts for neutral molecules on a much larger time- and length-scale, using the obtained densities of the first model as an input. The major finding was that primary RONS are dominantly generated from exited Ar species.

Next to the kINPen, 0D models are also employed to study the COST reference microplasma jet. Indeed, Gorbanev et al. [44] used such simulations to study the chemistry occurring inside the source, as well as during transport to a liquid surface. Moreover, a 3D fluid dynamics model was developed to assess the humidity induced in the effluent by evaporation of water from the liquid target. In a follow-up study [45], the 0D model was used to help unraveling reaction pathways leading to the formation of aqueous $ClO^-$. Schröter et al. [46] employed 0D simulations to study the effect of surface reaction probabilities of reactive neutral species (H, O, and OH). They found that the densities of low-mass species (e.g., H) can change significantly based on the choice of the surface reaction probabilities. This in turn offers potential for tailoring the composition of these plasmas by choosing different wall materials or plasma geometries.

Besides gas phase simulations, plasma–liquid modeling is also crucial for plasma medicine applications, for two reasons: (1) the growing interest in PTLs and (2) even in direct plasma treatment, biological samples are usually covered by a liquid medium. The chemical cocktail in the gas phase is not necessarily the same after the gas phase species are transferred to the liquid phase.

Plasma–liquid modeling has also been studied by 0D chemical kinetics and (mainly 2D) fluid dynamics models, for medical (as well as environmental) applications of plasmas. Locke et al. [47] wrote a general review paper on modeling of electrical discharges in liquid water, including streamer-like plasma channels, spark and arc discharges. van Gils et al. [48] presented a 0D chemical kinetics model for the liquid phase chemistry upon plasma treatment, assuming certain values for the fluxes of $O_3$, NO, and OH originating from the plasma, using an optimization algorithm, to find the values at which the end concentrations of nitrate and nitrite correspond to the values measured by ion chromatography in the plasma-treated liquid. They used the model to elucidate the bacterial inactivation mechanisms in the liquid phase. Likewise, Lukes et al. [49] performed a kinetic study of the aqueous-phase chemistry, and more specifically the formation of peroxynitrite in plasma-activated water, also to investigate the bactericidal effects of an air discharge in contact with water. Hamaguchi and coworkers [50] developed a 1D reaction–diffusion model, as well as a 0D model, to describe the behavior of reactive species in pure water (with dissolved $O_2$ and $N_2$ in equilibrium with air) exposed to an atmospheric pressure plasma. The model contained 37 species and 111 chemical reactions. They estimated the flux of RONS towards the liquid (i.e., the initial conditions for the 0D model of the liquid), based on their typical gas phase concentrations in a representative plasma medicine device, and on the Henry's constants, which determine how easily a species dissolves in liquid. Chen et al. [51] evaluated the plasma–liquid chemistry for a He/$O_2$ DBD with a 0D model, for treating biofilms and biological tissues. Later, the same group [52] developed a semi-1D model for plasma–biofilm interaction and for plasma–tissue interaction, based on a reactive penetration model for mass transfer of plasma species across the gas–liquid boundary. The liquid phase model contained 19 species and 84 chemical reactions.

The most comprehensive models for describing (DBD) plasma–liquid interaction, especially for the purpose of plasma for wound treatment, were developed by Kushner and coworkers. As mentioned before, Babaeva et al. developed a combined 0D–2D multi-species fluid model, called nonPDPSIM, for atmospheric pressure DBDs interacting with slightly wounded skin, focusing on a dry wound [21]. This model was later extended by Tian and Kushner [53, 54] to the interaction of DBD plasma with liquid-covered tissue. The water layer (with liquid phase chemistry) was represented by an extremely dense gas and separated from the actual gas phase (with gas phase chemistry) by a boundary layer. Transport of RONS across this boundary was also dictated by Henry's law. Also water evaporation from the liquid layer and photolysis of water molecules into $OH_{liq}$ (triggered by photons created in the plasma above the liquid layer) was taken into account. The authors also found that ROS are readily consumed by hydrocarbons in the liquid, thus affecting their concentration [53, 54]. The same group also studied the interaction of DBD filaments with wounds and blood plasma, focusing on the electric fields delivered to blood platelets and cells [24], as well as the interaction of a DBD treating liquid-covered tissue [55]. Lietz et al. [55, 56] studied plasma–liquid interactions by coupling two 0D models. In the first model, the gas phase chemistry of a pulsed DBD operating in ambient air is studied. The second model investigates the chemistry occurring in a liquid target. These two models are characterized by a separate set of chemical species (gaseous and aqueous) as well as a separate set of chemical reactions. The species in these two zones (gas and liquid) only interact through the interface between the zones by means of diffusion into or out of the liquid, which is based on Henry's law. By using two 0D models, elaborate chemistry sets can be investigated on large time scales, which allowed the authors to model 5000 consecutive pulses, followed by an afterglow period of 5 min. The effect of the applied voltage, gas flow rate, pulse repetition frequency, and the presence of organic molecules in the liquid was investigated [55, 56].

To the best of our knowledge, only a limited number of papers reported the interaction between a plasma jet and a liquid. Van Boxem et al. applied a 0D model to a liquid treated by the kINPen plasma jet, to elucidate the reaction mechanisms for the formation of $H_2O_2$ and $HNO_2$ in order to explain experimental observations [57]. Du et al. [58] developed a 1D drift–diffusion model for the mass transfer process from a plasma jet into the liquid. A few papers also reported 2D simulations for the interaction between a plasma jet and a liquid layer. Lindsay et al. [59] developed a 2D model to study the transport of reactive species produced by a helium plasma jet to liquid water, using 13 species and 23 reactions. Lietz and Kushner [18] studied the effect of $H_2O$ and $O_2$ admixtures in the feed gas of a helium plasma jet, by means of a 2D fluid dynamics model. Norberg et al. [60] investigated the influence of the pulse repetition frequency on the plasma treatment of a reactive liquid layer. Verlackt et al. [61] studied the transport and accumulation of plasma-generated species in aqueous solution by a 2D fluid dynamics model, considering 20 gas phase and 22 liquid species, and 57 and 42 reactions in the gas and liquid phase, respectively. This model was recently extended by Heirman et al. [3], by coupling it to a 0D model, in order to take advantage and avoid the limitations of both models (in terms of calculation time vs approximations). Furthermore, this model also showed simulation results after plasma treatment, to reveal the stability of RONS in the liquid after treatment. The 0D model included 91 gas phase and 32 liquid species, which interact in 1390 gas phase and 89 liquid phase reactions, while the 2D model included 21 gas phase and 25 liquid species, interacting in 56 gas phase and 52 liquid phase reactions. This clearly illustrates the difference between the number of species, and especially number of reactions, in gas phase and liquid between 2D and 0D models, as explained in the beginning of this section.

### 7.2.2 Atomic Scale Models for the Interaction of RONS with Biomolecules

Besides modeling the plasma devices and plasma–liquid interactions, studying the interaction of RONS with biomolecules, important in the context of plasma for cancer treatment, such as the phospholipid bilayer (PLB), DNA, and proteins, is also needed to obtain a better insight in the underlying mechanisms. The same applies to the effect of these interactions (e.g., oxidation) on structural changes in the biomolecules. For these purposes, atomic (or molecular) scale models must be developed, as macroscopic models will not provide the required information. A wide range of atomic/molecular scale methods can be applied, ranging from quantum mechanics, density functional theory (DFT), density functional tight binding (DFTB), classical reactive molecular dynamics (MD), and non-reactive MD, including all-atom, united-atom, and coarse-grained approaches (see below). These methods vary in accuracy and level of detail they can describe, as well as the time- and length-scales they can handle, as illustrated in Fig. 7.1.

Quantum mechanical (QM) calculations are the most accurate computational methods, as they are based on first principles. Various QM techniques exist, varying in their approach to solve the Schrödinger equation and in their level of approximations (see more details in [62]). For plasma medicine, DFT calculations are the most suitable approach, keeping in mind the required system sizes. They

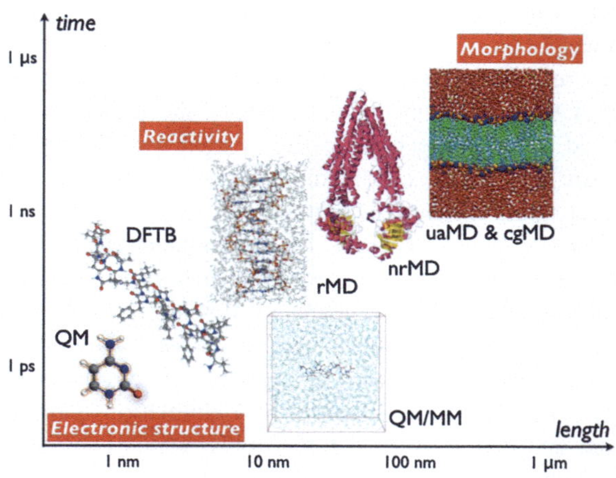


**Fig. 7.1** Overview of the computational methods that allow to obtain atomic (or molecular) level insight in the interaction of RONS with biomolecular systems, or the (longer-term) effects of these interactions, illustrating the length- and time-scales that can be handled. QM = quantum mechanics, DFTB = density functional-tight binding, QM/MM = quantum mechanics/molecular mechanics, rMD = reactive molecular dynamics, nrMD = non-reactive molecular dynamics, uaMD = united-atom molecular dynamics, cgMD = coarse-grained molecular dynamics. Adopted from [62] with permission

are more approximate than other QM methods, but still very time-consuming, able to handle system sizes only in the order of 100 atoms. In principle, they are static, but they can be integrated in MD simulations, where all atoms in the system are followed through space and time, by integrating their equations of motion. The forces acting on the atoms are obtained as the derivative of some suitable interatomic potential, which can thus be based on QM data. DFT-based MD calculations, also called "ab initio MD" (AIMD) can handle time scales in the order of picoseconds.

Somewhat larger systems can be handled with DFTB, which is an approximate DFT method, based on a Taylor series expansion of the DFT total energy expression [63]. DFTB can typically handle a few 1000 atoms on time scales of 10s of picoseconds. This method has been applied in the context of plasma medicine to study the interaction of ROS with the head group of the PLB [64], a specific protein (P-glycoprotein) [65] and peptides [66], as well as the behavior of O and OH in water [67].

Classical reactive MD simulations, in which the forces are not based on QM data but on classical fitting parameters, can typically handle much longer time scales and larger systems than DFT or DFTB calculations, ranging from 1 ps to 100 ns, and from $10^4$ to $10^6$ atoms, depending on the complexity of the interatomic potential. This (classical) potential is constructed with many parameters obtained by fitting against DFT calculations. The Brenner potential [68] and the ReaxFF potential [69] are two reactive potentials used for plasma medicine applications. Reactive MD simulations can describe bond breaking and formation, so they allow to study chemical reactions of RONS with biomolecules. This has been illustrated for the interaction of ROS with peptidoglycan [70, 71], lipid A [72], lipids [23, 73, 74], DNA [75, 76], a water layer [77], and simple organic molecules in water [78].

As opposed to reactive classical MD simulations, in non-reactive MD simulations, also called "molecular mechanics" (MM), the molecular connectivity in the system is fixed. Hence, the bond order of each bond does not have to be recalculated in every step, so they can handle system sizes and time scales two orders of magnitude larger than reactive MD simulations, hence, in the order of $10^6$–$10^8$ atoms, at time scales of 0.1 ns to 10 μs, for the so-called all-atom force fields. A drawback is that they cannot describe chemical reactions (as they do not consider bond breaking and formation), but as they can follow the system over a longer time scale, they can study conformational changes and stability of biomolecules.

In all-atom force fields, as the name suggests, all atoms in the system are treated separately. In addition, non-reactive MD simulations can also use the so-called united-atom and coarse-grained force fields, which can handle even larger system sizes (typically up to one order of magnitude larger), for the same time scales. United-atom force fields (e.g., [79]) treat all heavy atoms separately, but the H atoms bound to a C atom (e.g., in the apolar tails of phospholipids) are combined and treated as one (methyl or methylene) group. Hence, this yields a lower number of separate particles in the system, allowing to simulate larger systems. Some well-known non-reactive interatomic potentials are AMBER [80], CHARMM [81], and GROMOS [82]. Coarse-grained methods further reduce the number of particles in the system, and thus they allow even larger system sizes (or speed up the calculations), because the atoms comprising entire functional groups, i.e., typically, 3–5 heavy atoms with their H atoms, are represented by the so-called coarse-grained particles. The Martini force field [83] is an example of a coarse-grained potential.

Finally, it is also possible to combine the above methods in the so-called quantum mechanical/molecular mechanics (QM/MM) methods [84]. Indeed, it is clear that the various methods can handle different length- and time-scales (see Fig. 7.1), attributed to the accuracy and type of information they provide. In the QM/MM method, a small (chemically most relevant) part of the system (e.g., the active site of the biological system) will be described at the quantum chemical (electronic) level, while the surrounding embedding atoms and molecules will be treated at a classical (atomic) level.

For more details on these methods and examples of their simulation results for plasma medicine, we refer to [62, 85].

## 7.3 Typical Modeling Results

### *7.3.1 Gas Phase Simulations*

#### 0D Modeling Results

Figure 7.2 illustrates typical results obtained by a 0D model of a plasma jet. Using the pseudo-1D model approach discussed above, the evolution of species along the plasma axis can be obtained. These results illustrate, e.g., the difference in behavior of short- and long-lived RONS. The densities of short-lived RONS, such as OH or exited species, increase rapidly in the effluent when ambient air species enter the effluent. However, during transport to a target, these species recombine again into more stable, long-lived RONS, which are therefore generated gradually throughout the effluent.

Due to the low computational cost of such 0D models, the effect of different parameters on the generation of species can easily be investigated, including, e.g., the gas temperature at the nozzle exit, the flow rate of the feed gas, or the humidity of the ambient air. The effect of varying the flow rate of the feed gas in an atmospheric pressure argon plasma jet is illustrated in Fig. 7.3. This shows that altering one operational parameter can have different effects on different important species. Indeed, lowering the feed gas velocity yields an increase in the density of, e.g., $H_2O_2$, $HNO_2$, or $HNO_3$, whereas the density of NO, OH, and $O_2^-$ decreases.

#### 2D Modeling Results

Tian et al. [53] employed 2D simulations to study the interaction of DBDs in humid air with a thin water layer covering tissue. Such a model allows studying the spatio-temporal evolution of species in the discharge zone, as well as in the water

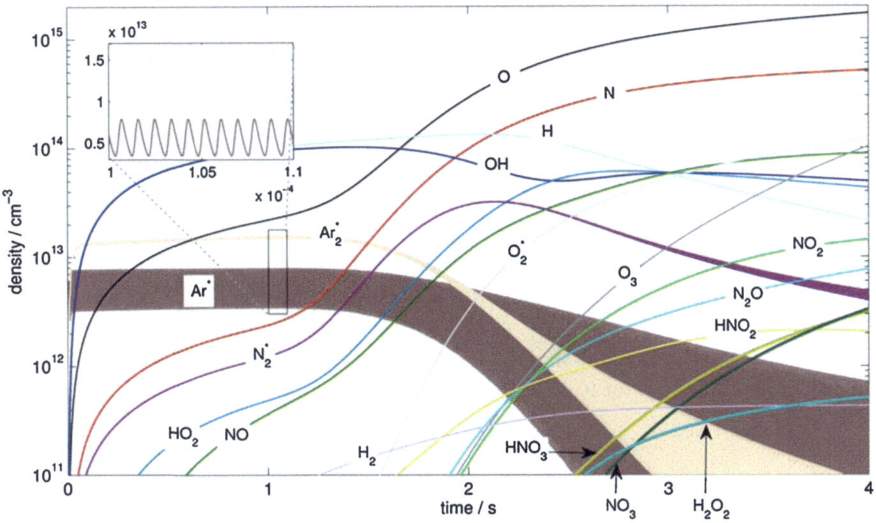


**Fig. 7.2** Calculated density profiles of different RONS in the effluent, obtained with a 0D model for the kINPen. Adopted from [43] with permission

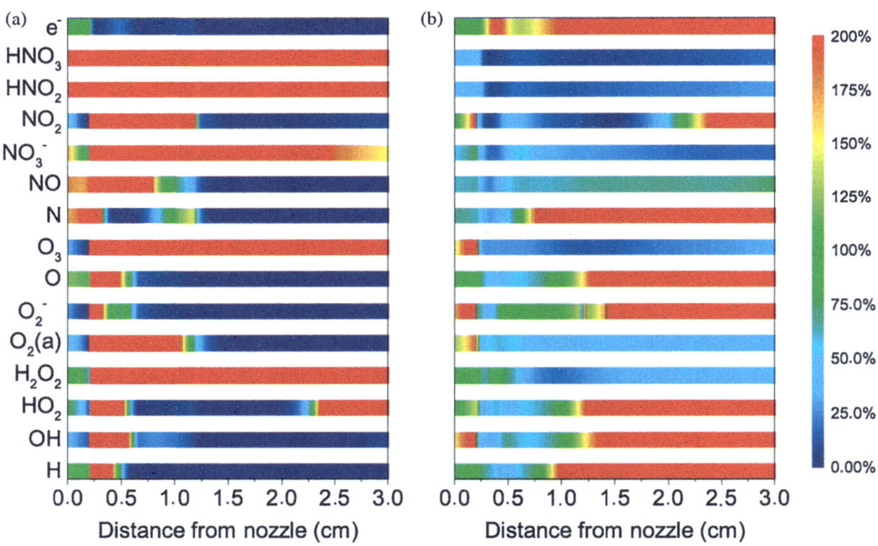


**Fig. 7.3** Calculated relative changes (%) in the density of different biomedically important RONS upon altering the flow rate of the feed gas, compared to a flow rate of 2 slm, as obtained from 0D modeling. In case (**a**), the flow rate is lowered to 1.5 slm, whereas in case (**b**), it is enhanced to 2.5 slm. Adopted from [2] with permission

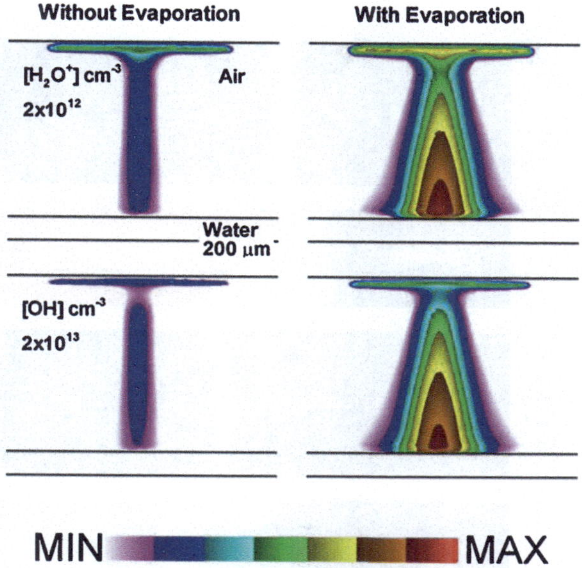


**Fig. 7.4** Effect of water evaporation on the density profiles of $H_2O^+$ ions and OH radicals during a filament incident on the water surface, as obtained from 2D modeling. Contours for $H_2O^+$ and OH are plotted on a log scale over three decades with the maximum density shown in each frame. Adopted from [53] with permission

layer, during consecutive filament discharges onto the liquid surface and afterglow periods. One of the important observations made in this study is the effect of water evaporation from the water layer covering the tissue. Indeed, when accounting for this evaporation, the densities of different important species changed significantly, as is illustrated in Fig. 7.4.

As mentioned before, Bourdon et al. [15] employed 2D simulations to study the propagation of the ionization front in a plasma gun setup. Using these models, they discovered that Penning reactions not only have an effect on the discharge dynamics, but also on its structure, as illustrated in Fig. 7.5. As Penning reactions produce ions and electrons, excluding these reactions makes the plasma column behind the ionization front less conductive, which leads to a slower propagation of the discharge in the tube.

## *7.3.2 Gas–Liquid Interactions and Liquid Phase Simulations*

### 0D Modeling Results

Lietz et al. [55] used a coupling of two 0D models (also called global models) to investigate the chemistry in the gas and liquid phase of a DBD device interacting with a water layer. As the computational cost of 0D models is much lower than for 2D models, the authors were able to study the chemistry in 5000 consecutive pulses, followed by an afterglow of several minutes. Figure 7.6 shows the accumulation of species in the water layer during the first 50 pulses (a–c), as well as the decay

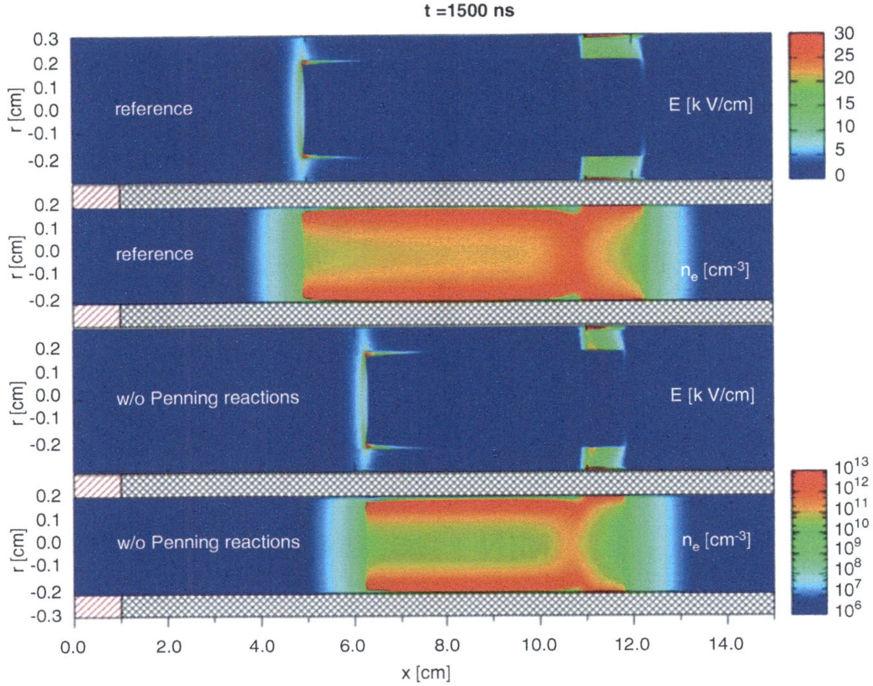


**Fig. 7.5** Cross-sectional views of the magnitude of the electric field and electron density, as obtained from 2D modeling, at 1500 ns, for an applied voltage of 16 kV, 1000 ppm of $N_2$, with a pre-ionization level $n_{\mathrm{init}} = 10^4\ \mathrm{cm}^{-3}$ and for a photo-ionization coefficient $A_{\mathrm{ph}} = 10^{-2}$, with the reference kinetic scheme presented in [15], as well as without two- and three-body Penning reactions. Adopted from [15] with permission

of some important RONS after treatment (d). After treatment, gas phase species continue to solvate, which serves as a source term of the aqueous species. Liquid reactions, on the other hand, lead to a depletion of these species.

In the same paper, Lietz et al. investigated the effect of hydrocarbons present in the liquid on the generation and stability of aqueous species. Their results clearly show that these hydrocarbons (e.g., peptidoglycan) have a big impact on the density of certain RONS, due to reactions occurring between plasma-generated RONS and hydrocarbons, which is shown in Fig. 7.7.

## 2D Modeling Results

2D fluid dynamics models also provide information on the flow profiles in both gas and liquid phase. This is illustrated in Fig. 7.8, for the kINPen plasma jet interacting with water, contained in a well of 2 mL, at a flow rate of 3 slm Ar [3]. A maximum velocity of 57 m/s inside the plasma jet was predicted, while the maximum gas

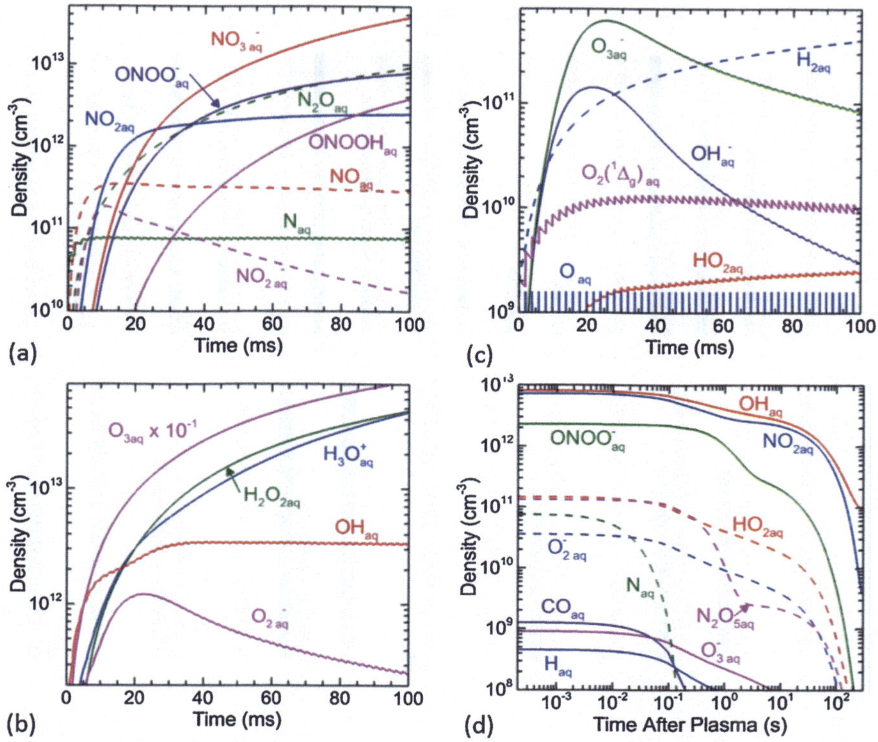


**Fig. 7.6** The accumulation of RNS (**a**) and ROS (**b**, **c**) during the first 50 pulses, as well as the decay of some important RONS after plasma treatment (**d**), as obtained from 0D modeling. The time refers to the time after the last pulse. Adopted from [55] with permission

velocity outside the jet is 34 m/s. When the gas flow reaches the liquid surface, it flows towards the edge of the well, causing a shear stress on the liquid surface. Upon hitting the edge of the well, the gas flow results in a vortex within the well, so that the gas flows back towards the afterglow. Because of the shear stress on the liquid surface, the upper layer of the liquid starts to move in the same direction as the gas, but with a lower velocity. The maximum velocity in the liquid (2.5 m/s) is reached near the edge of the well. The liquid movement results in another vortex in the liquid phase (see close-up in Fig. 7.8). Note that this vortex behavior is characteristic for the small volume of the liquid. In larger volumes, other vortex patterns can be observed [61].

The calculated gas phase densities and liquid phase concentrations of the three major RONS, i.e., $H_2O_2$, $HNO_2$, and $HNO_3$, obtained with the same model and at the same conditions, are plotted in Fig. 7.9, for 10 s of plasma treatment [3]. The profiles clearly follow the gas and liquid flow profiles of Fig. 7.8. Indeed, convection of species was found to be more important than diffusion. The model predicts that

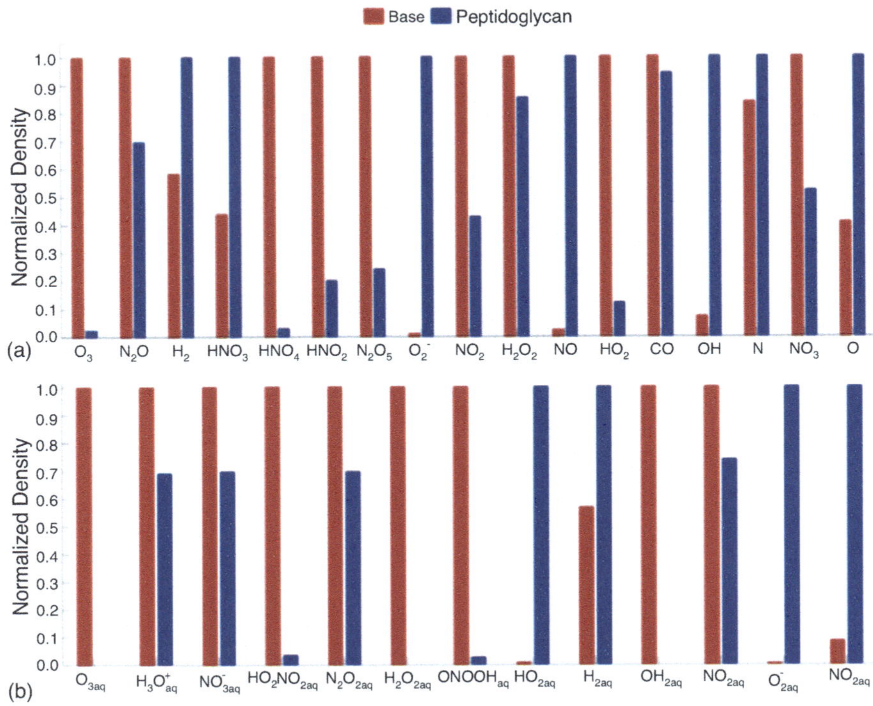


**Fig. 7.7** Calculated (**a**) gas phase and (**b**) liquid phase densities of different RONS after the 5000th discharge pulse, with and without the presence of peptidoglycan in the liquid. O, OH, $O_3$, and $H_2O_2$ are able to react directly with peptidoglycan in this mechanism, so their densities are zero in the liquid upon presence of peptidoglycan. Adopted from [55] with permission

the $H_2O_2$ density is at maximum underneath the plasma jet, just above the liquid surface, and its density drops to half its value towards the edge of the well and in the bulk gas. $HNO_2$ is only formed to some extent underneath the plasma jet, just above the liquid surface, but its density reaches its maximum towards the edge of the well. Finally, the $HNO_3$ density profile does not rise at all underneath the plasma jet. The $HNO_2$ and $HNO_3$ species accumulate in the gas phase inside the vortex.

Transport to the liquid phase is dictated by the Henry's constants, which are above 1 for $H_2O_2$, $HNO_2$, and $HNO_3$. Hence, their equilibrium is towards the liquid phase. This is clear from the difference between gas phase density just above the liquid surface (being small, because the species are transported towards the liquid) and the liquid phase concentration just below the liquid surface (being the highest in the first few nm, because the species enter from the gas phase); see close-ups in Fig. 7.9.

The highest concentration of $H_2O_2$ in the liquid was calculated to be ten times higher than that of $HNO_2$, which is ten times higher than that of $HNO_3$ (cf. different values in the color scales of Fig. 7.9).

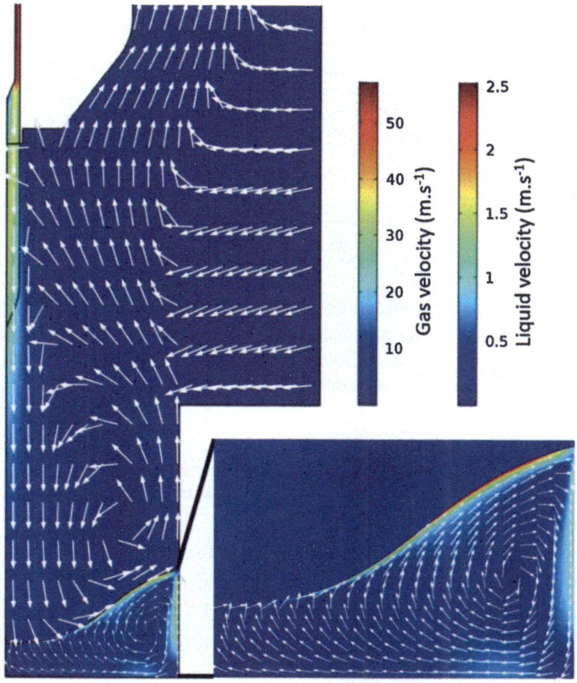


**Fig. 7.8** Calculated gas and liquid flow profiles, obtained with a 2D fluid dynamics model for the kINPen plasma jet interacting with water, contained in a well of 2 mL, at an inlet argon flow rate of 3 slm. The color scale gives the velocity magnitudes, while the arrows show the flow direction. Adopted from [3]—reproduced by permission of the PCCP Owner Societies

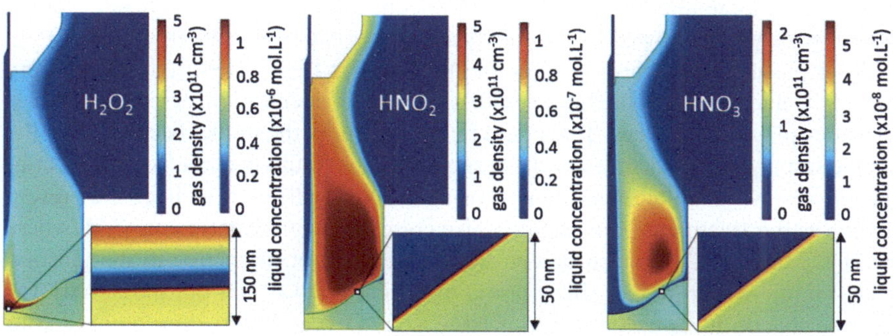


**Fig. 7.9** 2D plots of the gas phase densities and liquid phase concentrations of $H_2O_2$, $HNO_2$, and $HNO_2$, calculated with a 2D fluid dynamics model for the kINPen plasma jet interacting with water, contained in a well of 2 mL, at an inlet Ar flow rate of 3 slm, for 10 s of plasma treatment. Adopted from [3]—reproduced by permission of the PCCP Owner Societies

This model also calculated the behavior of the RONS in the liquid after plasma treatment, to check their stability, which is very important for the application of PTLs. It was predicted that only $H_2O_2$, $HNO_2$, $HNO_3$, $HO_2$, $O_3$, and ONOOH can accumulate (to some extent) in the liquid (the so-called long-lived liquid species), while the other RONS, such as OH, NO, and $NO_2$, are only important in the first

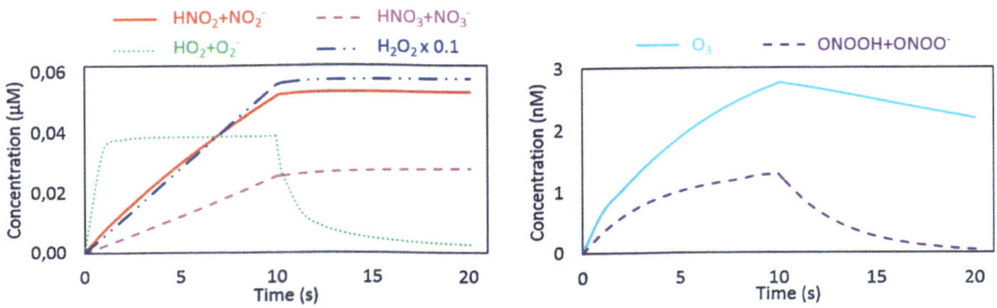


**Fig. 7.10** Volume-integrated liquid phase concentrations of the long-lived RONS, as a function of time during and after plasma treatment, calculated with a 2D fluid dynamics model for the kINPen plasma jet interacting with water, contained in a well of 2 mL, at an inlet argon flow rate of 3 slm, for 10 s of plasma treatment. Adopted from [3]—reproduced by permission of the PCCP Owner Societies

few nm below the liquid surface and have a lifetime around 10 ms ("short-lived species").

Figure 7.10 illustrates the volume-integrated liquid phase concentrations of the long-lived RONS, as a function of time, both during and after plasma treatment, for the same conditions as Figs. 7.8 and 7.9. During plasma treatment (10 s), the $H_2O_2$, $HNO_2$, and $HNO_3$ concentrations increase linearly with time, which is in agreement with experimental data [57]. The concentrations of $O_3$ and ONOOH rise less than linearly with treatment time, while the $HO_2$ concentration rises significantly in the first milliseconds, followed by a steady state till the end of plasma treatment.

After plasma treatment (i.e., after 10 s), obviously $H_2O_2$, $HNO_2$, and $HNO_3$ are the only stable species in the liquid, with a constant concentration, at least in the first 10 s after plasma treatment. This could be explained by the model because both the production and loss processes of these species in the liquid become negligible after plasma treatment, as their rates depend on short-lived radicals, which are not present anymore after treatment. This stability also correlates with experiments [57], where the $H_2O_2$ and $HNO_2$ concentrations were observed not to change in the plasma-treated liquid for up to 2 h after plasma treatment ($HNO_3$ was not measured). On the other hand, the concentrations of $HO_2$, ONOOH, and $NO_2$ drop quickly after plasma treatment, to virtually zero after 10 s, while the $O_3$ concentration decreases linearly. After 10 s, it reaches a value equal to 80% of its value at the end of plasma treatment. Extrapolating this linear drop to longer times would mean that it takes about 50 s for the $O_3$ concentration to drop to negligible values. This drop in $O_3$ concentration can be explained by transport of $O_3$ back to the gas phase, because its Henry's constant is lower than 1; in order words, its equilibrium points towards the gas phase.

Note that the calculated liquid phase concentrations of $H_2O_2$, $HNO_2$, and $HNO_3$ in this model are significantly underestimated, compared to experiments [3]. This was attributed to the static liquid interface defined in the model. It was argued that a turbulently moving liquid layer would deplete the liquid phase concentrations in

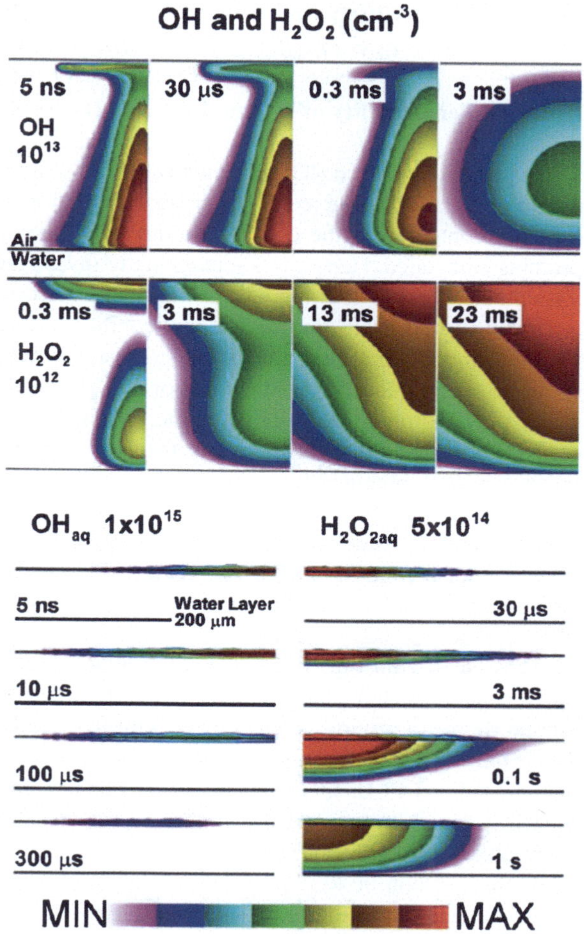


**Fig. 7.11** Time evolution of the OH and $H_2O_2$ densities in the (top) gas gap and (bottom) 200 μm water layer, as obtained from 2D fluid dynamics simulations. The densities of OH and $OH_{aq}$ are shown during the first pulse and interpulse afterglow. $H_2O_2$ and $H_2O_{2aq}$ densities are shown accumulating after each pulse (0, 10 ms, 20 ms) and during the terminal afterglow. The contours are plotted on a 3-decade log scale with the maximum values noted in each frame. Adopted from [53] with permission

the interface, so that the gas-to-liquid equilibrium shifts towards the liquid, which will enhance the liquid phase concentrations, but this hypothesis will need to be investigated further.

Tian et al. [53] also investigated the transport of RONS from the gas phase into a liquid layer upon treatment of a water layer with a DBD device. Figure 7.11 illustrates the time evolution of the OH and $H_2O_2$ densities in the (top) gas gap and (bottom) 200 μm water layer. These results again clearly illustrate the dual behavior of (1) short-lived species (such as OH), which are lost in the liquid interface, versus (2) long-lived species (such as $H_2O_2$), which accumulate in the bulk of the liquid and can reach targets below the water layer. The generation of $OH_{aq}$ was attributed to both diffusion and solvation processes, as well as photo-dissociation at the surface of the water by VUV produced by the plasma.

### 7.3.3 Permeability of RONS and Pore Formation Through Native and Oxidized Phospholipid Bilayers (PLBs)

**Different Permeability of Hydrophilic and Hydrophobic RONS**

When plasma-produced RONS come in contact with cells, either directly or through a liquid layer covering cells, or upon application of PTLs to cells, they first come in contact with the cell membrane. The latter is composed of a PLB with proteins embedded. Hence, it is interesting to investigate the permeability of RONS through the PLB. In addition, as plasma-produced RONS can oxidize the PLB, it is interesting to compare the RONS permeability through both native and oxidized PLBs. Although the cell membrane consists of both lipids and proteins, each contributing for about 50% to the mass of the cell membrane, the PLB plays a crucial role in the structure of the cell membrane, so it makes sense, in first instance, to study the RONS permeability through only the PLB. Figure 7.12 depicts the free energy profiles (FEPs), obtained from umbrella sampling (US) MD simulations, of various ROS and RNS across both native and 50% oxidized PLBs, assuming oxidation of the lipid tails into aldehyde (see details in [86]). The structure of the PLB is drawn in pale color behind the FEPs. The center of the PLB is at $z = 0$, the head groups are around $z = \pm\, 2$ nm, and beyond this distance is the water phase surrounding the PLB.

As is clear from Fig. 7.12a, when the hydrophilic ROS (i.e., OH, $HO_2$, $H_2O_2$) move from the water phase to the PLB center, their FEP first reaches a minimum around the head groups, in agreement with [87], followed by a steep rise towards the center, thus creating a clear energy barrier when they cross the PLB. In case of the oxidized PLB, this free energy barrier is greatly reduced (see Fig. 7.12b). Indeed, upon oxidation, the hydrophilicity of the PLB rises, resulting in a larger permeability for hydrophilic ROS.

As the hydrophilic ROS prefer to reside close to the head groups, it is interesting to study oxidation of the head groups of the PLB, which was performed by DFTB [64]. While $HO_2$ and $H_2O_2$ molecules do not seem to react with the head groups and only show weak attractive non-bonded interactions, OH radicals do react with the head groups, resulting in detachment of some parts in the lipids, which reduces the lipid order and enhances the membrane fluidity, in agreement with experiments [64]. The latter might allow RONS to penetrate more easily through the PLB, causing further lipid tail (per)oxidation, which might result in pore formation (see next section). Note that OH radicals react with virtually all biomolecules and thus, oxidation of the head groups will only occur if the OH radicals are generated nearby the cell membrane, due to their small diffusion distance.

As is obvious from Fig. 7.12c, d, the hydrophobic ROS (such as $O_2$ and $O_3$) and RNS (e.g., NO, $NO_2$, and $N_2O_4$) show a completely different behavior, with very low permeation barriers around the PLB head groups and minima in the center (compared to the water phase). Indeed, these species prefer to reside in the lipid tail region, where they can cause lipid (per)oxidation. Upon oxidation of the PLB,

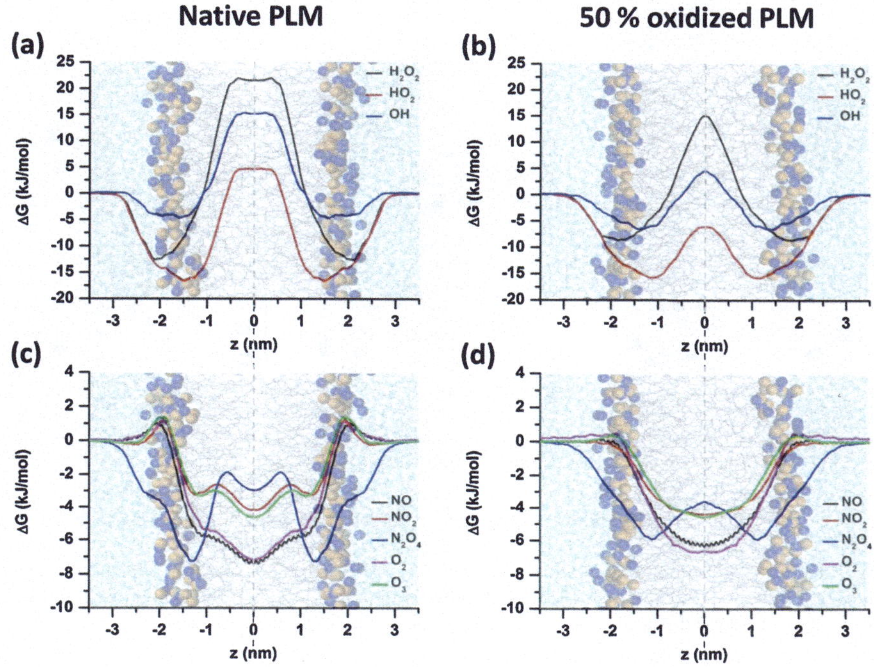


**Fig. 7.12** Free energy profiles (FEPs) of the hydrophilic (**a**, **b**) and hydrophobic (**c**, **d**) ROS and RNS, across native and 50% oxidized PLBs. The PLB structure is drawn in pale color at the background, to indicate the position of the water layer, head groups, and lipid tails

the FEPs of these hydrophobic ROS and RNS become a bit smoother, due to the higher membrane fluidity [86], but otherwise they do not change significantly. Note that hydrophilic RNS, like $HNO_2$, $HNO_3/NO_3^-$, and ONOOH, might behave the same as the hydrophilic ROS, but their permeability across the PLB could not yet be investigated, as no force fields are available.

These simulation results are in good qualitative correlation with experiments, where the permeability of hydrophobic RONS (NO and $O_2$) was found to be 3–6 orders of magnitude higher than the permeability of hydrophilic ROS ($H_2O_2$) [88, 89]. Obviously, pores or aquaporin (AQP) channels will be needed for the active transport of hydrophilic ROS in and out of the cell, as will be illustrated in the next sections, while hydrophobic RONS may permeate through the cell membrane even in the absence of AQP channels and pores.

## Effect of Oxidation on Pore Formation in the PLB

As hydrophilic RO(N)S cannot easily penetrate through the PLB, caused by their high free energy barriers, it is interesting to study pore formation in the cell

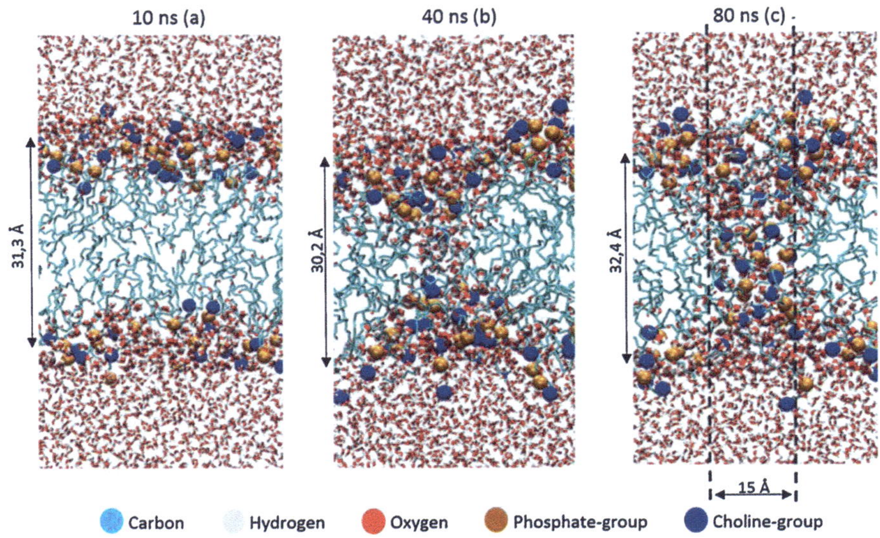


**Fig. 7.13** Snapshot of MD simulations, at 10 ns (**a**), 40 ns (**b**), and 80 ns (**c**), illustrating pore formation in a model system of a PLB with 100% oxidation. A pore with diameter of ca. 15 Å is formed in (**c**). Adopted from [91]—Published by The Royal Society of Chemistry

membrane after lipid (per)oxidation. For this purpose, some lipid (per)oxidation products, based on data from literature [90], were added to the model systems of a native PLB, with concentrations varying between 0 and 100% (see details in [91]).

Non-reactive MD simulations revealed that, upon oxidation, the thickness of the PLB drops, followed by a rise when the oxidation reaches 100%, and this was attributed to pore formation [91]. Figure 7.13 illustrates such a pore formation for a PLB with 100% lipid oxidation. The initial conformation (after 10 ns) does not show water defects (Fig. 7.13a), while after 40 ns (Fig. 7.13b), there is a significant amount of water in the center of the PLB, and after 80 ns (Fig. 7.13c), a pore is formed with diameter of 15 Å, which might allow RONS to penetrate and reach the cell interior. Figure 7.13 also indicates the thickness of the PLB: as pore formation allows water to enter the PLB, this yields swelling, resulting in a somewhat thicker bilayer (cf. Fig. 7.13c vs Fig. 7.13a, b). Other MD simulations in literature on oxidized PLBs [92–94] also reported an overall rise in membrane permeability [92], a change in the lipid mobility [93], as well as pore formation and bilayer disintegration [94] upon lipid oxidation.

### Synergistic Effect of Plasma Oxidation and Electric Field on Pore Formation

Besides plasma oxidation, electric fields can also cause pore formation, which is applied in the so-called electroporation [95–97]. As some biomedical plasma sources generate electric fields, with values in the order of a few up to 100 kV/cm (see, e.g., [98–100]), these electric fields might also contribute to pore formation in the cell membrane, and perhaps there is even synergy with plasma-induced lipid oxidation.

Figure 7.14 indeed illustrates that lipid oxidation yields a drop in the electric field threshold needed for pore formation in the PLB, as well as a shorter average pore formation time. More details can be found in [101]. Note that the pore formation time can fluctuate considerably, resulting in large error bars in the figure, but the latter cannot be reduced by increasing the number of simulations, because pore formation is a stochastic process. Nevertheless, it is clearly visible that oxidation of the lipid tails in the PLB facilitates pore formation, by lowering the threshold electric field, as well as the pore formation time.

We need to remark that the applied electric fields in electroporation simulations are typically much stronger than the fields used in electroporation or plasma medicine [98–100]. Indeed, the latter typically vary between 0.1 and 100 kV/cm (corresponding to 0.01 and 10 mV/nm). However, it should be noted that the macroscopic field applied in experiments is not at all equivalent to the field experienced by the cell membrane (and thus applied in MD simulations), so the actual electric field values should not be directly compared, as explained in

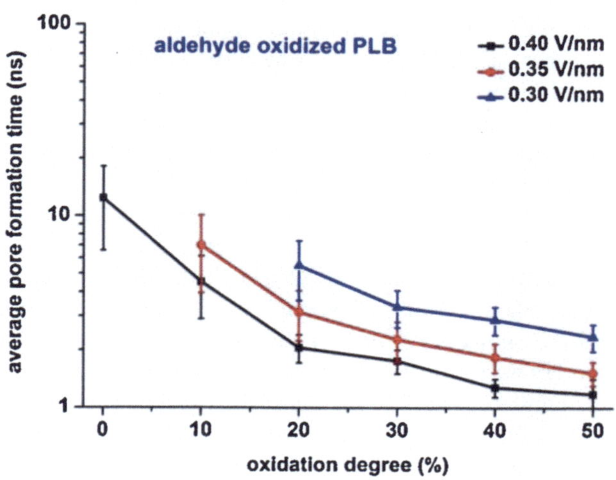


**Fig. 7.14** Average pore formation time for three different electric field values, as a function of the oxidation degree of the PLB, for lipid oxidation into aldehydes. Adopted from [101] with permission

detail in [101]. In addition, it is not possible to directly relate the average pore formation times obtained in MD simulations to the experimental pore formation kinetics, as also explained in [101]. Nevertheless, the trends of pore formation times for different values of electric fields and oxidation degrees shown in Fig. 7.14 qualitatively illustrate the synergistic effect of plasma oxidation and electric fields on the permeability of the cell membrane.

### Stabilizing Effect of Cholesterol in the PLB: Possible Explanation for the Selectivity of Plasma for Cancer Treatment

It is known that the cholesterol-to-phospholipid ratio in the cell membrane of some cancer cells (e.g., leukemic cells) is lower than in their normal counterparts (such as lymphocytes) [102]. Therefore, it is interesting to study the effect of the presence of cholesterol in the cell membrane, in varying concentrations, on the FEPs of ROS (both hydrophilic and hydrophobic), to reveal whether this can provide an explanation for the selectivity of plasma towards cancer cells vs. normal cells [103].

In Fig. 7.15 (upper part) the FEP of $H_2O_2$ is plotted for various cholesterol concentrations. Higher cholesterol concentrations obviously enhance the free energy barrier height and width and also result in local free energy minima in the center of the PLB, which will make the penetration of $H_2O_2$ towards the intracellular environment more difficult, even if it would succeed to penetrate into the PLB center. Other hydrophilic ROS, such as OH and $HO_2$, showed the same behavior [103].

Figure 7.15 (lower part) illustrates the FEP of $O_2$, which exhibits a minimum in the center (cf. also Fig. 7.12). However, upon increasing cholesterol concentrations, some extra free energy barriers are formed at about 1 nm from the center, which is due to the presence of the bulky sterol rings. These extra barriers will limit the probability of lipid (per)oxidation of the lipid tails, and hence pore formation. As some cancer cells have less cholesterol in their cell membrane than normal cells, this might explain why RONS can more easily reach the cell interior of cancer cells, and hence, the selective action of plasma treatment.

As indicated above, the FEPs of the hydrophilic RO(N)S are too high for permeation, even without the presence of cholesterol, indicating the need for pore formation to allow these hydrophilic RO(N)S to enter the cell. Hence, it is also interesting to study the effect of cholesterol on pore formation. It was found that cholesterol fractions above 15% do not give rise to increasing cell membrane fluidity upon oxidation, and even for 100% oxidation degree, no pore formation was observed [91]. Indeed, Fig. 7.16 illustrates the water density in the center of the PLB in case of 100% lipid oxidation, as a function of the cholesterol fraction in the PLB. A cholesterol fraction up to 11% yields a significant water density, resulting from pore formation, while for higher cholesterol fractions, the water density drops to negligible values, indicating that no pores are formed. Hence, RONS should be able to penetrate more easily through the membrane of cells with lower cholesterol fraction, such as cancer cells, resulting in oxidative stress inside the cell, possibly

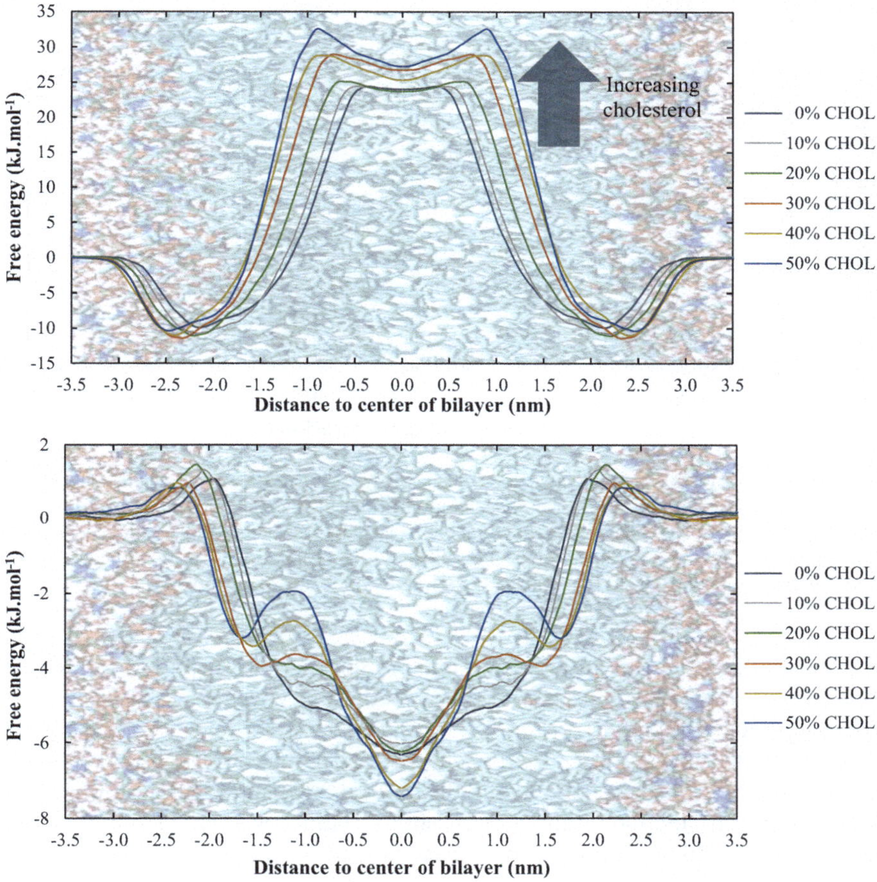


**Fig. 7.15** FEPs of $H_2O_2$ (upper part) and $O_2$ (lower part) across the PLB, for various cholesterol concentrations in the cell membrane. Adopted from [103] with permission

inducing apoptosis of cancer cells, while this effect might not happen in normal cells, because the cholesterol in their cell membrane inhibits pore formation. Hence, this might be one of the explanations of the selectivity of plasma towards cancer cells.

### Permeation of RONS Through Aquaporins (AQPs) Vs PLB: Possible Explanation for the Selectivity of Plasma for Cancer Treatment

Another possible explanation for the selectivity of plasma towards cancer cells vs. normal cells is the higher expression of AQPs in the cell membrane of cancer cells. Indeed AQPs are known as $H_2O_2$ channels [104–107]. It is thus interesting

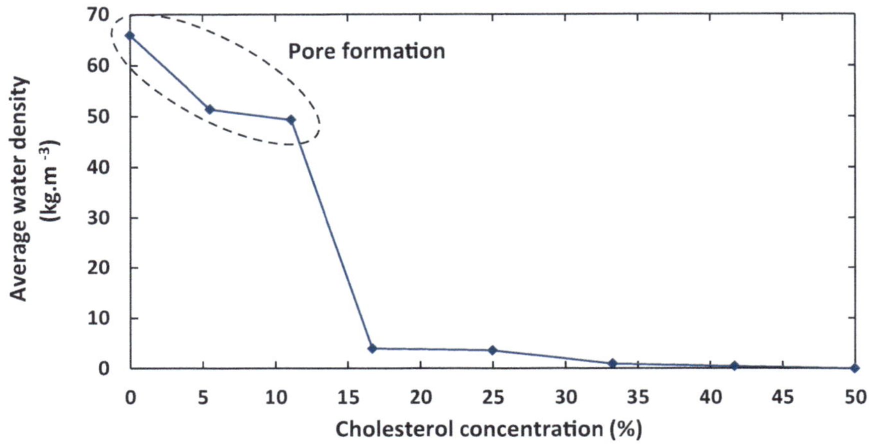


**Fig. 7.16** Calculated average water density in the center of the PLB, for model systems with 100% oxidation, as a function of cholesterol concentration in the PLB, indicating that pore formation occurs more easily in cell membranes containing less cholesterol, which is typical for cancer cells. This might be one of the explanations of the selectivity of plasma treatment for cancer cells vs normal cells. Adopted from [91]—Published by The Royal Society of Chemistry

to compare the FEPs and the diffusion rate profiles of $H_2O_2$ across AQP and the PLB, as well as the resulting permeability coefficient, which can all be obtained from MD simulations [108].

As illustrated in Fig. 7.17, the free energy barrier for $H_2O_2$ transport through AQP1 (which is one model system of AQP) is ca. three times lower than through the PLB, and the resulting permeability coefficient was found to be more than two orders higher through AQP than through the PLB, i.e., 2.57 cm/s vs. $6.62 \times 10^{-3}$ cm/s. We can thus conclude that AQP creates a more favorable path for $H_2O_2$ permeation, which might also explain the selectivity of plasma towards cancer cells.

In [109] the permeation of other RONS, both hydrophilic (OH) and hydrophobic ($NO_2$ and NO) through AQP was studied, and it was found that these RONS can all penetrate across the pores of AQP, with a lower permeation free energy barrier for OH and NO than for $H_2O_2$ and $NO_2$, suggesting that these radicals may have easier access to the pore interior and interact with the amino acid residues of AQP. Furthermore, oxidation of both the phospholipids and AQP (i.e., sulfenylation of $Cys_{191}$) seemed to slightly enhance the free energy barrier for $H_2O_2$ and $NO_2$ permeation, while for OH and NO no clear effect was observed [109].

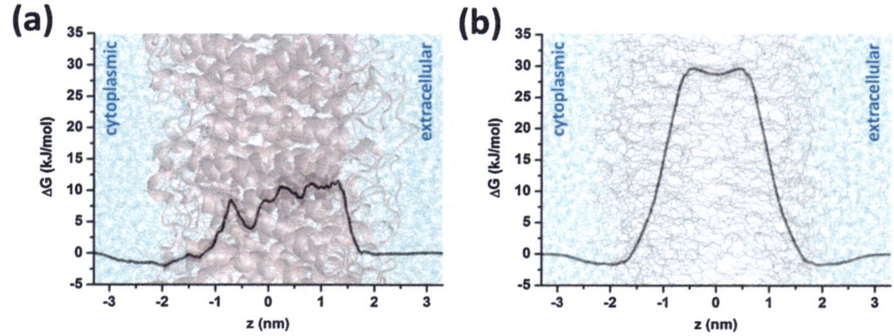


**Fig. 7.17** FEPs of $H_2O_2$ across AQP1 (**a**) and the PLB (**b**). The cytoplasmic and extracellular water layers, the AQP protein, and lipid tails are shown in the background, as light blue, pink and gray colors, respectively. The associated standard deviations of the FEPs are shown in gray

### *7.3.4 Interaction of RONS with DNA: Single and Double Strand Breaks*

Besides studying the behavior of the PLB, the interaction of RONS with DNA might also be important in the context of cancer treatment. This has been investigated with classical reactive MD simulations [76]. As mentioned in Sect. 7.2.2, this method is less time-consuming than DFT and DFTB, so it can handle somewhat larger model systems, although the latter are still quite limited (order of a few 10,000 atoms). Hence, DNA as a whole cannot be simulated, because the structure is far too large. However, it is possible to select some representative fragments, as a model system. This is illustrated in Fig. 7.18, for a small part of a DNA double helix, containing 12 base pairs, surrounded by water molecules.

When studying the interaction of OH radicals with this DNA model system, two main types of reactions were observed, i.e., (i) H-abstraction from a (primary or secondary) amine in the nucleic bases and (ii) OH-addition on the purine ring of the nucleic bases (see details in [76]).

The H-abstractions result in a radical that may give rise to some more (intermolecular) H-abstractions, which might affect the H-bridge structure between both DNA strands, possibly leading to single strand breaks (SSBs). When combined with a second SSB at the opposite strand in close vicinity, this might result in a double strand break (DSB), cleaving the DNA double helix, which might activate pro-apoptotic factors. It is indeed known from literature [110] that when both strands are broken within 20–57 base pairs by SSBs, DSB will eventually occur when enough H-bridges between both helices are broken. This could, however, not be observed in the MD simulations, because they are limited to a certain time scale, due to the high computational cost.

The OH-addition on the purine ring of the nucleic bases, more specifically at the C-8′ position of dAMP and dGMP, yields 8-hydroxy-purine adduct radicals

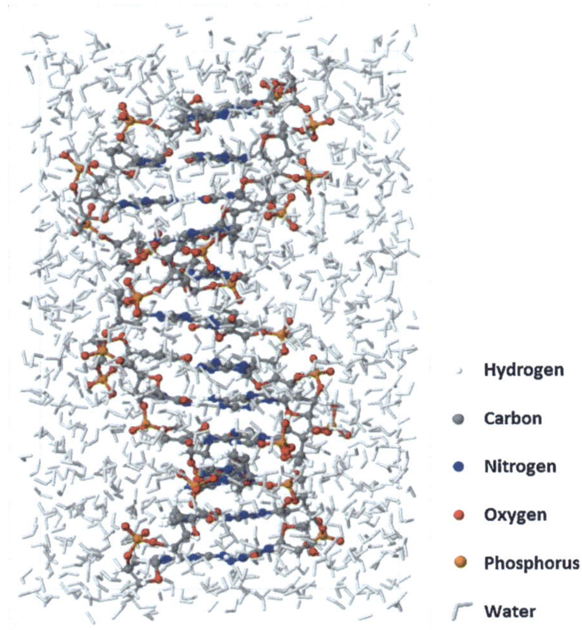


**Fig. 7.18** Part of a DNA double helix, consisting of 12-base pairs surrounded by water molecules, used as model system for the interaction with OH radicals in reactive MD simulations. Adopted from [76] with permission

(8-OH-Ade$^{\bullet}$ or 8-OH-Gua$^{\bullet}$), as illustrated in Fig. 7.19 for dGMP. This reaction is the first step towards 8-oxo-guanine (8-O-Gua) and 2,6-diamino-4-hydroxy-5-formamidopyrimidine (FapydG) formation [111], which are markers for oxidative stress in cells [110–113]. This may affect biochemical pathways within the affected cell, e.g., introduction of DNA mutations or inhibition of gene expressions, possibly leading to apoptosis. Again, however, because of the high computational cost of the MD simulations, the simulated time was not long enough to observe these further reactions.

### 7.3.5 *Effect of (Plasma) Oxidation on Proteins*

As third type of biomolecules, proteins also play a crucial role in the context of cancer. In [114] non-reactive MD and molecular docking simulations, as well as binding free energy calculations, were performed to investigate the structural conformation and binding affinity of human epidermal growth factor (hEGF) to its receptor (hEGFR) under oxidative stress. hEGF is one of the important signaling proteins playing a role in both wound healing and cancer treatment. The simulations showed that a low amount of oxidation has minor effect on the binding affinity of hEGF with its receptor, while a higher level of oxidation causes less interaction with its receptor. This can be deduced from Fig. 7.20, illustrating the alignments

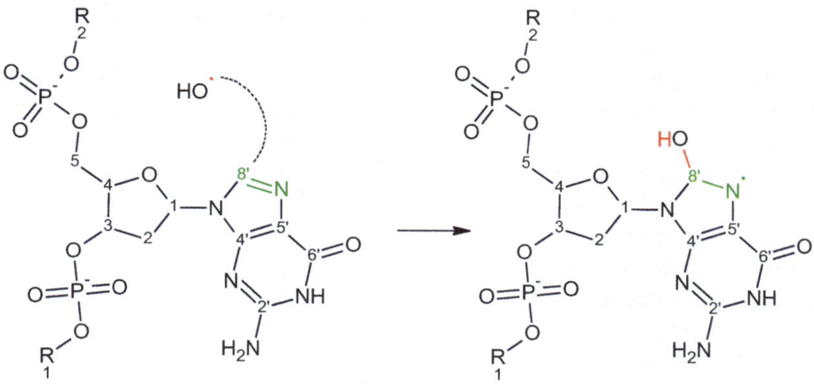

**Fig. 7.19** Schematic illustration of OH-addition on C-8′ of DGP, yielding the formation of an 8-OH-guanine adduct radical, as predicted by MD simulations. R1 and R2 correspond to the preceding and following nucleotides

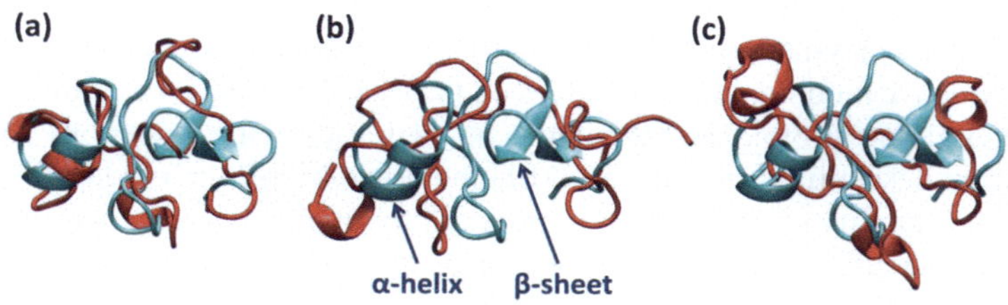


**Fig. 7.20** Alignment of the oxidized hEGF structures (red), at mild (**a**), intermediate (**b**), and strong (**c**) oxidation, with native hEGF (cyan), illustrating more conformational changes upon higher oxidation of hEGF, compared to the native hEGF. Adopted from [114] with permission

of the oxidized structures with the native hEGF. Indeed, the structure with mild oxidation has a better alignment with the native hEGF, whereas the structures with more enhanced oxidation show significant deviations. Furthermore, the helical structures also start to form at random sites, which is more visible in the strongest oxidized structure. Thus, we can conclude that a higher level of oxidation may lead to completely new conformations in the hEGF protein (see Fig. 7.20).

Molecular docking and binding free energy simulations furthermore predicted that these structural modifications, and corresponding conformational changes, affect the binding free energies of hEGF with its receptor (hEGFR) [114]. Indeed, the binding free energy of native hEGF with its receptor was calculated to be −105.15 kcal/mol, while the binding free energies for the oxidized hEGF structures were found to be −71.21, −63.85, and −49.49 kcal/mol, for the mildly, intermediate, and strongly oxidized structures, respectively, thus indicating that higher oxidation causes a lower interaction.

The fact that mild oxidation does not significantly affect the binding affinity of hEGF with its receptor means that most likely, it does not strongly influence the signaling pathways in a cell, and thereby the cell proliferation, which might explain why short plasma treatment times are beneficial for chronic wound healing. On the other hand, the reduced alignment and interaction upon a higher level of oxidation of hEGF may cause a disturbance of the cell signaling pathways, ultimately leading to disruption of the cell proliferation. The latter might explain why longer plasma treatment times result in inhibiting cancer cell proliferation and even cancer cell death.

Although not in the context of plasma for cancer treatment, some other studies focused on the effect of oxidation on proteins by means of computer simulations. For instance, in [115] classical MD simulations were carried out to study the structure and dynamics of the oxidized (i.e., carbonylated) headpiece domain of villin. The modification of certain residues in this protein (i.e., seven lysine, arginine, and proline) resulted in a destabilization and unfolding of the protein structure, especially at high levels of carbonylation. This was attributed to the disruption of specific stabilizing elements, i.e., proline kinks and salt bridges, and a change of surface hydrophobicity [115]. The same group also examined the effect of oxidation on the stability and dimerization of superoxide dismutase 1 (SOD1), a key antioxidant enzyme in human cells [116]. The MD simulation results revealed that the oxidative modification of residues on the dimer interface (i.e., carbonylation of lysine, arginine, proline, and threonine, as well as cysteine oxidation) may lead to significant destabilization in the SOD1 homodimer. Moreover, oxidation can also decrease the stability of SOD1 monomer, resulting in a partial local unfolding and subsequent increase of the aggregation propensity [116]. Finally, MD simulations were also applied to investigate the role of post-translational modifications (PTMs), namely side chain hydroxylations, *C*- and *N*-methylations, on the stability of $\beta^{6.3}$-helical structure of the polytheonamide B (pTB) peptide [117], by reverting the modified residues to their precursors. It was shown that the reversion of the hydroxylations and *C*-methylations represents insignificant or even no effect on the structure. On the other hand, the reversion of the *N*-methylations leads to the complete loss of the β-helical conformation [117].

In general, the effects of PTMs (e.g., methylation, phosphorylation, glycosylation, and acetylation) on the protein structure have been widely studied. However, it is also important to examine the effect of oxidation (and nitrosylation induced by RNS) on the protein stability and its binding affinity, especially in the context of plasma for cancer treatment, which is till now only scarcely investigated. Indeed, such studies would help to unravel the molecular level mechanisms of cancer cell death induced by plasma, as well as by other cancer therapies based on oxidation.

## 7.4 Conclusion and Future Challenges

This chapter gave an overview of the different types of modeling approaches that can be applied to study plasma for cancer treatment. This includes both macro-scale models, i.e., 0D chemical kinetics and 1D/2D fluid dynamics models of the most common plasma sources and their interaction with liquids, and atomic/molecular scale models, to study the interaction of plasma-produced RONS with biomolecules important in the context of plasma for cancer treatment, such as the phospholipid bilayer (PLB), DNA, and proteins, as well as the effect of these interactions (e.g., oxidation) on structural changes in the biomolecules.

Subsequently, typical calculation results of these models were presented, to illustrate the type of information that can be obtained from the different models.

Several different 0D chemical kinetics models and 1D/2D fluid dynamics models have been developed for different types of plasma sources and their interaction with liquids. They typically provide information on the important RONS present inside the plasma, the effluent, and/or the liquid. 0D models focus on a detailed chemistry, as they can handle a large number of species and chemical reactions with limited computational cost, while 1D/2D (or in principle 3D) fluid dynamics simulations typically consider a more limited chemistry (due to calculation time), but can be used to model specific plasma devices, and provide spatial information, e.g., gas and liquid flow profiles, species concentration profiles in the gas and liquid phase (including transport of RONS from the interface to the bulk), and chemical pathways for the various species in gas and liquid phase.

Furthermore, we showed several calculation results of the atomic/molecular scale simulations, for instance, to gain more insight in the permeability of various RONS across both native and oxidized PLBs, as model systems for the cell membrane, as well as in the synergistic effect of plasma oxidation and electric fields on pore formation, the effect of cholesterol present in the cell membrane on the permeability and on the chance for pore formation, and the different permeability of $H_2O_2$ (as representative RONS) across AQP channels vs. the PLB, in order to contribute to a better understanding of the selectivity of plasma for cancer treatment.

Hydrophobic RONS are shown to penetrate quite easily across both native and oxidized PLBs, while hydrophilic ROS experience too high energy barriers. Although the latter are reduced upon oxidation of the PLB, they are still too high, indicating the need for specific protein channels (e.g., AQPs) or pores created by plasma oxidation and/or electric fields, to allow their penetration into the cytoplasm, eventually to cause oxidative damage. In addition, the calculations reveal that the lower cholesterol fraction and the higher AQP expression in the cell membrane of various types of cancer cells might explain the selectivity of plasma towards cancer cells vs. normal cells.

Although these simulations give more insight in the formation and reactivity/stability of RONS inside plasma and in plasma-treated liquids, and how these RONS can penetrate through the cell membrane, either by passive transport (for hydrophobic RONS) or through pores or AQP channels (for hydrophilic RONS), more research will be needed to elucidate all the underlying mechanisms of plasma for cancer treatment. For instance, the simulations for RONS penetration through cell membranes are only applied to PLB model systems, while in reality the cell membrane is more complex and contains more proteins besides AQP, such as antiporters, catalase, etc., which can also be important in the context of plasma for cancer treatment. In addition, it would be interesting to study the cell membrane permeability of other RONS, e.g., hydrophilic RNS, like $HNO_2$, $HNO_3/NO_3^-$, and ONOOH. They were not investigated up to now, because no accurate force fields are available yet for these simulations. This obviously shows the need to develop such force fields, to obtain a more comprehensive picture of the behavior of all possible RONS in the cell membrane. Furthermore, the various RONS might create a myriad of different lipid oxidation (and nitration) products in the cell membrane, so more knowledge is crucial on these different products formed, and how they affect the biophysical properties of the cell membrane, and thus its function. Up to now, simulations of oxidized membranes were based on rather simplified descriptions of membrane composition, while lipid peroxidation might lead to liquid ordered–liquid disordered phase separation in membranes [118], and this might favor pore formation. Thus, it would be interesting to account for phase-separated membranes.

We also showed examples of the interaction of RONS with DNA, as well as of the effect of plasma-induced oxidation on the structural conformation of proteins, more specifically for hEGF, playing a role in both wound healing and cancer therapy. Of course, many more proteins play a role in cancer therapy and could also be the subject of future investigations.

Specifically, it would be interesting to investigate the interaction between immune cell proteins and cancer cell proteins, under native conditions and oxidative stress (as could be induced by plasma therapy), which would be of great interest for plasma cancer immunotherapy. Protein–protein interactions have indeed been computationally studied in many different fields (e.g., [119, 120]). In particular, some MD simulations have already been reported for the molecular level functions of immune checkpoint proteins. For instance, Du et al. investigated the high affinity of PD-1 mutant by MD simulations, which helps to design engineered PD-1 mutants to modulate the PD-1/PD-L1 interaction [121]. Likewise, the binding of major histocompatibility complex (MHC) with different peptides and their receptors has also been studied by MD simulations [122]. In general, the effect of mutations in proteins and drugs that block the binding of immune checkpoint proteins to receptors has been studied for PD-1 and PD-L1, CD47 and SIRPα, and CD86 and CTLA-4 [123, 124]. However, to the best of our knowledge, the interaction between immune cell and cancer cell proteins under oxidative stress, as induced by plasma, has not yet been investigated by computer simulations, showing the need for such type of modeling, for a better insight in plasma for cancer immunotherapy.

Besides the above atomic/molecular level simulations, it would also be very interesting to study the cell signaling pathways, which might also explain the selectivity of plasma for cancer therapy [125]. Indeed, experiments showed that two apoptosis-inducing signaling pathways that are inhibited by catalase in the extracellular compartment can be reactivated when catalase in the cell membrane of cancer cells is inactivated. As plasma might be able to inactivate catalase, it might induce cancer cell death by reactivation of these signaling pathways. It would be interesting to obtain a better understanding of this apoptosis-inducing mechanism and the role of catalase inactivation. Such mechanisms can of course not be simulated by molecular level simulations, but it requires reaction kinetics modeling, including also spatial dynamics. Such modeling is very challenging and would need to be built up step-by-step, but it would be very rewarding to gain further insight in the underlying mechanisms of the selective action of plasma towards cancer cells.

Finally, to validate the model predictions, especially those at the molecular level, dedicated experiments should be designed, under very controlled conditions, generating, for instance, only a beam of OH radicals, instead of a complex mixture of RONS and other plasma effects, and for well-defined model systems of biomolecules, gradually mimicking the more complex tissues. Such experiments are set up in various labs, for instance, for studying the separate and synergistic effects of plasma-generated radicals and UV/VUV photons at the cellular and molecular level for various kinds of biomolecules, or experiments with simple model systems for the cell membrane, based on synthetic phospholipid membrane vesicles or liposomal model membranes (e.g., [126–137]). We believe that the combination of such experiments and modeling is crucial for gaining a deeper understanding of the underlying mechanisms of plasma for cancer treatment.

**Acknowledgements** We acknowledge the financial support from the Research Foundation—Flanders (FWO; grant numbers: 1200219 N and 11U5416N). The computational work was carried out using the Turing HPC infrastructure at the CalcUA core facility of the Universiteit Antwerpen (UA), a division of the Flemish Supercomputer Center VSC, funded by the Hercules Foundation, the Flemish Government (department EWI), and the UA. We are also very thankful to R. Cordeiro for the very interesting discussions.

## References

1. J. Van der Paal, Generation, transport and molecular interactions of reactive species in plasma medicine, PhD-dissertation, University of Antwerp, 2019
2. W. Van Gaens, A. Bogaerts, Kinetic modelling for an atmospheric pressure argon plasma jet in humid air. J. Phys. D Appl. Phys. **46**, 275201 (2013)
3. P. Heirman, W. Van Boxem, A. Bogaerts, Reactivity and stability of plasma-generated oxygen and nitrogen species in buffered water solution: a computational study. Phys. Chem. Chem. Phys. **21**, 12881–12894 (2019)
4. G.V. Naidis, Modelling of streamer propagation in atmospheric-pressure helium plasma jets. J. Phys. D Appl. Phys. **43**, 402001 (2010)
5. G.V. Naidis, Modelling of plasma bullet propagation along a helium jet in ambient air. J. Phys. D Appl. Phys. **44**, 215203 (2011)
6. J.P. Boeuf, L.L. Yang, L.C. Pitchford, Dynamics of a guided streamer ('plasma bullet') in a helium jet in air at atmospheric pressure. J. Phys. D Appl. Phys. **46**, 015201 (2013)
7. D. Breden, K. Miki, L.L. Raja, Self-consistent two-dimensional modeling of cold atmospheric-pressure plasma jets/bullets. Plasma Sources Sci. Technol. **21**, 034011 (2012)
8. M. Yousfi, O. Eichwald, N. Merbahi, N. Jomaa, Analysis of ionization wave dynamics in low-temperature plasma jets from fluid modeling supported by experimental investigations. Plasma Sources Sci. Technol. **21**, 045003 (2012)
9. X.Y. Liu, X.K. Pei, X.P. Lu, D.W. Liu, Numerical and experimental study on a pulsed-dc plasma jet. Plasma Sources Sci. Technol. **23**, 035007 (2014)

10. Z. Xiong, E. Robert, V. Sarron, J.M. Pouvesle, M.J. Kushner, Atmospheric-pressure plasma transfer across dielectric channels and tubes. J. Phys. D Appl. Phys. **46**, 155203 (2013)
11. Z. Xiong, M.J. Kushner, Atmospheric pressure ionization waves propagating through a flexible high aspect ratio capillary channel and impinging upon a target. Plasma Sources Sci. Technol. **21**, 034001 (2012)
12. Z. Xiong, E. Robert, V. Sarron, J.M. Pouvesle, M.J. Kushner, Dynamics of ionization wave splitting and merging of atmospheric-pressure plasmas in branched dielectric tubes and channels. J. Phys. D Appl. Phys. **45**, 275201 (2012)
13. N.Y. Babaeva, M.J. Kushner, Interaction of multiple atmospheric-pressure micro-plasma jets in small arrays: He/O2 into humid air. Plasma Sources Sci. Technol. **23**, 015007 (2014)
14. G.V. Naidis, On the ring-shaped structure of helium plasma jets. IEEE Trans. Plasma Sci. **43**, 733–736 (2015)
15. A. Bourdon, T. Darny, F. Pechereau, et al., Numerical and experimental study of the dynamics of a μs helium plasma gun discharge with various amounts of N2 admixture. Plasma Sources Sci. Technol. **25**, 035002 (2016)
16. C. Lazarou, C. Anastassiou, C. Topala, et al., Numerical simulation of capillary helium and helium−oxygen atmospheric pressure plasma jets: propagation dynamics and interaction with dielectric. Plasma Sources Sci. Technol. **27**, 105007 (2018)
17. P. Viergas, F. Péchereau, A. Bourdon, Numerical study on the time evolutions of the electric field in helium plasma jets with positive and negative polarities. Plasma Sources Sci. Technol. **27**, 025007 (2018)
18. A.M. Lietz, M.J. Kushner, Molecular admixtures and impurities in atmospheric pressure plasma jets. J. Appl. Phys. **124**, 153303 (2018)
19. L. Chang, L. Nie, Y. Xian, X. Lu, The effect of seed electrons on the repeatability of atmospheric pressure plasma plume propagation. II. Modeling. Phys. Plasmas **23**, 123513 (2016)
20. Y. Sakiyama, D.B. Graves, H.W. Chang, T. Shimizu, G.E. Morfill, Plasma chemistry model of surface microdischarge in humid air and dynamics of reactive neutral species. J. Phys. D Appl. Phys. **45**, 425201 (2012)
21. N.Y. Babaeva, M.J. Kushner, Reactive fluxes delivered by dielectric barrier discharge filaments to slightly wounded skin. J. Phys. D Appl. Phys. **46**, 025401 (2013)
22. H.W. Lee, G.Y. Park, Y.S. Seo, Y.H. Im, S.B. Shim, H.J. Lee, Modelling of atmospheric pressure plasmas for biomedical applications. J. Phys. D Appl. Phys. **44**, 053001 (2011)
23. N.Y. Babaeva, N. Ning, D.B. Graves, M.J. Kushner, Ion activation energy delivered to wounds by atmospheric pressure dielectric-barrier discharges: sputtering of lipid-like surfaces. J. Phys. D Appl. Phys. **45**, 115203 (2012)
24. N.Y. Babaeva, W. Tian, M.J. Kushner, The interaction between plasma filaments in dielectric barrier discharges and liquid covered wounds: electric fields delivered to model platelets and cells. J. Phys. D Appl. Phys. **47**, 235201 (2014)
25. G.V. Naidis, Modelling of OH production in cold atmospheric-pressure He–H2O plasma jets. Plasma Sources Sci. Technol. **22**, 035015 (2013)
26. D.X. Liu, P. Bruggeman, F. Iza, M.Z. Rong, M.G. Kong, Global model of low-temperature atmospheric-pressure He + H2O plasmas. Plasma Sources Sci. Technol. **19**, 025018 (2010)
27. J. Waskoenig, K. Niemi, N. Knake, L.M. Graham, S. Reuter, V. Schulz-vonder Gathen, T. Gans, Atomic oxygen formation in a radio-frequency driven micro-atmospheric pressure plasma jet. Plasma Sources Sci. Technol. **19**, 045018 (2010)
28. K. McKay, D.X. Liu, M.Z. Rong, F. Iza, M.G. Kong, Generation and loss of reactive oxygen species in low-temperature atmospheric-pressure RF He + O2 + H2O plasmas. J. Phys. D Appl. Phys. **45**, 172001 (2012)
29. T. Murakami, K. Niemi, T. Gans, D. O'Connell, W.G. Graham, Afterglow chemistry of atmospheric-pressure helium–oxygen plasmas with humid air impurity. Plasma Sources Sci. Technol. **23**, 025005 (2014)

30. T. Murakami, K. Niemi, T. Gans, D. O'Connell, W.G. Graham, Interacting kinetics of neutral and ionic species in an atmospheric-pressure helium–oxygen plasma with humid air impurities. Plasma Sources Sci. Technol. **22**, 045010 (2013)
31. T. Murakami, K. Niemi, T. Gans, D. O'Connell, W.G. Graham, Chemical kinetics and reactive species in atmospheric pressure helium–oxygen plasmas with humid-air impurities. Plasma Sources Sci. Technol. **22**, 015003 (2013)
32. W.J.M. Brok, M.D. Bowden, J. van Dijk, J.J.A.M. van der Mullen, G.M.W. Kroesen, Numerical description of discharge characteristics of the plasma needle. J. Appl. Phys. **98**, 013302 (2005)
33. Y. Sakiyama, D.B. Graves, Finite element analysis of an atmospheric pressure RF-excited plasma needle. J. Phys. D Appl. Phys. **39**, 3451 (2006)
34. Y. Sakiyama, D.B. Graves, Corona-glow transition in the atmospheric pressure RF-excited plasma needle. J. Phys. D Appl. Phys. **39**, 3644 (2006)
35. Y. Sakiyama, D.B. Graves, Non-thermal atmospheric RF plasma in one-dimensional spherical coordinates. IEEE Trans. Plasma Sci. **35**, 1279 (2007)
36. Y. Sakiyama, D.B. Graves, Nonthermal atmospheric rf plasma in one-dimensional spherical coordinates: asymmetric sheath structure and the discharge mechanism. J Appl. Phys. **101**, 073306 (2007)
37. Y. Sakiyama, D.B. Graves, Neutral gas flow and ring-shaped emission profile in non-thermal RF-excited plasma needle discharge at atmospheric pressure. Plasma Sources Sci. Technol. **18**, 025022 (2009)
38. W. Van Gaens, A. Bogaerts, Reaction pathways of biomedically active species in an Ar plasma jet. Plasma Sources Sci. Technol. **23**, 035015 (2014)
39. S. Zhang, W. Van Gaens, B. Van Gessel, S. Hofmann, E. van Veldhuizen, A. Bogaerts, P. Bruggeman, Spatially resolved ozone densities and gas temperatures in a time modulated RF driven atmospheric pressure plasma jet: an analysis of the production and destruction mechanisms. J Phys. D Appl. Phys. **46**, 205202 (2013)
40. W. Van Gaens, P.J. Bruggeman, A. Bogaerts, Numerical analysis of the NO and O generation mechanism in a needle-type plasma jet. New J. Phys. **16**, 063054 (2014)
41. W. Van Gaens, S. Iséni, A. Schmidt-Bleker, K.-D. Weltmann, S. Reuter, A. Bogaerts, Numerical analysis of the effect of nitrogen and oxygen admixtures on the chemistry of an argon plasma jet operating at atmospheric pressure. New J. Phys. **17**, 033003 (2015)
42. A. Schmidt-Bleker, J. Winter, S. Iseni, M. Dünnbier, K.D. Weltmann, S. Reuter, Reactive species output of a plasma jet with a shielding gas device - combination of FTIR absorption spectroscopy and gas phase modelling. J. Phys. D Appl. Phys. **47**, 145201 (2014)
43. A. Schmidt-Bleker, J. Winter, A. Bösel, S. Reuter, K.-D. Weltmann, On the plasma chemistry of a cold atmospheric argon plasma jet with shielding gas device. Plasma Sources Sci. Technol. **25**, 015005 (2016)
44. Y. Gorbanev, C.C.W. Verlackt, S. Tinck, E. Tuenter, K. Foubert, P. Cos, A. Bogaerts, Combining experimental and modelling approaches to study the sources of reactive species induced in water by the COST RF plasma jet. Phys. Chem. Chem. Phys. **20**, 2797–2808 (2018)
45. Y. Gorbanev, J. Van der Paal, W. Van Boxem, S. Dewilde, A. Bogaerts, Reaction of chloride anion with atomic oxygen in aqueous solutions: can cold plasma help in chemistry research? Phys. Chem. Chem. Phys. **21**, 4117–4121 (2019)
46. S. Schröter, R.A. Gibson, M.J. Kushner, T. Gans, D. O'Connell, Numerical study of the influence of surface reaction probabilities on reactive species in an rf atmospheric pressure plasma containing humidity. Plasma Phys. Control. Fusion **60**, 014035 (2018)
47. B.R. Locke, S.M. Thagard, Analysis and review of chemical reactions and transport processes in pulsed electrical discharge plasma formed directly in liquid water. Plasma Chem. Plasma Process. **32**, 875 (2012)
48. C.A.J. van Gils, S. Hofmann, B.K.H.L. Boekema, R. Brandenburg, P.J. Bruggeman, Mechanisms of bacterial inactivation in the liquid phase induced by a remote RF cold atmospheric pressure plasma jet. J Phys. D. Appl. Phys. **46**, 175203 (2013)

49. P. Lukes, E. Dolezalova, I. Sisrova, M. Clupek, Aqueous-phase chemistry and bactericidal effects from an air discharge plasma in contact with water: evidence for the formation of peroxynitrite through a pseudo-second-order post-discharge reaction of H2O2 and HNO2. Plasma Sources Sci. Technol. **23**, 015019 (2014)
50. S. Hamaguchi, K. Ikuse, T. Kanazawa, Generation of free radicals in liquid by atmospheric-pressure plasmas and its application to biology and medicine. JPS Conf. Proc. **1**, 015055 (2014)
51. C. Chen, D.X. Liu, Z.C. Liu, A.J. Yang, H.L. Chen, G. Shama, M.G. Kong, A model of plasma-biofilm and plasma-tissue interactions at ambient pressure. Plasma Chem. Plasma Process. **34**, 403 (2014)
52. D.X. Liu, Z.C. Liu, C. Chen, A.J. Yang, D. Li, M.Z. Rong, H.L. Chen, M.G. Kong, Aqueous reactive species induced by a surface air discharge: heterogeneous mass transfer and liquid chemistry pathways. Sci. Rep. **6**, 23737 (2016)
53. W. Tian, M.J. Kushner, Influence of excitation pulse duration of dielectric barrier discharges on biomedical applications. J. Phys. D Appl. Phys. **47**, 165201 (2014)
54. W. Tian, M.J. Kushner, Long-term effects of multiply pulsed dielectric barrier discharges in air on thin water layers over tissue: stationary and random streamers. J. Phys. D Appl. Phys. **48**, 494002 (2015)
55. A.M. Lietz, M.J. Kushner, Air plasma treatment of liquid covered tissue: long timescale chemistry. J. Phys. D Appl. Phys. **49**, 425204 (2016)
56. A.M. Lietz, M.J. Kushner, Corrigendum: air plasma treatment of liquid covered tissue: long timescale chemistry (2016 J. Phys. D: Appl. Phys. 49 425204). J. Phys. D Appl. Phys. **50**, 119501 (2017)
57. W. Van Boxem, J. Van der Paal, Y. Gorbanev, S. Vanuytsel, E. Smits, S. Dewilde, A. Bogaerts, Anti-cancer capacity of plasma-treated PBS: effect of chemical composition on cancer cell cytotoxicity. Sci. Rep. **7**, 16478 (2017)
58. J. Du, Z. Liu, C. Bai, L. Li, Y. Zhao, L. Wang, J. Pan, Concentration distributions and reaction pathways of species in the mass transfer process from atmospheric pressure plasma jet to water. Eur. Phys. J. D **72**, 179 (2018)
59. A. Lindsay, C. Anderson, E. Slikboer, S. Shannon, D. Graves, Momentum, heat, and neutral mass transport in convective atmospheric pressure plasma-liquid systems and implications for aqueous targets. J. Phys. D Appl. Phys. **48**, 424007 (2015)
60. S.A. Norberg, G.M. Parsey, A.M. Lietz, E. Johnsen, M.J. Kushner, Atmospheric pressure plasma jets onto a reactive water layer over tissue: pulse repetition rate as a control mechanism. J. Phys. D Appl. Phys. **52**, 015201 (2019)
61. C.C.W. Verlackt, W. Van Boxem, A. Bogaerts, Transport and accumulation of plasma generated species in aqueous solution. Phys. Chem. Chem. Phys. **20**, 6845–6859 (2018)
62. A. Bogaerts, N. Khosravian, J. Van der Paal, C.C.W. Verlackt, M. Yusupov, B. Kamaraj, E.C. Neyts, Multi-level molecular modeling for plasma medicine. J. Phys. D Appl. Phys. **49**, 054002 (2016)
63. M. Elstner, D. Porezag, G. Jungnickel, J. Elsner, M. Haugk, T. Frauenheim, S. Suhai, G. Seifert, Self-consistent-charge density-functional tight-binding method for simulations of complex materials properties. Phys. Rev. B **58**, 7260–7268 (1998)
64. M. Yusupov, K. Wende, S. Kupsch, E.C. Neyts, S. Reuter, A. Bogaerts, Effect of head group and lipid tail oxidation in the cell membrane revealed through integrated simulations and experiments. Sci. Rep. **7**, 5761 (2017)
65. N. Khosravian, B. Kamaraj, E.C. Neyts, A. Bogaerts, Structural modification of P-glycoprotein induced by OH radicals: insights from atomistic simulations. Sci. Rep. **6**, 19466 (2016)
66. C.C.W. Verlackt, W. Van Boxem, D. Dewaele, F. Lemière, F. Sobott, J. Benedikt, E.C. Neyts, A. Bogaerts, Mechanisms of peptide oxidation by hydroxyl radicals: insight at the molecular scale. J. Phys. Chem. C **121**, 5787–5799 (2017)
67. C.C.W. Verlackt, E.C. Neyts, A. Bogaerts, Atomic scale behavior of oxygen-based radicals in water. J. Phys. D Appl. Phys. **50**, 11LT01 (2017)

68. D.W. Brenner, Empirical potential for hydrocarbons for use in simulating the chemical vapor deposition of diamond films. Phys. Rev. B **42**, 9458–9471 (1990)
69. A.C.T. van Duin, S. Dasgupta, F. Lorant, W.A. Goddard, ReaxFF: a reactive force field for hydrocarbons. J. Phys. Chem. A **105**, 9396–9409 (2001)
70. M. Yusupov, E.C. Neyts, U. Khalilov, R. Snoeckx, A.C.T. van Duin, A. Bogaerts, Atomic scale simulations of plasma species interacting with bacterial cell walls. New J. Phys. **14**, 093043 (2012)
71. M. Yusupov, A. Bogaerts, S. Huygh, S. Snoeckx, A.C.T. van Duin, E.C. Neyts, Plasma-induced destruction of bacterial cell wall components: a reactive molecular dynamics simulation. J. Phys. Chem. C **117**, 5993–5998 (2013)
72. M. Yusupov, E.C. Neyts, C.C. Verlackt, U. Khalilov, A.C.T. van Duin, A. Bogaerts, Inactivation of the endotoxic biomolecule lipid a by oxygen plasma species: a reactive molecular dynamics study. Plasma Process. Polym. **12**, 162–171 (2015)
73. J. Van der Paal, S. Aernouts, A.C.T. van Duin, E.C. Neyts, A. Bogaerts, Interaction of O and OH radicals with a simple model system for lipids in the skin barrier: a reactive molecular dynamics simulation for plasma medicine. J. Phys. D Appl. Phys. **46**, 395201 (2013)
74. J. Van der Paal, C.C. Verlackt, M. Yusupov, E.C. Neyts, A. Bogaerts, Structural modification of the skin barrier by OH radicals: a reactive molecular dynamics study for plasma medicine. J. Phys. D Appl. Phys. **48**, 155202 (2015)
75. R.M. Abolfath, P.K. Biswas, R. Rajnarayanam, T. Brabec, R. Kodym, L. Papiez, Multiscale QM/MM molecular dynamics study on the first steps of guanine damage by free hydroxyl radicals in solution. J. Phys. Chem. A **116**, 3940–3945 (2012)
76. C.C.W. Verlackt, E.C. Neyts, T. Jacob, D. Fantauzzi, M. Golkaram, Y.-K. Shin, A.C.T. van Duin, A. Bogaerts, Atomic-scale insight in the interactions between hydroxyl radicals and DNA in solution using the ReaxFF reactive force field. New J. Phys. **17**, 103005 (2015)
77. M. Yusupov, E.C. Neyts, P. Simon, G. Bergiyorov, R. Snoeckx, A.C.T. van Duin, A. Bogaerts, Reactive molecular dynamics simulations of oxygen species in a liquid water layer of interest for plasma medicine. J. Phys. D Appl. Phys. **47**, 025205 (2014)
78. N. Khosravian, A. Bogaerts, S. Huygh, M. Yusupov, E.C. Neyts, How do plasma-generated OH radicals react with biofilm components? Insights from atomic scale simulations. Biointerphases **10**, 029501 (2015)
79. O. Berger, O. Edholm, F. Jähnig, Molecular dynamics simulations of a fluid bilayer of dipalmitoylphosphatidylcholine at full hydration, constant pressure, and constant temperature. Biophys. J. **72**, 2002–2013 (1997)
80. W.D. Cornell, P. Cieplak, C.I. Bayly, I.R. Gould, K.M. Merz Jr., D.M. Ferguson, D.C. Spellmeyer, T. Fox, J.W. Caldwell, P.A. Kollman, A second generation force field for the simulation of proteins, nucleic acids, and organic molecules. J. Am. Chem. Soc. **117**, 5179–5197 (1995)
81. W. Yu, X. He, K. Vanommeslaeghe, A.D. MacKerell Jr., Extension of the CHARMM general force field to sulfonyl-containing compounds and its utility in biomolecular simulations. J. Comput. Chem. **33**, 2451 (2012)
82. W.F. van Gunsteren, H.J.C. Berendsen, *Groningen Molecular Simulation (GROMOS) Library Manual* (Biomos, Groningen, 1987), p. 221. http://www.gromos.net/gromos87/GROMOS87_manual.pdf
83. S.J. Marrink, H.J. Risselada, S. Yefimov, D.P. Tieleman, A.H. De Vries, The MARTINI force field: coarse grained model for biomolecular simulations. J. Phys. Chem. B **111**, 7812–7824 (2007)
84. J. Åqvist, A. Warshel, Simulation of enzyme reactions using valence bond force fields and other hybrid quantum/classical approaches. Chem. Rev. **93**, 2523–2544 (1993)
85. E.C. Neyts, M. Yusupov, C.C. Verlackt, A. Bogaerts, Computer simulations of plasma-biomolecule and plasma-tissue interactions for a better insight in plasma medicine. J. Phys. D Appl. Phys. **47**, 293001 (2014)

86. J. Razzokov, M. Yusupov, R.M. Cordeiro, A. Bogaerts, Atomic scale understanding of the permeation of plasma species across native and oxidized membranes. J. Phys. D Appl. Phys. **51**, 365203 (2018)
87. R.M. Cordeiro, Reactive oxygen species at phospholipid bilayers: distribution, mobility and permeation. Biochim. Biophys. Acta Biomembr. **1838**, 438–444 (2014)
88. M.N. Möller, Q. Li, J.R. Lancaster, A. Denicola, Acceleration of nitric oxide autoxidation and nitrosation by membranes. IUBMB Life **59**, 243–248 (2007)
89. W.K. Subczynski, M. Lomnicka, J.S. Hyde, Permeability of nitric oxide through lipid bilayer membranes. Free Radic. Res. **24**, 343–349 (1996)
90. A. Reis, M.R.M. Domingues, F.M.L. Amado, A.J.V. Ferrer-Correia, P. Domingues, Separation of peroxidation products of diacyl-phosphatidylcholines by reversed-phase liquid chromatography-mass spectrometry. Biomed. Chromatogr. **19**, 129–137 (2005)
91. J. Van der Paal, E.C. Neyts, C.C.W. Verlackt, A. Bogaerts, Effect of lipid peroxidation on membrane permeability of cancer and normal cells subjected to oxidative stress. Chem. Sci. **7**, 489–498 (2016)
92. J. Wong-Ekkabut, Z. Xu, W. Triampo, I.-M. Tang, D.P. Tieleman, L. Monticelli, Effect of lipid peroxidation on the properties of lipid bilayers: a molecular dynamics study. Biophys. J. **93**, 4225–4236 (2007)
93. L. Beranova, L. Cwiklik, P. Jurkiewicz, M. Hof, P. Jungwirth, Oxidation changes physical properties of phospholipid bilayers: fluorescence spectroscopy and molecular simulations. Langmuir **26**, 6140–6144 (2010)
94. L. Cwiklik, P. Jungwirth, Massive oxidation of phospholipid membranes leads to pore creation and bilayer disintegration. Chem. Phys. Lett. **486**, 99–103 (2010)
95. P.T. Vernier, M.J. Ziegler, Nanosecond field alignment of head group and water dipoles in electroporating phospholipid bilayers. J. Phys. Chem. B **111**, 12993–12996 (2007)
96. M. Casciola, M. Tarek, A molecular insight into the electro-transfer of small molecules through electropores driven by electric fields. Biochim. Biophys. Acta. Biomembr. **1858**, 2278–2289 (2016)
97. S.J. Marrink, A.H. de Vries, D.P. Tieleman, Lipids on the move: simulations of membrane pores, domains, stalks and curves. Biochim. Biophys. Acta Biomembr. **1788**, 149–168 (2009)
98. A.M. Hirst, F.M. Frame, M. Arya, N.J. Maitland, D. O'Connell, Low temperature plasmas as emerging cancer therapeutics: the state of play and thoughts for the future. Tumor Biol. **37**, 7021–7031 (2016)
99. E. Robert, T. Darny, S. Dozias, S. Iseni, J.-M. Pouvesle, New insights on the propagation of pulsed atmospheric plasma streams: from single jet to multi jet arrays. Phys. Plasmas **22**, 122007 (2015)
100. A. Begum, M. Laroussi, M.R. Pervez, Atmospheric pressure He-air plasma jet: breakdown process and propagation phenomenon. AIP Adv. **3**, 062117 (2013)
101. M. Yusupov, J. Van der Paal, E.C. Neyts, A. Bogaerts, Synergistic effect of electric field and lipid oxidation on the permeability of cell membranes. Biochim. Biophys. Acta Gen. **1861**, 839–847 (2017)
102. M. Shinitzky, Membrane fluidity in malignancy adversative and recuperative. Biochim. Biophys. Acta Rev. Cancer **738**, 251–261 (1984)
103. J. Van der Paal, C. Verheyen, E.C. Neyts, A. Bogaerts, Hampering effect of cholesterol on the permeation of reactive oxygen species through phospholipids bilayer: possible explanation for plasma cancer selectivity. Sci. Rep. **7**, 39526 (2017)
104. P.S. Hole, J. Zabkiewicz, C. Munje, Z. Newton, L. Pearn, P. White, N. Marquez, R.K. Hills, A.K. Burnett, A. Tonks, R.L. Darley, Overproduction of NOX-derived ROS in AML promotes proliferation and is associated with defective oxidative stress signaling. Blood **122**, 3322–3330 (2013)
105. M.C. Papadopoulos, S. Saadoun, Key roles of aquaporins in tumor biology. Biochim. Biophys. Acta Biomembr. **1848**, 2576–2583 (2015)
106. R.M. Cordeiro, Molecular dynamics simulations of the transport of reactive oxygen species by mammalian and plant aquaporins. Biochim. Biophys. Acta Gen. **1850**, 1786–1794 (2015)

107. D. Yan, A. Talbot, N. Nourmohammadi, J.H. Sherman, X. Cheng, M. Keidar, Toward understanding the selective anticancer capacity of cold atmospheric plasma - a model based on aquaporins (review). Biointerphases **10**, 040801 (2015)
108. M. Yusupov, D. Yan, R.M. Cordeiro, A. Bogaerts, Atomic scale simulation of $H_2O_2$ permeation through aquaporin: toward the understanding of plasma-cancer treatment. J. Phys. D Appl. Phys. **51**, 125401 (2018)
109. M. Yusupov, J. Razzokov, R.M. Cordeiro, A. Bogaerts, Transport of reactive oxygen and nitrogen species across aquaporin: a molecular level picture. Oxidative Med. Cell Longev. **2019**, 2930504 (2019)
110. C. Shao, M. Saito, Z. Yu, Formation of single- and double-strand breaks of pBR322 plasmid irradiated in the presence of scavengers. Radiat. Environ. Biophys. **38**, 105 (1999)
111. M. Dizdaroglu, P. Jaruga, Mechanisms of free radical-induced damage to DNA. Free Radic. Res. **46**, 382 (2012)
112. J. Cadet, T. Douki, J.-L. Ravanat, P. Di Mascio, Sensitized formation of oxidatively generated damage to cellular DNA by UVA radiation. Photochem. Photobiol. Sci. **8**, 903 (2009)
113. N.R. Jena, P.C. Mishra, Mechanisms of formation of 8-oxoguanine due to reactions of one and two OH• radicals and the H2O2 molecule with guanine: a quantum computational study. J. Phys. Chem. B **109**, 14205 (2005)
114. M. Yusupov, J.-W. Lackmann, J. Razzokov, S. Kumar, K. Stapelmann, A. Bogaerts, Impact of plasma oxidation on structural features of human epidermal growth factor. Plasma Process. Polym. **15**, e1800022 (2018)
115. D. Petrov, B. Zagrovic, Microscopic analysis of protein oxidative damage: effect of carbonylation on structure, dynamics, and aggregability of villin headpiece. J. Am. Chem. Soc. **133**, 7016–7024 (2011)
116. D. Petrov, X. Daura, B. Zagrovic, Effect of oxidative damage on the stability and dimerization of superoxide dismutase 1. Biophys. J. **110**, 1499–1509 (2016)
117. A. Renevey, R. Sereina, The importance of N-methylations for the stability of the $\beta^{6.3}$-helical conformation of polytheonamide B. Eur. Biophys. J. **46**, 363–374 (2017)
118. C.K. Haluska, M.S. Baptista, A.U. Fernades, A.P. Schroder, C.M. Marques, R. Itri, Photo-activated phase separation in giant vesicles made from different lipid mixture. Biochim. Biophys. Acta Biomembr. **1818**, 666–672 (2012)
119. D.E. Scott et al., Small molecules, big targets: drug discovery faces the protein-protein interaction challenge. Nat. Rev. Drug Discov. **15**, 533–550 (2016)
120. T. Kortemme et al., Computational redesign of protein-protein interaction specificity. Nat. Struct. Mol. Biol. **11**, 371–379 (2004)
121. J. Du et al., The design of high affinity human PD-1 mutants by using molecular dynamics simulations (MD). Cell. Commun. Signal. **16**, 25 (2018)
122. B. Knapp et al., Current status and future challenges in T-cell receptor/peptide/MHC molecular dynamics simulations. Brief. Bioinform. **16**, 1035 (2015)
123. K. Weiskopf et al., Engineered SIRPα variants as immunotherapeutic adjuvants to anticancer antibodies. Science **341**, 88–91 (2013)
124. J.Y. Lee et al., Structural basis of checkpoint blockade by monoclonal antibodies in cancer immunotherapy. Nat. Commun. **7**, 13354 (2016)
125. G. Bauer, D.B. Graves, Mechanisms of selective antitumor action of cold atmospheric plasma-derived reactive oxygen and nitrogen species. Plasma Process. Polym. **13**, 1157–1178 (2016)
126. J.-W. Lackmann, S. Schneider, E. Edengeiser, F. Jarzina, S. Brinckmann, E. Steinborn, M. Havenith, J. Benedikt, J.E. Bandow, Photons and particles emitted from cold atmospheric-pressure plasma inactivate bacteria and biomolecules independently and synergistically. J. R. Soc. Interface **10**, 20130591 (2013)
127. T.-Y. Chung, N. Ning, J.W. Chu, D.B. Graves, E. Bartis, J. Seog, G.S. Oehrlein, Plasma deactivation of endotoxic biomolecules: vacuum ultraviolet photon and radical beam effects on Lipid A. Plasma Process. Polym. **10**, 167–180 (2013)
128. E.A.J. Bartis, D.B. Graves, J. Seog, G.S. Oehrlein, Atmospheric pressure plasma treatment of lipopolysaccharide in a controlled environment. J. Phys. D Appl. Phys. **46**, 312002 (2013)

129. E.A.J. Bartis, C. Barrett, T.Y. Chung, N. Ning, J.W. Chu, D.B. Graves, J. Seog, G.S. Oehrlein, Deactivation of lipopolysaccharide by Ar and $H_2$ inductively coupled low-pressure plasma. J. Phys. D Appl. Phys. **47**, 045202 (2014)
130. J.H. Park, N. Kumar, D.H. Park, M. Yusupov, E.C. Neyts, C.C.W. Verlackt, A. Bogaerts, M.H. Kang, H.S. Uhm, E.H. Choi, P. Attri, A comparative study for the inactivation of multidrug resistance bacteria using dielectric barrier discharge and nano-second pulsed plasma. Sci. Rep. **5**, 13849 (2015)
131. M. Marschewski, J. Hirschberg, T. Omairi, O. Hofft, W. Viol, S. Emmert, W. Maus-Friedrichs, Electron spectroscopic analysis of the human lipid skin barrier: cold atmospheric plasma-induced changes in lipid composition. Exp. Dermatol. **21**, 921–925 (2012)
132. E. Takai, T. Kitamura, J. Kuwabara, S. Ikawa, S. Yoshizawa, K. Shiraki, H. Kawasaki, R. Arakawa, K. Kitano, Chemical modification of amino acids by atmospheric-pressure cold plasma in aqueous solution. J. Phys. D Appl. Phys. **47**, 285403 (2014)
133. G.S. Madugundu, J. Cadet, J.R. Wagner, Hydroxyl-radical-induced oxidation of 5-methylcytosine in isolated and cellular DNA. Nucleic Acids Res. **42**, 7450–7460 (2014)
134. S.-H. Hong, E.J. Szili, A.T.A. Jenkins, R.D. Short, Ionized gas (plasma) delivery of reactive oxygen species (ROS) into artificial cells. J. Phys. D Appl. Phys. **47**, 362001 (2014)
135. E.J. Szili, J.W. Bradley, R.D. Short, A 'tissue model' to study the plasma delivery of reactive oxygen species. J. Phys. D Appl. Phys. **47**, 152002 (2014)
136. M.U. Hammer, E. Forbrig, S. Kupsch, K.-D. Weltmann, S. Reuter, Influence of plasma treatment on the structure and function of lipids. Plasma Med. **3**, 97–114 (2013)
137. J. Van der Paal, S.-H. Hong, M. Yusupov, N. Gaur, J.-S. Oh, R.D. Short, E.J. Szili, A. Bogaerts, Cell membrane response to oxidative stress for anticancer therapies: an experimental and computational study. Phys. Chem. Chem. Phys. **21**(35), 19327–19341 (2019)

# Chapter 8
# Immunology in Plasma Cancer Treatment

**Sander Bekeschus, Georg Bauer, and Vandana Miller**

**Contents**



**Abstract** Cancer is a devastating and life-threatening disease. For decades, the emphasis of investigations into therapeutic modalities has focused on drug development and delivery. However, it soon became evident that the direct killing of cancer cells via anticancer therapies is only one side of the coin. The other side is the notion that the immune system can recognize and eliminate tumors in a highly specific manner. Nevertheless, cancer patients' immunity often needs help in doing so, leading to the creation of checkpoint antibody-based immunotherapies. These were awarded the Nobel Prize for Medicine or Physiology in 2018 and have set the stage for a multifaceted view of methods to promote anticancer immunity. Among the novel tools to do so are medical gas plasma devices and treatment regimens. While in principle gentle to the tissue with no necrotic or thermal harm being

S. Bekeschus (✉)
Department of ZIK Plasmatis, Leibniz Institute for Plasma Science and Technology (INP), Greifswald, Germany
e-mail: sander.bekeschus@inp-greifswald.de

G. Bauer
Institute of Virology, Medical Center - University of Freiburg, Freiburg, Germany

V. Miller
Department of Microbiology and Immunology, Drexel University College of Medicine, Philadelphia, PA, USA



M. Keidar (ed.), *Plasma Cancer Therapy*, Springer Series on Atomic, Optical, and Plasma Physics 115, https://doi.org/10.1007/978-3-030-49966-2_8

imposed, this reactive oxygen species-triggered tumor cell death was recently shown to have pro-immunogenic properties. The latest results and findings in this field are the subject of this book chapter along with a brief lay-reader-explanation of cancer and immunology.

## 8.1 Introduction

### *8.1.1 Cancer*

In a few words, cancer is a disease of uncontrolled cell growth, often accompanied by gain/loss of specific cell functions. The expansion of cancer tissue becomes an acute medical problem once it grows into vital regions of the body, leading to organ failure that is eventually fatal if left untreated. This problem is especially compounded when metastasizing tumor cells are migrating within the body and cause tumor growth at distant sites. As almost all types of cells can proliferate, tumor entities are known for each of these cell types. This includes tumors of immune cells (leukemia), the brain, the gut, the skin, and many others. In the US alone, about 600,000 cancer-related deaths are projected to occur [1]. This number is fast approaching the number of deaths from heart and cardiovascular disease, making cancer the second leading cause of death in western countries. To combat this disease and its devastating consequences, several different types of therapies have emerged in the last decades. The most commonly used approach is antineoplastic drugs. The major classes of anticancer drugs include antimetabolites, alkylating agents, hormones, antimetabolites, and natural products. Based on clinical trials usually involving thousands of patients, the type of cancer, disease stage, and the other anticancer therapies interaction dictate the specific type of drug to be used. Each drug has a specific treatment scheme, antitumor efficacy, and side effects. In addition to drugs, anticancer therapies based on principles of physics constitute the second important pillar of anticancer therapy.

In contrast to drugs that are usually given systemically to act throughout the whole body, physical approaches are applied locally at the tumor site. Abstractly, the process of locally excising tumor mass (surgery) belongs to such a regimen. Other more elaborate techniques have been developed in the past century. These include radiotherapy, photodynamic therapy, cryoablation, electroporation, high hydrostatic pressure, and hyperthermic therapy. Cold physical plasma is another emerging concept using the basic principles of physics to target cancer cells [2]. At the same time, there are a large number of targeted biologic therapeutics using, for instance, antibodies to block growth factor receptors needed by the cancer cells for their proliferation. All these approaches, however, were initially rooted in the idea that directly tackling the tumor cells is the prime and single motivation of anticancer therapy. Traditional oncology studies neglected the groundbreaking work of immunologists in the field of cancer for a long time, which included the finding that the immune system is a significant contributor to antitumor efficacy, both experimentally and clinically. The modern view, which seems to be realistic, is based on the concept that most ways of tumor treatment (except surgery) require the concomitant or subsequent activity of the immune system for complete tumor regression [3–7].

### 8.1.2 Immunology

The final action of the immune system is based on a cellular fraction (cellular immunity) and a noncellular fraction (humoral immunity). Humoral immunity consists of the complement system and antibodies. These effectors play a pivotal role in fighting infectious diseases; their role in antitumor responses is, however, considered to be of less importance when compared to cellular immunity that consists of immune cells, called leukocytes. There are a large number of leukocyte subtypes. The primary distinction is made between cells of the innate and cells of the adaptive immune system. The innate immune system's role is to respond quickly to a "danger" situation that signifies a shift from "normal", such as a wound or infection. To perform their action, innate immune cells are equipped with several different "preconfigured" receptor types that can sense evolutionary-conserved "alarm" signals. This includes, for example, bacterial lipids and proteins that are mandatory for most strains to multiply, nucleic acids obligatory for the replication of viruses, and proteins and molecules of the host only exposed upon severe damage of cells and tissues. In such situations, the innate immune cells known as phagocytes (macrophages, dendritic cells, and neutrophils) can ingest such material to remove the debris as well as the infectious agents. Moreover, a particular subset of lymphocytes, natural killer (NK) cells, can detect virus-infected and severely stressed cells to induce their cell death.

A principal trait of the immune system is the extensive cross talk between cells responsible for innate and adaptive immunity. For instance, dendritic cells and macrophages act as antigen presenting cells (APC) and present parts of their ingested material (e.g., peptide sequences digested from proteins), also called antigens, to lymphocytes, such as T-cells. The role of B-cells, the lymphocytes subtype responsible for generating antibodies specific against known and new antigens, will not be discussed in this chapter because their specific role has only just been begun to be studied in cancer [8]. Both B-cells and T-cells have mechanisms in place to "shuffle" their receptor specificity in a random fashion. This way, the adaptive immune system can respond against virtually any antigen. In practice, this means that if a synthetic protein not known in nature is injected together with immunostimulating agents into an immune-competent host such as mice, the animals soon will develop antibodies and T-cells specifically recognizing parts of this protein. It would not be an exaggeration to state that this potent mechanism of defense was vital to the survival and development of vertebrates, including humans. This is based on the fact the pathogens, such as bacteria (replication time: in the order of minutes) and viruses, quickly change their "molecular appearance", rendering them potentially invisible to the innate immune system. Humans only replicate every 20–30 years, putting the "survival of the fittest" (best receptor repertoire) idea *ad absurdum* when it comes to time in racing the deadly pathogens in terms of protein variation and its recognition. Hence, adaptive immunity provides the life-saving flexibility of recognizing and eliminating any novel antigens appearing in the body. It is now vital to understand that not only pathogens change their proteins to evade immunity, a process counteracted by adaptive immunity, but also tumor cells acquire genetic changes through their evolution. As a consequence of generic mutations in the DNA, tumor cells compromise "normal cell" signaling and metabolic pathways, eventually leading to the promotion of their growth unchecked. It is inevitable in this process that changes in the DNA sequences are copied as changed RNA sequences, and hence translated to changes in the protein and finally the peptide sequence. Because T-cells are theoretically capable of recognizing any "new" protein and peptide sequence introduced to the body, these cells are also able to eliminate cancer cells specifically. This is one principle of cancer immunology. It must be emphasized that each tumor entity as well as each patient-specific tumor has individual quality and quantity of acquired DNA mutations that can be somewhat predictive in estimating the impact of antitumor T-cell responses [9].

### *8.1.3 Cancer Immunology*

To generate antitumor T-cell responses, several prerequisites must be met. During the evolution of tumor cells, many cells die because of lack of nutrition, hypoxia, fatal genetic mutations, and/or NK cell-mediated lysis of stressed cells. Phagocytes such as dendritic cells (DCs) are constantly patrolling the tissues in search of pathogens and dead cells and debris to remove. The uptake of dead tumor cells always is expertly facilitated by DCs but the consequences of such uptake largely depend on the context in the tissue. If the uptake takes place in a noninflammatory environment, the material will be removed essentially without activating or priming the DC in an inflammatory fashion. Therefore, no tumor-specific adaptive immunity occurs. If, however, the uptake takes place in the context of inflammatory mediators, the DCs will be primed, increase the expression of T-cell co-stimulatory molecules, and the DCs will migrate to the draining lymph node to present the phagocytosed material to T-cells promoting their activation and proliferation [10–14]. Inflammatory stimuli may be microbe-associated molecule patterns (MAMPs) present during infection or damage-associated molecular patterns (DAMPs) present during a pro-inflammatory type of cell death. Hence, in cancer immunology, it is not imperative if a tumor cell dies but how (in which context) it dies as only the inflammatory type of cell death promotes antitumor T-cell immunity. The list of potential DAMPs in cell death is long, with a few molecules such as adenosine triphosphate (ATP) and calreticulin (CRT) having been investigated extensively [13]. Hence, in brief, the immunogenic type of cancer cell death (ICD) characterized by DAMP release by dying tumor cells facilitates the uptake of tumor cells by DCs. The latter then become primed and migrate to the draining lymph node, where they spur the activation of antitumor T-cells. The caveat here is that these are the patient's own body cells and uncontrolled T cell activation would be likely to produce bystander damage of normal cells. To combat this and to return the body to homeostasis after the danger has been dealt with, the body has several safeguard mechanisms in place to prevent overshooting of T-cell response targeted against the body's cells. All cell types in the body carry receptors and ligands that abrogate aberrant T-cell activation as a means of protection. This mechanism, also called tumor cell evasion, is excessively utilized by cancer cells to escape antitumor immunity [15].

In 2018, the Nobel Prize in Medicine or Physiology was awarded for the discovery of checkpoint immunotherapy evolving from 30 years of extensive research. The idea is based on the fact that most cancer patients have tumor-specific T-cells, but inhibitory ligands and receptors arising from the tumor cells halt their toxic action. If such inhibitory signaling is interfered with by antibodies that either block the ligand or the receptor, antitumor T-cells can unleash their full cytotoxic potential on tumors, i.e., overcoming the inhibition of T cells. The elegance of this concept lies in the idea that such mechanism provides systemic immune-protection against tumors, including metastases located at distance sites of the body (except immuno-privileged tissues being void or low in T-cell infiltration). As we speak, more than two thousand ongoing clinical trials are testing the safety and efficacy of known and new antibodies aimed at potentiating antitumor T-cell responses in either mono or combination treatment regimens together with radio or chemotherapy.

Another approach in cancer immunology is the ex vivo generation of autologous antitumor T-cells before their subsequent injection into the patient. This can be either "natural" T-cells grown in large numbers ex vivo and targeting several tumor-specific antigens or genetically engineered T-cells (called CAR T-cells) with a single type of T-cell receptor that targets a dominant antigen derived from bioinformatics' prediction of the most promising patient-specific tumor mutation. While these approaches are very elegant and provide a high level of specificity in terms of precision oncology, they come at very high manufacturing costs, prohibitive for their general application to the millions of cancer patients worldwide. The same is valid for generating antitumor DCs loaded ex vivo with patient-specific tumor material before their reinjection into the patient. While the idea of providing maximum antigen-presentation driving antitumor T-cell immunity is compelling, high costs and the missing integration of the concept of immune checkpoints and ICD have resulted in decades of low clinical efficacy [16].

Macrophages are another critical subset of innate immune cells in cancer. This is not necessarily due to their tumor-toxic action that has been often reported but is mechanistically still underexplored. Instead, macrophages can differentiate into different subtypes that either support (M2 or tumor-associated macrophages, TAM) or suppress (M1 macrophages) tumor growth. Several types of cancers, for instance, pancreatic cancer [17] and glioblastoma [18], consist of up to 50% macrophages and heavily exploit their growth-supporting characteristics, including anti-inflammatory mediators and promotion of angiogenesis [19]. Several targeted approaches are being investigated to either abrogate M2 or promote M1 macrophage polarization in tumors, eventually depriving the cancer cells of growth promotion [20–22]. A prime example of not including immunology aspects in cancer therapy is the use of gemcitabine in pancreatic cancer. This drug not only is of low therapeutic efficacy and comes with tremendous side effects in the patients [23] but also promotes monocyte infiltration and M2-macrophage polarization into the tumor [24], leading to an immunosuppressive tumor microenvironment (TME).

## 8.2 Plasma Onco-Immunology

### *8.2.1 The Concept*

In the light of the concepts mentioned above, several opportunities lie in the application of a local antitumor treatment regimen. It needs to be stated that this does not only apply to plasma but also other physical treatment modalities targeting tumor cells locally [25]. It is generally accepted that many (but not all) cancer patients had some existing antitumor immunity before they are diagnosed with cancer. An effective therapy would work in conjunction with and support such immunity by qualitative and quantitative means [26, 27]. Qualitative changes include the polarization (e.g., m1 or M2 macrophages) and activity status of immune cells in the TME or influxing into the TME. Quantitative means would affect the number of leukocytes and the diversity of the T-cell receptor repertoire available for the attack on the tumor. Cold physical plasma-assisted onco-therapy can be effective at many levels for these and the sequence and benefit of events includes:

1. Local damage of tumor cells,
2. Release of DAMPs and induction of ICD,
3. Promotion of uptake of tumor material into phagocytes,
4. Increase in co-stimulatory potential and migration of DCs to lymph nodes,
5. Elevation of the number and diversity of antitumor cytotoxic T-cells systemically,
6. Suppression of inhibitory cell types in the TME, including cancer-associated fibroblasts, TAMs, and regulatory T-cells,
7. Promotion of pro-immunogenic markers and secreted molecules in live tumor cells to further stimulate immunity,
8. Increasing infiltration of anticancer immune cells in the TME,
9. Immune-mediated growth retardation of the primary tumor together with distant metastases,
10. Self-promoting enhancement of free tumor antigen and antitumor immunity.

Selective antitumor action of plasma or plasma-conditioned media thereby would ensure a specific immunological response, whereas nonselective cell killing by plasma (through inadequate doses) might enharbor the danger to trigger unwanted autoimmune effects. In principle, autoimmune effects have been linked to excessive

ROS production in the literature [28–30]. However, this often relates to chronic ROS exposure, as seen in chronic inflammation while plasma-mediated ROS exposure is a rather short event. Along these lines, van der Linde and colleagues have shown in an animal model that repeated plasma treatment was not able to trigger autoimmune events in vivo when compared against a positive control, at least for the kINPen MED argon plasma jet [31].

### 8.2.2 In Vitro Studies of Plasma-Assisted Immunostimulation

Two different scenarios are present when evaluating the effect of cold physical plasma or plasma-conditioned liquids concerning leukocytes in vitro. The first scenario is the effect of plasma on immune cells alone. The second scenario involves the effects and cytotoxic action of immune cells in the coculture conditions with tumor cells. The second scenario also involves the induction of ICD in cancer cells following plasma treatment.

The effect of plasma on leukocyte has been reviewed recently [32]. It is conceivable to hypothesize that plasma may be able to stimulate T-cell activation. However, this is not the case. T-cells are highly sensitive to plasma treatment and die at very short exposure times [33–35]. The same is true for NK cells and B-cells. There are several T-cell subtypes and plasma treatment eradicates memory T-cells to a greater extent compared to naïve T-cells [36]. This is in line with findings in redox biology using hydrogen peroxide [37]. Moreover, specific stimulation of T-cells following plasma treatment could not be found [38], which is reassuring regarding the autoimmune side effects potentially arising with accidentally activated cells [39]. So far, there is no report on regulatory T-cells in plasma medicine but it can be assumed that this subtype is extraordinarily robust toward oxidative stress-induced cell death as previous findings in redox research implicate [40]. Nevertheless, proteomic studies on leukocyte microparticles identified specific responses in response to plasma treatment, involving DAMP release and changes of the microparticle content [41]. Another route of immunomodulation with plasma is its effect on myeloid cell differentiation [42]. It was previously demonstrated that plasma changes the surface marker expression of bone-marrow-derived macrophages [43]. It induces cellular and morphological as well as migratory changes in human monocytes [44], which came with a differentiation-like reaction in monocytic cell lines and primary monocytes [45]. The latter study also provided evidence of an increased cytotoxic response of plasma-conditioned monocytes/macrophages in tumor cells. This corroborates previous results with murine macrophages and tumor cells [46], which was further confirmed in studies using human monocytes or monocyte-derived macrophages [47–49]. The latter two studies also provided evidence of ICD induction with plasma treatment. Similar findings were made in murine melanoma cells [50, 51]. The direct stimulation of these cells in coculture with primary murine splenic macrophages and DCs was modest, however [52]. A prominent ICD induction was also observed in murine colon cancer cells with both jet and

DBD plasma as well as plasma-conditioned phosphate-buffered saline [53–55]. With the latter treatment regimen and using human pancreatic cancer cells, an ICD induction, as well as an enhanced uptake by human monocyte-derived DCs, was described [3]. By contrast, the modestly toxic plasma-conditioned liquid did not trigger the induction of CRT in eight human cancer cell lines [4] while prominently modulating the expression of several immunostimulatory and immuno-inhibitory cell surface marker receptors on these cell lines [5]. This study identified CD112 as a putative typical immunomodulatory indicator on plasma-treated human tumor cells, not explicitly succumbing to cell death and ICD.

### 8.2.3 Studies on Plasma-Assisted Immunostimulation in Cancer In Vivo

There are relatively few studies addressing antitumor immunity in vivo. An extended ICD response in tissues in vivo along with an increased influx of DCs and T-cell activation was found using a DBD plasma in murine colon cancer [54]. These results were replicated using plasma-conditioned phosphate-buffered saline in vivo, not only effectively decreasing intraperitoneal carcinomatosis but also T-cell activation upon restimulation with tumor cells ex vivo [55]. The same DBD as for the colon model was recently also shown to be effective in melanoma [51], which is also the case with a jet plasma system [56]. Mizuno and colleagues were the first to show that plasma effectively induces antitumor immunity using a syngeneic model of melanoma and nanosecond-pulsed DBD treatment in vivo. They observed an increased influx of $CD8^+$ cytotoxic T-cells into the TME [6, 7]. An increased T-cell influx was also observed in a syngeneic murine model of pancreatic peritoneal carcinomatosis treated with repeated injections of plasma-conditioned cell culture medium [57, 58]. This study also provided evidence of a massive macrophage influx into the tissue, possibly a consequence of apoptosis. Of note, there was neither enhancement in the polarization of macrophages toward an M2 phenotype nor an increase of DCs in this model, while at the same time, the immunogenicity of the approach was confirmed using CRT as lead ICD-marker. Finally, there is little but notable evidence of changes of the leukocyte infiltrate in the TME of a head and neck cancer patient repeatedly exposed to the certified medical product kINPen MED used for palliation in the patient [59].

### 8.2.4 Mechanism of Plasma-Driven Antitumor Immunity

It can be predicted that despite its precise mechanism, multiple concerted actions, and intercalating synergistic mechanisms, cold physical plasma and plasma-conditioned liquid-triggered apoptosis induction in tumor cells is most likely not sufficient to cause complete tumor regression in vivo. Therefore, the chance for survival of remaining tumor cells will most likely increase. This might lead to a subpopulation of tumor cells with the potential to start a new round of tumor formation immediately, or after a period of dormancy. Based on the importance, as well as the efficacy of immunogenic cell death [12, 60–63], it is reasonable to postulate that ROS/RNS-mediated cell death after cold physical plasma and plasma-conditioned liquid application will lead to successful tumor regression only if it also triggers ICD. This then might be the final powerful step in cold physical plasma and plasma-conditioned liquid-mediated antitumor therapy. It is an interesting question whether the induction of cell death by cold physical plasma or plasma-conditioned liquid-derived ROS/RNS leads to immunogenicity of the dying cells, or whether ROS/RNS trigger cell death and immunogenicity in parallel. According to assumption # 1, the acquisition of immunogenicity would be a consequence of induction of cell death, whereas according to assumption # 2, the two processes would be coincident processes that are triggered by either the same molecular species or different species out of the group of ROS/RNS. This question is in the focus of current research. Bekeschus et al. [53] report that upon plasma treatment CT26 colon carcinoma cells express CRT, one of the classical markers of ICD, on their membranes. Calreticulin was detected on the dead as well as live cells, up to 72 h after plasma treatment. This finding indicates that the expression of calreticulin in the colon carcinoma cells was triggered by plasma, without necessarily requiring preceding or parallel apoptosis induction. This finding might open specific approaches to enhance the immunological control of tumors. In the same study, Bekeschus et al. [53] detected mitochondrial $^1O_2$ 4 h after treatment of the tumor cells with plasma. At this time point, the mitochondria in apoptotic cells were already damaged. It seems that this response is a *consequence* rather than a *cause* of cell death, as during execution of the mitochondrial pathway of apoptosis cytochrome c is released from the mitochondria. This causes a breakdown of the respiratory chain with massive production and $O_2^{\bullet}$ . These are dismutated to $H_2O_2$ by mitochondrial SOD. As cytochrome c also exhibits peroxidase activity and can generate HOCl, using $H_2O_2$ as substrate (Bauer, unpublished result), within and in the vicinity of depolarized mitochondria efficient $^1O_2$ generation can be expected to take place through $HOCl/H_2O_2$ interaction under these conditions [64]. According to the kinetic analysis and functional inhibitor studies [65–68], $^1O_2$ generation late in the process seemed to play no role in the execution of apoptosis and the bystander-like communication between the cells. Instead, a very short phase (less than 30 min) of $^1O_2$ generation from plasma components and by the targeted cells was shown to have functional relevance to trigger the process and to inactivate catalase. In the study by Bekeschus, the detection of primary and secondary $^1O_2$ close to the membrane of the tumor cells might have either been prevented by the scavenging function of membrane-associated catalase, or their plasma source contained sufficient atomic oxygen to establish direct HOCl-mediated apoptosis induction, which is independent of $^1O_2$ [69].

Most researchers in the field of ICD are focusing on the classical DAMPs as markers for this process [12, 60–63]. Established noncanonical mechanisms of immunogenic cell death, such as the HOCl-mediated enhancement of tumor-specific cytotoxic T cell response [70–75], are presently not in the focus of plasma research. As HOCl is one of the central elements of intercellular ROS-mediated apoptosis-inducing signaling after the action of cold physical plasma or plasma-conditioned liquid [65, 67], it might be promising to study the role of HOCl in cold physical plasma or plasma-conditioned liquid-mediated immunogenicity in parallel to the canonical pathway.

## References

1. R.L. Siegel, K.D. Miller, A. Jemal, Cancer statistics, 2019. CA Cancer J. Clin. **69**, 7–34 (2019). https://doi.org/10.3322/caac.21551
2. A. Privat-Maldonado, A. Schmidt, A. Lin, K.D. Weltmann, K. Wende, A. Bogaerts, S. Bekeschus, Ros from physical plasmas: redox chemistry for biomedical therapy. Oxidative Med. Cell. Longev. **2019**, 1 (2019). https://doi.org/10.1155/2019/9062098
3. J. Van Loenhout, T. Flieswasser, L. Freire Boullosa, J. De Waele, J. Van Audenaerde, E. Marcq, J. Jacobs, A. Lin, E. Lion, H. Dewitte, M. Peeters, S. Dewilde, F. Lardon, A. Bogaerts, C. Deben, E. Smits, Cold atmospheric plasma-treated pbs eliminates immunosuppressive pancreatic stellate cells and induces immunogenic cell death of pancreatic cancer cells. Cancers **11**, 1597 (2019). https://doi.org/10.3390/cancers11101597
4. S. Bekeschus, E. Freund, K. Wende, R.K. Gandhirajan, A. Schmidt, Hmox1 upregulation is a mutual marker in human tumor cells exposed to physical plasma-derived oxidants. Antioxidants **7**, 151 (2018). https://doi.org/10.3390/antiox7110151
5. J. Moritz, H.-R. Metelmann, S. Bekeschus, Physical plasma treatment of 8 human cancer cell lines demarcates upregulation of cd112 as a common immunomodulatory response element. IEEE Transactions on Radiation and Plasma Medical Sciences **4**, 343–349 (2020). https://doi.org/10.1109/trpms.2019.2936790
6. K. Mizuno, Y. Shirakawa, T. Sakamoto, H. Ishizaki, Y. Nishijima, R. Ono, Plasma-induced suppression of recurrent and reinoculated melanoma tumors in mice. IEEE Trans. Radiat. Plasma Med. Sci. **2**, 353–359 (2018). https://doi.org/10.1109/trpms.2018.2809673
7. K. Mizuno, K. Yonetamari, Y. Shirakawa, T. Akiyama, R. Ono, Anti-tumor immune response induced by nanosecond pulsed streamer discharge in mice. J. Phys. D. Appl. Phys. **50**, 12LT01 (2017). https://doi.org/10.1088/1361-6463/aa5dbb
8. P. Tsou, H. Katayama, E.J. Ostrin, S.M. Hanash, The emerging role of b cells in tumor immunity. Cancer Res. **76**, 5597–5601 (2016). https://doi.org/10.1158/0008-5472.CAN-16-0431
9. T.N. Schumacher, R.D. Schreiber, Neoantigens in cancer immunotherapy. Science **348**, 69–74 (2015). https://doi.org/10.1126/science.aaa4971
10. M. Khalili, L. Daniels, A. Lin, F.C. Krebs, A.E. Snook, S. Bekeschus, W.B. Bowne, V. Miller, Non-thermal plasma-induced immunogenic cell death in cancer. J. Phys. D. Appl. Phys. **52**, 423001 (2019). https://doi.org/10.1088/1361-6463/ab31c1
11. A.D. Garg, L. Galluzzi, L. Apetoh, T. Baert, R.B. Birge, J.M. Bravo-San Pedro, K. Breckpot, D. Brough, R. Chaurio, M. Cirone, A. Coosemans, P.G. Coulie, D. De Ruysscher, L. Dini, P. de Witte, A.M. Dudek-Peric, A. Faggioni, J. Fucikova, U.S. Gaipl, J. Golab, et al., Molecular and translational classifications of damps in immunogenic cell death. Front. Immunol. **6**, 588 (2015). https://doi.org/10.3389/fimmu.2015.00588
12. G. Kroemer, L. Galluzzi, O. Kepp, L. Zitvogel, Immunogenic cell death in cancer therapy. Annu. Rev. Immunol. **31**, 51–72 (2013). https://doi.org/10.1146/annurev-immunol-032712-100008

13. L. Galluzzi, A. Buque, O. Kepp, L. Zitvogel, G. Kroemer, Immunogenic cell death in cancer and infectious disease. Nat. Rev. Immunol. **17**, 97–111 (2017). https://doi.org/10.1038/nri.2016.107
14. K.N. Balogh, D.J. Templeton, J.V. Cross, Macrophage migration inhibitory factor protects cancer cells from immunogenic cell death and impairs anti-tumor immune responses. PLoS One **13**, e0197702 (2018). https://doi.org/10.1371/journal.pone.0197702
15. A. Abdollahi, J. Folkman, Evading tumor evasion: current concepts and perspectives of anti-angiogenic cancer therapy. Drug Resist. Updat. **13**, 16–28 (2010). https://doi.org/10.1016/j.drup.2009.12.001
16. K. Palucka, J. Banchereau, Cancer immunotherapy via dendritic cells. Nat. Rev. Cancer **12**, 265–277 (2012). https://doi.org/10.1038/nrc3258
17. L.I. Partecke, C. Gunther, S. Hagemann, C. Jacobi, M. Merkel, M. Sendler, N. van Rooijen, A. Kading, D. Nguyen Trung, E. Lorenz, S. Diedrich, F.U. Weiss, C.D. Heidecke, W. von Bernstorff, Induction of m2-macrophages by tumour cells and tumour growth promotion by m2-macrophages: a quid pro quo in pancreatic cancer. Pancreatology **13**, 508–516 (2013). https://doi.org/10.1016/j.pan.2013.06.010
18. S.M. Pyonteck, L. Akkari, A.J. Schuhmacher, R.L. Bowman, L. Sevenich, D.F. Quail, O.C. Olson, M.L. Quick, J.T. Huse, V. Teijeiro, M. Setty, C.S. Leslie, Y. Oei, A. Pedraza, J. Zhang, C.W. Brennan, J.C. Sutton, E.C. Holland, D. Daniel, J.A. Joyce, Csf-1r inhibition alters macrophage polarization and blocks glioma progression. Nat. Med. **19**, 1264–1272 (2013). https://doi.org/10.1038/nm.3337
19. M.R. Galdiero, E. Bonavita, I. Barajon, C. Garlanda, A. Mantovani, S. Jaillon, Tumor associated macrophages and neutrophils in cancer. Immunobiology **218**, 1402–1410 (2013). https://doi.org/10.1016/j.imbio.2013.06.003
20. E. Bolli, K. Movahedi, D. Laoui, J.A. Van Ginderachter, Novel insights in the regulation and function of macrophages in the tumor microenvironment. Curr. Opin. Oncol. **29**, 55–61 (2017). https://doi.org/10.1097/CCO.0000000000000344
21. I. Rhee, Diverse macrophages polarization in tumor microenvironment. Arch. Pharm. Res. **39**, 1588–1596 (2016). https://doi.org/10.1007/s12272-016-0820-y
22. B. Ruffell, N.I. Affara, L.M. Coussens, Differential macrophage programming in the tumor microenvironment. Trends Immunol. **33**, 119–126 (2012). https://doi.org/10.1016/j.it.2011.12.001
23. D. Ciliberto, C. Botta, P. Correale, M. Rossi, M. Caraglia, P. Tassone, P. Tagliaferri, Role of gemcitabine-based combination therapy in the management of advanced pancreatic cancer: a meta-analysis of randomised trials. Eur. J. Cancer **49**, 593–603 (2013). https://doi.org/10.1016/j.ejca.2012.08.019
24. S.K. Deshmukh, N. Tyagi, M.A. Khan, S.K. Srivastava, A. Al-Ghadhban, K. Dugger, J.E. Carter, S. Singh, A.P. Singh, Gemcitabine treatment promotes immunosuppressive microenvironment in pancreatic tumors by supporting the infiltration, growth, and polarization of macrophages. Sci. Rep. **8**, 12000 (2018). https://doi.org/10.1038/s41598-018-30437-2
25. I. Adkins, J. Fucikova, A.D. Garg, P. Agostinis, R. Spisek, Physical modalities inducing immunogenic tumor cell death for cancer immunotherapy. Onco. Targets. Ther. **3**, e968434 (2014). https://doi.org/10.4161/21624011.2014.968434
26. S. Bekeschus, R. Clemen, H.-R. Metelmann, Potentiating anti-tumor immunity with physical plasma. Clin. Plasma Med. **12**, 17–22 (2018). https://doi.org/10.1016/j.cpme.2018.10.001
27. S. Bekeschus, P. Favia, E. Robert, T. von Woedtke, White paper on plasma for medicine and hygiene: future in plasma health sciences. Plasma Process. Polym. **16**, 1800033 (2019). https://doi.org/10.1002/ppap.201800033
28. H. Ahsan, A. Ali, R. Ali, Oxygen free radicals and systemic autoimmunity. Clin. Exp. Immunol. **131**, 398–404 (2003). https://doi.org/10.1046/j.1365-2249.2003.02104.x
29. W.L. Suarez-Pinzon, C. Szabo, A. Rabinovitch, Development of autoimmune diabetes in nod mice is associated with the formation of peroxynitrite in pancreatic islet beta-cells. Diabetes **46**, 907–911 (1997). https://doi.org/10.2337/diab.46.5.907

30. B.T. Kurien, R.H. Scofield, Autoimmunity and oxidatively modified autoantigens. Autoimmun. Rev. **7**, 567–573 (2008). https://doi.org/10.1016/j.autrev.2008.04.019
31. J. van der Linde, K.R. Liedtke, R. Matthes, A. Kramer, C.-D. Heidecke, L.I. Partecke, Repeated cold atmospheric plasma application to intact skin does not cause sensitization in a standardized murine model. Plasma Med. **7**, 383–393 (2017). https://doi.org/10.1615/PlasmaMed.2017019167
32. S. Bekeschus, C. Seebauer, K. Wende, A. Schmidt, Physical plasma and leukocytes - immune or reactive? Biol. Chem. **400**, 63–75 (2018). https://doi.org/10.1515/hsz-2018-0224
33. S. Bekeschus, J. Kolata, A. Muller, A. Kramer, K.-D. Weltmann, B. Broker, K. Masur, Differential viability of eight human blood mononuclear cell subpopulations after plasma treatment. Plasma Med. **3**, 1–13 (2013). https://doi.org/10.1615/PlasmaMed.2014008450
34. S. Bekeschus, T. von Woedtke, A. Kramer, K.-D. Weltmann, K. Masur, Cold physical plasma treatment alters redox balance in human immune cells. Plasma Med. **3**, 267–278 (2013). https://doi.org/10.1615/PlasmaMed.2014011972
35. L. Bundscherer, S. Bekeschus, H. Tresp, S. Hasse, S. Reuter, K.-D. Weltmann, U. Lindequist, K. Masur, Viability of human blood leukocytes compared with their respective cell lines after plasma treatment. Plasma Med. **3**, 71–80 (2013). https://doi.org/10.1615/PlasmaMed.2013008538
36. S. Bekeschus, K. Rödder, A. Schmidt, M.B. Stope, T. von Woedtke, V. Miller, A. Fridman, K.-D. Weltmann, K. Masur, H.-R. Metelmann, K. Wende, S. Hasse, Cold physical plasma selects for specific t helper cell subsets with distinct cells surface markers in a caspase-dependent and nf-κb-independent manner. Plasma Process. Polym. **13**, 1144–1150 (2016). https://doi.org/10.1002/ppap.201600080
37. A. Takahashi, M.G. Hanson, H.R. Norell, A.M. Havelka, K. Kono, K.J. Malmberg, R.V. Kiessling, Preferential cell death of cd8+ effector memory (ccr7-cd45ra-) t cells by hydrogen peroxide-induced oxidative stress. J. Immunol. **174**, 6080–6087 (2005)
38. S. Bekeschus, K. Masur, J. Kolata, K. Wende, A. Schmidt, L. Bundscherer, A. Barton, A. Kramer, B. Broker, K.D. Weltmann, Human mononuclear cell survival and proliferation is modulated by cold atmospheric plasma jet. Plasma Process. Polym. **10**, 706–713 (2013). https://doi.org/10.1002/ppap.201300008
39. S. Sakaguchi, N. Sakaguchi, J. Shimizu, S. Yamazaki, T. Sakihama, M. Itoh, Y. Kuniyasu, T. Nomura, M. Toda, T. Takahashi, Immunologic tolerance maintained by cd25+ cd4+ regulatory t cells: their common role in controlling autoimmunity, tumor immunity, and transplantation tolerance. Immunol. Rev. **182**, 18–32 (2001). https://doi.org/10.1034/j.1600-065x.2001.1820102.x
40. D. Mougiakakos, C.C. Johansson, R. Jitschin, M. Bottcher, R. Kiessling, Increased thioredoxin-1 production in human naturally occurring regulatory t cells confers enhanced tolerance to oxidative stress. Blood **117**, 857–861 (2011). https://doi.org/10.1182/blood-2010-09-307041
41. S. Bekeschus, J. Moritz, A. Schmidt, K. Wende, Redox regulation of leukocyte-derived microparticle release and protein content in response to cold physical plasma-derived oxidants. Clin. Plasma Med. **7-8**, 24–35 (2017). https://doi.org/10.1016/j.cpme.2017.07.001
42. M.L. Semmler, S. Bekeschus, M. Schäfer, T. Bernhardt, T. Fischer, K. Witzke, C. Seebauer, H. Rebl, E. Grambow, B. Vollmar, J.B. Nebe, H.-R. Metelmann, T. von Woedtke, S. Emmert, L. Boeckmann, Molecular mechanisms of the efficacy of cold atmospheric pressure plasma (cap) in cancer treatment. Cancers **12**, 269 (2020)
43. S. Bekeschus, L. Scherwietes, E. Freund, K.R. Liedtke, C. Hackbarth, T. von Woedtke, L.-I. Partecke, Plasma-treated medium tunes the inflammatory profile in murine bone marrow-derived macrophages. Clin. Plasma Med. **11**, 1–9 (2018). https://doi.org/10.1016/j.cpme.2018.06.001
44. S. Bekeschus, A. Schmidt, L. Bethge, K. Masur, T. von Woedtke, S. Hasse, K. Wende, Redox stimulation of human thp-1 monocytes in response to cold physical plasma. Oxidative Med. Cell. Longev. **2016**, 5910695 (2016). https://doi.org/10.1155/2016/5910695

45. E. Freund, J. Moritz, M. Stope, C. Seebauer, A. Schmidt, S. Bekeschus, Plasma-derived reactive species shape a differentiation profile in human monocytes. Appl. Sci. **9**, 2530 (2019). https://doi.org/10.3390/app9122530
46. N.K. Kaushik, N. Kaushik, B. Min, K.H. Choi, Y.J. Hong, V. Miller, A. Fridman, E.H. Choi, Cytotoxic macrophage-released tumour necrosis factor-alpha (tnf-alpha) as a killing mechanism for cancer cell death after cold plasma activation. J. Phys. D. Appl. Phys. **49**, 084001 (2016). https://doi.org/10.1088/0022-3727/49/8/084001
47. N.K. Kaushik, N. Kaushik, M. Adhikari, B. Ghimire, N.N. Linh, Y.K. Mishra, S.J. Lee, E.H. Choi, Preventing the solid cancer progression via release of anticancer-cytokines in co-culture with cold plasma-stimulated macrophages. Cancers **11**, 842 (2019). https://doi.org/10.3390/cancers11060842
48. A. Lin, B. Truong, A. Pappas, L. Kirifides, A. Oubarri, S.Y. Chen, S.J. Lin, D. Dobrynin, G. Fridman, A. Fridman, N. Sang, V. Miller, Uniform nanosecond pulsed dielectric barrier discharge plasma enhances anti-tumor effects by induction of immunogenic cell death in tumors and stimulation of macrophages. Plasma Process. Polym. **12**, 1392–1399 (2015). https://doi.org/10.1002/ppap.201500139
49. A. Lin, B. Truong, S. Patel, N. Kaushik, E.H. Choi, G. Fridman, A. Fridman, V. Miller, Nanosecond-pulsed dbd plasma-generated reactive oxygen species trigger immunogenic cell death in a549 lung carcinoma cells through intracellular oxidative stress. Int. J. Mol. Sci. **18**, 966 (2017). https://doi.org/10.3390/ijms18050966
50. S. Bekeschus, K. Rodder, B. Fregin, O. Otto, M. Lippert, K.D. Weltmann, K. Wende, A. Schmidt, R.K. Gandhirajan, Toxicity and immunogenicity in murine melanoma following exposure to physical plasma-derived oxidants. Oxidative Med. Cell. Longev. **2017**, 4396467 (2017). https://doi.org/10.1155/2017/4396467
51. A. Lin, Y. Gorbanev, J. De Backer, J. Van Loenhout, W. Van Boxem, F. Lemiere, P. Cos, S. Dewilde, E. Smits, A. Bogaerts, Non-thermal plasma as a unique delivery system of short-lived reactive oxygen and nitrogen species for immunogenic cell death in melanoma cells. Adv. Sci. **6**, 1802062 (2019). https://doi.org/10.1002/advs.201802062
52. K. Rödder, J. Moritz, V. Miller, K.-D. Weltmann, H.-R. Metelmann, R. Gandhirajan, S. Bekeschus, Activation of murine immune cells upon co-culture with plasma-treated b16f10 melanoma cells. Appl. Sci. **9**, 660 (2019). https://doi.org/10.3390/app9040660
53. S. Bekeschus, A. Mueller, V. Miller, U. Gaipl, K.-D. Weltmann, Physical plasma elicits immunogenic cancer cell death and mitochondrial singlet oxygen. IEEE Trans. Radiat. Plasma Med. Sci. **2**, 138–146 (2018). https://doi.org/10.1109/trpms.2017.2766027
54. A.G. Lin, B. Xiang, D.J. Merlino, T.R. Baybutt, J. Sahu, A. Fridman, A.E. Snook, V. Miller, Non-thermal plasma induces immunogenic cell death in vivo in murine ct26 colorectal tumors. Onco. Targets. Ther. **7**, e1484978 (2018). https://doi.org/10.1080/2162402X.2018.1484978
55. E. Freund, K.R. Liedtke, J. van der Linde, H.R. Metelmann, C.D. Heidecke, L.I. Partecke, S. Bekeschus, Physical plasma-treated saline promotes an immunogenic phenotype in ct26 colon cancer cells in vitro and in vivo. Sci. Rep. **9**, 634 (2019). https://doi.org/10.1038/s41598-018-37169-3
56. S. Bekeschus, R. Clemen, F. Nießner, S.K. Sagwal, E. Freund, A. Schmidt, Medical gas plasma jet technology targets murine melanoma in an immunogenic fashion. Adv. Sci. **7**, 1903438 (2020). https://onlinelibrary.wiley.com/doi/full/10.1002/advs.201903438
57. K.R. Liedtke, E. Freund, C. Hackbarth, C.-D. Heidecke, L.-I. Partecke, S. Bekeschus, A myeloid and lymphoid infiltrate in murine pancreatic tumors exposed to plasma-treated medium. Clin. Plasma Med. **11**, 10–17 (2018). https://doi.org/10.1016/j.cpme.2018.07.001
58. K.R. Liedtke, S. Bekeschus, A. Kaeding, C. Hackbarth, J.P. Kuehn, C.D. Heidecke, W. von Bernstorff, T. von Woedtke, L.I. Partecke, Non-thermal plasma-treated solution demonstrates antitumor activity against pancreatic cancer cells in vitro and in vivo. Sci. Rep. **7**, 8319 (2017). https://doi.org/10.1038/s41598-017-08560-3
59. H.-R. Metelmann, C. Seebauer, V. Miller, A. Fridman, G. Bauer, D.B. Graves, J.-M. Pouvesle, R. Rutkowski, M. Schuster, S. Bekeschus, K. Wende, K. Masur, S. Hasse, T. Gerling, M. Hori, H. Tanaka, E. Ha Choi, K.-D. Weltmann, P.H. Metelmann, D.D. Von Hoff, et al., Clinical

experience with cold plasma in the treatment of locally advanced head and neck cancer. Clin. Plasma Med. **9**, 6–13 (2018). https://doi.org/10.1016/j.cpme.2017.09.001
60. L. Apetoh, A. Tesniere, F. Ghiringhelli, G. Kroemer, L. Zitvogel, Molecular interactions between dying tumor cells and the innate immune system determine the efficacy of conventional anticancer therapies. Cancer Res. **68**, 4026–4030 (2008). https://doi.org/10.1158/0008-5472.CAN-08-0427
61. D.R. Green, T. Ferguson, L. Zitvogel, G. Kroemer, Immunogenic and tolerogenic cell death. Nat. Rev. Immunol. **9**, 353–363 (2009). https://doi.org/10.1038/nri2545
62. D.V. Krysko, A.D. Garg, A. Kaczmarek, O. Krysko, P. Agostinis, P. Vandenabeele, Immunogenic cell death and damps in cancer therapy. Nat. Rev. Cancer **12**, 860–875 (2012). https://doi.org/10.1038/nrc3380
63. A.D. Garg, P. Agostinis, ER stress, autophagy and immunogenic cell death in photodynamic therapy-induced anti-cancer immune responses. Photochem. Photobiol. Sci. **13**, 474–487 (2014). https://doi.org/10.1039/c3pp50333j
64. A.M. Held, D.J. Halko, J.K. Hurst, Mechanisms of chlorine oxidation of hydrogen-peroxide. J. Am. Chem. Soc. **100**, 5732–5740 (1978). https://doi.org/10.1021/ja00486a025
65. G. Bauer, The synergistic effect between hydrogen peroxide and nitrite, two long-lived molecular species from cold atmospheric plasma, triggers tumor cells to induce their own cell death. Redox Biol. **26**, 101291 (2019). https://doi.org/10.1016/j.redox.2019.101291
66. G. Bauer, Intercellular singlet oxygen-mediated bystander signaling triggered by long-lived species of cold atmospheric plasma and plasma-activated medium. Redox Biol. **26**, 101301 (2019). https://doi.org/10.1016/j.redox.2019.101301
67. G. Bauer, D. Sersenova, D.B. Graves, Z. Machala, Cold atmospheric plasma and plasma-activated medium trigger rons-based tumor cell apoptosis. Sci. Rep. **9**, 14210 (2019). https://doi.org/10.1038/s41598-019-50291-0
68. G. Bauer, D. Sersenova, D.B. Graves, Z. Machala, Dynamics of singlet oxygen-triggered, rons-based apoptosis induction after treatment of tumor cells with cold atmospheric plasma or plasma-activated medium. Sci. Rep. **9**, 13931 (2019). https://doi.org/10.1038/s41598-019-50329-3
69. G. Bauer, Hocl-dependent singlet oxygen and hydroxyl radical generation modulate and induce apoptosis of malignant cells. Anticancer Res. **33**, 3589–3602 (2013)
70. C.L. Chiang, J.A. Ledermann, A.N. Rad, D.R. Katz, B.M. Chain, Hypochlorous acid enhances immunogenicity and uptake of allogeneic ovarian tumor cells by dendritic cells to cross-prime tumor-specific t cells. Cancer Immunol. Immunother. **55**, 1384–1395 (2006). https://doi.org/10.1007/s00262-006-0127-9
71. C.L. Chiang, J.A. Ledermann, E. Aitkens, E. Benjamin, D.R. Katz, B.M. Chain, Oxidation of ovarian epithelial cancer cells by hypochlorous acid enhances immunogenicity and stimulates t cells that recognize autologous primary tumor. Clin. Cancer Res. **14**, 4898–4907 (2008). https://doi.org/10.1158/1078-0432.CCR-07-4899
72. C.L. Chiang, G. Coukos, L.E. Kandalaft, Whole tumor antigen vaccines: where are we? Vaccines **3**, 344–372 (2015). https://doi.org/10.3390/vaccines3020344
73. Z.M. Prokopowicz, F. Arce, R. Biedron, C.L. Chiang, M. Ciszek, D.R. Katz, M. Nowakowska, S. Zapotoczny, J. Marcinkiewicz, B.M. Chain, Hypochlorous acid: a natural adjuvant that facilitates antigen processing, cross-priming, and the induction of adaptive immunity. J. Immunol. **184**, 824–835 (2010). https://doi.org/10.4049/jimmunol.0902606
74. R. Biedron, M.K. Konopinski, J. Marcinkiewicz, S. Jozefowski, Oxidation by neutrophils-derived hocl increases immunogenicity of proteins by converting them into ligands of several endocytic receptors involved in antigen uptake by dendritic cells and macrophages. PLoS One **10**, e0123293 (2015). https://doi.org/10.1371/journal.pone.0123293
75. R. Zhou, W.J. Huang, C. Ma, Y. Zhou, Y.Q. Yao, Y.X. Wang, L.T. Gou, C. Yi, J.L. Yang, Hocl oxidation-modified ct26 cell vaccine inhibits colon tumor growth in a mouse model. Asian Pac. J. Cancer Prev. **13**, 4037–4043 (2012). https://doi.org/10.7314/apjcp.2012.13.8.4037

# Chapter 9
# Adaptive Plasma and Machine Learning

**Taeyoung Lee and Michael Keidar**

## Contents



**Abstract** Cold Atmospheric Plasma (CAP) jets provide a unique combination of reactive oxygen species, reactive nitrogen species, photons, and electric fields that exhibit desirable properties of triggering cell death pathway, selectively for cancer cells. However, the effects of CAP on cancer cells vary substantially depending on the particular type of cancer cell under treatments as well as various properties of plasma jet, such as gas composition, discharge voltage, and treatment duration. On the other hand, artificial intelligence has demonstrated remarkable capabilities in decision-making under uncertainties. Adaptive plasma in conjunction with artificial intelligence could lead to breakthroughs in autonomous, personalized cancer treatments. This chapter presents mathematical modeling of plasma cancer treatment, and summarizes the recent results in adaptive learning plasma.

T. Lee (✉)
Mechanical and Aerospace Engineering, The George Washington University, Washington, DC, USA
e-mail: tylee@gwu.edu

M. Keidar
Mechanical and Aerospace Engineering, School of Engineering and Applied Science, The George Washington University, Washington, DC, USA
e-mail: keidar@gwu.edu

 
M. Keidar (ed.), *Plasma Cancer Therapy*, Springer Series on Atomic, Optical, and Plasma Physics 115, https://doi.org/10.1007/978-3-030-49966-2_9

## 9.1 Introduction

Cold atmospheric plasma (CAP) jet is formed by ionization of noble gases, such as helium and argon, initiated when the gas jet flows through a high electric field.

The recent progress has led to a generation of CAP with the corresponding ion temperature close to the room temperature in a laboratory setting [1], and it is also referred to as non-thermal plasma jet or non-equilibrium plasma jet [2].

CAP jet has attracted much attention in the past decade due to its potential application in cancer therapy: it triggers cell death pathway in cancer cells while leaving normal cells unharmed. Various researchers have pointed that its therapeutic effects are related with the reactive oxygen and nitrogen species (RONS), including atomic nitrogen and oxygen, hydroxyl (OH), singlet delta oxygen, superoxide, and nitric oxide (NO) [3–7].

In particular, it is shown that the CAP jet eliminates cancer cells in vitro selectively with minimal damage to healthy cells, and furthermore, it significantly reduces tumor size in vivo [8]. The selectivity is suggested to be from the synergistic effect of selective diffusion of RONS into tumor cells and the high basic RONS level in tumor cells [9, 10]. As the level of RONS is already increased in cancer cells, the additional RONS delivered by CAP causes leads to death pathways especially for cancer cells. More specifically, these cause damage in DNA, causing cell cycle arrest in G2/M or programmed cell death referred to as apoptosis.

Despite success with various in vitro and in vivo experiments, there are several challenges in reliable CAP cancer treatment [11]. First, the therapeutic effect of the CAP jet is susceptible to the variability of plasma parameters (such as discharge voltage, flow rate, and frequency) and exogenous disturbances (such as temperature, target properties, and gas composition of surrounding environment) [12–14]. Second, different types of cancers exhibit different responses when exposed to the same CAP treatment conditions. Change of cancer type, properties of the medium in the device, and the treating duration exposure by plasma jet can drastically influence the plasma characteristics and their effects on the viability of cancer cells [15]. Third, the underlying biological mechanisms of novel therapy approaches have not been fully understood and guidelines on how to schedule these therapies may need to be established. Due to the complexity in clinical trials, the scheduling of treatments is often guided by exhaustive and expensive trial-and-error approaches.

One of the unique features of CAP compared to other sources of reactivity is the ability to rapidly change the reactive species production. As such, it is possible to adjust the CAP parameters such that the reactive species delivered to cells are customized in real-time according to the cancer cell response. Based on these, the idea of adaptive plasma for medical application was proposed recently in [16, 17]. More specifically, the objective is to develop an autonomous, self-adaptive therapeutic system that determines the parameters in the device generating plasma, after diagnosing the response of the cancer cells to CAP in a particular patience under treatment.

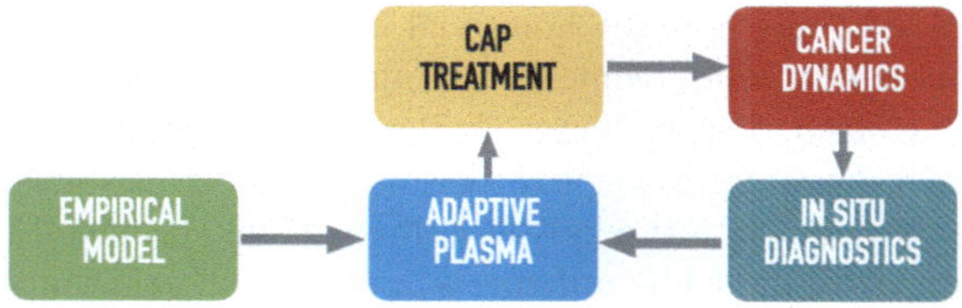


**Fig. 9.1** Schematics of adaptive plasma

The component of the adaptive plasma framework is illustrated in Fig. 9.1. First the prior knowledge of the cancer cell response to CAP is represented in the modeling block. This corresponds to our best projection for the behavior of cancer in the future, which can be constructed by a mathematical model of cancer cell dynamics or prior experiences in CAP cancer treatments. Then, according to the model, the adaptive plasma system is designed to plan the parameters to generate CAP for a specific treatment goal. Next, the actual cancer cell response is diagnosed in real-time, and according to the discrepancy between the empirical model and the in situ diagnostics, the adaptive plasma system learns the characteristics of the particular cancer cell under treatments and adjusts the treatment conditions accordingly.

This framework essentially corresponds to a real-time feedback mechanism in control system engineering or a decision-making process in machine learning. However, there is a unique challenge in developing the adaptive plasma system particularly due to the complexities of cancer cell dynamics. The conventional control systems rely on a mathematical model of the dynamic system, which is often formulated as an ordinary differential equation that describes the time evolution of its state, derived according to the first principles. While there have been adaptive control techniques to handle uncertainties in the mathematical model, or robust control to mitigate the effects of unknown disturbances, they focus on a certain class of parametric uncertainties or additive disturbance perturbing the given dynamics. The cancer cell response to CAP involves various complicated processes in biochemistry under a spectrum of time scales, and as such, it is not feasible to construct a mathematical model with sufficient reliability and simplicity required for real-time implementation of such control systems.

On the other hand, reinforcement learning has been successfully applied to sequential decision-making for a Markov process [18]. Reinforcement learning aims to construct a so-called action-value function to evaluate the projected outcome of each possible action, in selecting the optimal action at the current time. While this approach is comparable to the conventional optimal control, the distinct feature is that the action-value function is constructed in situ based on the prior experiences, possibly avoiding the need to construct any dynamic model in prior. However, successful implementation of reinforcement learning often requires numerous trials before the value function converges, and the number of trials exponentially increases as the complexities of the problem is intensified. It is impossible to administer multiple trials for cancer patient in clinics. Even conducting several in vitro experiments requires nontrivial efforts and costs. As such, it is not practical to implement reinforcement learning directly to the proposed adaptive plasma system.

Considering the above challenges of control system engineering and reinforcement learning in adaptive plasma, the most reasonable choice is integrating both approaches in a synergistic fashion. More specifically, we propose to construct an empirical model by utilizing a limited set of experimental data or by characterizing the overall behaviors from the current knowledge of treatment mechanism. Starting from this empirical model, learning control techniques can be designed to generate self-adaptive CAP treatments that adjust both of the treatment parameters and the empirical model in real-time. This can be further extended to reinforcement learning that does not require extensive pre-training.

Finally, our approach should be distinguished from model predictive control and machine learning presented in [19–21], which are designed to regulate treatment conditions of a device producing CAP jet, such as substrate temperature, plasma current, and power with no consideration on the actual cellular and tissue response. In contrast, we focus on adjusting plasma treatment conditions adaptive to the cellular response of cancer.

This chapter is organized as follows. In the first part, we present two mathematical models, namely an oxidative DNA stress model and an empirical model that represent the dynamic response of cancer cells to CAP. Next, three control approaches based on adaptive learning control and reinforcement learning are presented.

## 9.2 Mathematical Model of Plasma Cancer Treatment

As discussed above, it is implausible to derive a dynamic model describing the cancer cell responses to CAP from first principles. However, a dynamic representing our current knowledge of the corresponding mechanism will be critical for the successful implementation of adaptive plasma, in terms of scheduling an initial treatment plan to be updated or reducing the amount of data required for reinforcement learning. This section presents two particular attempts: an analytical model based on the oxidative DNA stress caused by CAP, and an empirical model constructed by a limited set of experiments.

### *9.2.1 Oxidative DNA Stress Model*

The cell cycle is a series of phases that a cell goes through when it is divided into two daughter cells [22], which happen aggressively in cancer cells. It is composed of the first growing phase $G_1$, the synthesis $S$ to replicate DNA, the second growing phase $G_2$, and the mitosis phase $M$ where the cell is literally divided. To guarantee successful cell division over these delicate processes, there are several mechanisms to ensure its proper progression, referred to as cell cycle check points. In particular, multiple check points are involved in the transition from $G_2$ to $M$ to examine DNA for the possible damage or incomplete replication. The presence of DNA damage detected in these check points prevents the cells from transitioning into the next phase for division, causing the cell cycle arrest, which eventually leads to a programmed cell death called apoptosis.

In [23, 24], the following hypotheses are presented to explain the effects of CAP on the cell cycle:

- the plasma treatment causes oxidative DNA stress at the most vulnerable $S$ phase;
- the increased damage in DNA results in $G_2/M$ cell cycle arrest and apoptosis.

This proposition describes the effects of CAP reasonably without excessive complication. Here, we present a mathematical model representing the above effects of CAP on the cell cycle.

### Cell Cycle Dynamics

To formulate the above hypotheses mathematically, we first present a cell cycle model that specifies the population density of cells with respect to the stress level at each cell cycle. In other words, we specify the distribution of the oxidative stress for the cells going through a specific cell cycle.

More specifically, let $x \in [0, \infty)$ be the level of oxidative DNA stress, represented by a positive real number. We assume $x = 0$ represents no stress, and the stress is more intense as $x$ becomes greater. The density of cells at time $t$ for the specific stress level $x$ is denoted by $g_1(t, x)$, $s(t, x)$, and $g_2(t, x) \in [0, \infty)$, respectively, for the cell cycle $G_1$, $S$, and $G_2/M$. We do not distinguish $M$ from $G_2$, as the conventional flow cytometry is not able to separate those two cycles. However, the proposed model is readily extended to four cell cycles.

According to the presented density model, cell population at the stress interval $[x, x + dx]$ is given by $g_1(t, x)dx$ for the $G_1$ phase, and the total cell population at $G_1$ is given by

$$G_1(t) = \int_0^\infty g_1(t, x)dx.$$

Similarly, the population for $S$ and $G_2/M$ are given by

$$S(t) = \int_0^\infty s(t, x)dx, \quad G_2(t) = \int_0^\infty g_2(t, x)dx.$$

The proposed mathematical model of the cell cycle with oxidative stress is given by

$$\frac{\partial g_1(t,x)}{\partial t} = 2k_2(x)g_2(t,x) - k_1 g_1(t,x), \tag{9.1}$$

$$\frac{\partial s(t,x)}{\partial t} = -v(u)\frac{\partial s(t,x)}{\partial x} + d(u)\frac{\partial^2 s(t,x)}{\partial x^2} + k_1 g_1(t,x) - k_s s(t,x), \tag{9.2}$$

$$\frac{\partial g_2(t,x)}{\partial t} = k_s s(t,x) - k_2(x)g_2(t,x) - \mu_2(x)a(t,x;T_a), \tag{9.3}$$

where the cells subject to apoptosis is given by

$$a(t,x;T_a) = k_s \exp(-k_2(x)T_a)s(t-T_a,x), \tag{9.4}$$

for a time delay $T_a > 0$, and the boundary condition for (9.2) is

$$d(u)\frac{\partial s(t,0)}{\partial x} - v(u)s(t,0) = 0. \tag{9.5}$$

Here, the parameters $k$ denote the rate of transition from one cycle to the next one. For example, $k_1$ is the rate of transition from $G_1$ to $S$, and $k_s$ is the rate of transition from $S$ to $G_2/M$. In (9.1), the population density at $G_1$ is reduced by the rate of $k_1 g_1$ leaving $G_1$, and it is increased by the rate $2k_2 g_2$, where the factor 2 implies that the cells leaving $G_2/M$ at the rate of $k_2 g_2$ are divided. In other words, as there is cell division in the $M$ phase, the influx rate to $G_1$, namely $2k_2$, is twice of the efflux rate $k_2$ from $G_2/M$. Next, in (9.2), the first two terms represent the effects of CAP denoted by $u$, describing the increase and the diffusion of the stress over the $S$ phase. Finally, (9.3) is for the $G_2/M$ phase, where the last term $\mu_2$ corresponds to the effects of the cell cycle arrest and the apoptosis, which are discussed below. See Fig. 9.2 for an illustration of the above dynamics model.

## Effects of CAP Treatments

Now, we discuss how the presented mathematical model reflects the hypotheses on the cell cycle dynamics. Equations (9.1)–(9.3) are defined such that the following three effects are formulated mathematically: the increase of the oxidative stress at $S$; $G_2/M$ cell cycle arrest; the resulting apoptosis.

### Increased Oxidative DNA Stress in $S$

As discussed above, it is considered that CAP treatment increases the oxidative DNA stress at $S$. This is modeled by the overall shift and the diffusion of the stress at (9.2) as follows. Assume the intensity or the gas composition of the plasma treatment, such as the voltage and the helium gas flow rate, is parameterized by $u \in \mathbb{R}^m$. The treatment is modeled by the terms $v(u) \in \mathbb{R}$ and $d(u) \in \mathbb{R}$ in (9.2)

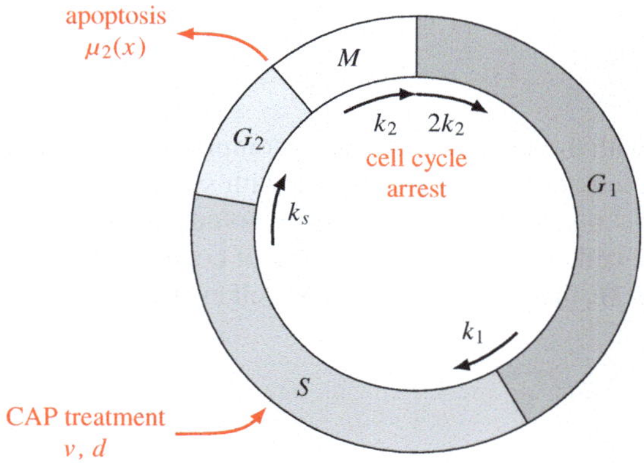


**Fig. 9.2** The effects of CAP treatment to the cell cycle dynamics: under nominal cancer growths, the cells go through the phase $G_1$, $S$, and $G_2/M$ at the exponential rates of $k_1$, $k_s$, and $k_2$, respectively. Due to the cell divisions, the influx of $G_1$ is twice of the efflux of $G_2/M$. As illustrated by red colors, it is assumed that CAP treatments cause oxidative DNA stress at $S_2$, which result in $G_2/M$ cell cycle arrest represented by reduction of $k_2$, and apoptosis modeled by $\mu_2$

that correspond to the advection or shift of the stress and the diffusion of the stress, respectively. In other words, the distribution of the stress will shift toward higher $x$ with the rate $v(u)$, and it will be smoothed according to $d(u)$, and the rates of the advection and diffusion are considered to be dependent of the treatment conditions parameterized by $u$. The given boundary condition (9.5) ensures that no stress is created arbitrarily at $x = 0$.

### Cell Cycle Arrest

In the presented cell cycle model, the cell division is represented by the fact that the rate of influx to $G_1$, namely $2k_2$, is twice to that of the efflux from $G_2/M$. The effects of the increased cell stress at $S$ to the other cell cycle are accounted by formulating the cell division rate $k_2$ and the death rate $\mu_2$ of $G_2/M$ as a function of stress $x$ so that only the cells with lower stress go through the mitosis to enter $G_1$, and the other cells with higher stress are destroyed.

More explicitly, consider the step function $\rho(x)$ defined in the appendix, that is, a smooth function satisfying (9.31). For constants $k_{2_0} \in \mathbb{R}$ and $0 < x_0 < x_1 \in \mathbb{R}$, the cell division rate is defined as

$$k_2(x) = k_{2_0}(1 - \rho(x; x_0, x_1)). \tag{9.6}$$

According to the property of the step function $\rho$ summarized at (9.31),

$$k_2(x) = k_{2_0}, \quad x \leq x_0,$$

$$k_2(x) < k_{2_0}, \quad x_0 < x < x_1,$$
$$k_2(x) = 0, \quad x \geq x_1.$$

As such, the cell with the stress lower than $x_0$ completes the cell division at the fixed rate $k_{2_0}$, and proceeds to $G_1$. For other cells with stress greater than $x_0$, either the cell division rate is discounted for $x > x_0$ or no cell division occurs for $x \geq x_1$. Consequently, the cells with higher stress cannot complete the cell division to $G_1$, and remain at $G_2/M$, thereby causing $G_2/M$ cell cycle arrest.

Apoptosis

For mathematical modeling of apoptosis caused by the above cell cycle arrest, the death rate at $G_2/M$, namely $\mu_2(x)$, is defined as a function of the stress as follows:

$$\mu_2(x) = \mu_{2_0}\rho(x; x_2, x_3), \tag{9.7}$$

for constants $\mu_{2_0}, x_2, x_3 \in \mathbb{R}$ satisfying $x_2 \leq x_3$. Therefore,

$$\mu_2(x) = 0, \quad x \leq x_2,$$
$$\mu_2(x) < \mu_{2_0}, \quad x_2 < x < x_3,$$
$$\mu_2(x) = \mu_{2_0}, \quad x \geq x_3.$$

This implies that no apoptosis occurs when $x \leq x_2$, and the rate is increased to $\mu_{2_0}$ for $x \geq x_3$. As such, the cells with higher stress go through apoptosis.

It is further assumed that the apoptosis is completed with the time delay of $T_a > 0$. The main motivation of introducing the time delay is the experimental results showing that the apoptosis occurs a certain period after CAP treatment. As such, the cells subject to apoptosis at $t$ had entered $G_2/M$ at $t - T_a$ with the rate of $k_s s(t - T_a, x)$, and they have evolved according to

$$\frac{\partial a(\tau, x; T_a)}{\partial \tau} = -k_2(x)a(\tau, x; T_a),$$

for $\tau \in [t - T_a, t]$, with the boundary condition $a(t - T_a, x; T_a) = k_s s(t - T_a, x)$. This is identical to (9.3) with $k_s = \mu_2 = 0$.

The above equation is linear, and it yields an explicit solution given by

$$a(t, x; T_a) = \exp(-k_2(x)T_a)k_s s(t - T_a, x). \tag{9.8}$$

The above expression can be implemented even when $t \leq T_a$, assuming $s(t, x) = s(t, 0)$ for $t < 0$.

**Special Case: Low-Stress**

Here we consider a special case when the oxidative stress is sufficiently low. This corresponds to the case where the cancer cells proliferate naturally without CAP treatments. More explicitly, we assume $x \leq \min\{x_0, x_2\}$ so that

$$k_2(x) = k_{2_0}, \quad \mu_2(x) = 0.$$

In other words, all of the cells complete the cell division and there is no apoptosis. Throughout this subsection, as it is independent of $x$, $k_2(x)$ is denoted by $k_2(x) = k_2$ for convenience.

From (9.1) and (9.3), it is straightforward to show

$$\dot{G}_1(t) = 2k_2 G_2(t) - k_1 G_1(t),$$
$$\dot{G}_2(t) = k_2 S(t) - k_2 G_2(t).$$

Also, integrating (9.2) with respect to $x$,

$$\begin{aligned}\dot{S}(t) &= \frac{d}{dt}\int_0^\infty s(t,x)dx = \int \frac{\partial s(t)}{\partial t} dx \\ &= \int \frac{\partial}{\partial x}\left(-vs + d\frac{\partial s}{\partial x}\right) + k_1 G_1(t) - k_s S(t) \\ &= -vs + d\frac{\partial s}{\partial x}\bigg|_0^\infty + k_1 G_1(t) - k_s S(t).\end{aligned}$$

Therefore, for given boundary condition (9.5) of zero flux, this reduces to

$$\dot{S}(t) = k_1 G_1(t) - k_s S(t).$$

As such, the dynamics of the cell population is given by the following linear time-invariant system:

$$\begin{bmatrix}\dot{G}_1 \\ \dot{S} \\ \dot{G}_2\end{bmatrix} = \begin{bmatrix}-k_1 & 0 & 2k_2 \\ k_1 & -k_s & 0 \\ 0 & k_s & -k_2\end{bmatrix}\begin{bmatrix}G_1 \\ S \\ G_2\end{bmatrix}. \tag{9.9}$$

The characteristic equation of the above system matrix is given by

$$\lambda^3 + \sum_{i \in \mathcal{I}} k_i \lambda^2 + \frac{1}{2}\sum_{i,j \in \mathcal{I}, i \neq j} k_i k_j \lambda - k_1 k_s k_2 = 0,$$

with $\mathcal{I} = \{1, s, 2\}$.

One can show that there is one positive real eigenvalue, namely $\lambda$. Suppose the initial condition is chosen such that $[G_1(0), S(0), G_2(0)]$ is parallel to the corresponding eigenvector of the positive read eigenvalue, i.e., $(A - \lambda I)[G_1(0), S(0), G_2(0)]^T = 0$. Then, the corresponding solution of (9.9) is given by

$$G_1(t) = e^{\lambda t} G_1(0), \quad S(t) = e^{\lambda t} S(0), \quad G_2(t) = e^{\lambda t} G_2(0). \tag{9.10}$$

Let $C(t) = G_1(t) + S(t) + G_2(t)$ be the total cell population. The above implies

$$C(t) = e^{\lambda t} C(0). \tag{9.11}$$

Therefore, in the proposed model when the oxidative stress is sufficiently low, all of the cell population at each cell cycle and the total cell population grow exponentially with the same rate $\lambda$. Also the ratio of each cell cycle to the total cell population remains unchanged.

Furthermore, the parameters $k_1, k_s, k_2$ can be determined by the cell cycle ratio and the exponential growth factor. More specifically, let $f_1, f_s, f_2 \in [0, 1]$ be the ratio of the cell population at each cycle, i.e.,

$$f_1 = \frac{G_1}{G_1 + S + G_2}, \quad f_s = \frac{S}{G_1 + S + G_2}, \quad f_2 = \frac{G_2}{G_1 + S + G_2}. \tag{9.12}$$

Then, it is straightforward to show $k_1, k_s, k_2$ are given explicitly as

$$k_1 = \frac{2 - f_1}{f_1}\lambda, \quad k_s = \frac{f_2 + 1}{f_s}\lambda, \quad k_2 = \frac{\lambda}{f_2}. \tag{9.13}$$

As such, the above parameters can be easily identified by the flow cytometry of untreated cancer cells.

## Numerical Example

Several numerical examples are presented. Throughout this section, the unit of time is hours if unspecified. The cell cycle ratio is chosen as $f_1 = 0.5$, $f_s = 0.3$, and $f_2 = 0.2$, and the overall growth rate is chosen as $\lambda = \frac{\log 2}{24}$, which represents that the doubling time is 24 h. From (9.13), the corresponding cell cycle transition rates are given by

$$k_1 = 0.0866, \quad k_s = 0.1155, \quad k_{2_0} = 0.1444.$$

The apoptosis rate is $\mu_{2_0} = 1$, and the delay is $T_a = 12$. The parameters to define the step function in $k_2(x)$ and $\mu_2(x)$ are given by $x_0 = 0.6$, $x_1 = 1.4$, $x_2 = 0.7$, and $x_3 = 1.3$. For CAP treatment, the rate of advection and diffusion is selected as $v = 30$ and $d = 3$, and when there is no treatment, they are changed to zero.

The initial conditions are chosen as

$$g_1(0,x) = f_1 \frac{2}{\sqrt{2\pi}\sigma} \exp\left(-\frac{x^2}{2\sigma^2}\right), \quad s(0,x) = \frac{f_s}{f_1} g_1(0,x),$$
$$g_2(0,x) = \frac{f_2}{f_1} g_1(0,x),$$

with $\sigma = 0.2$. In other words, the stress is distributed according to the Gaussian distribution, and scaled according to the cell cycle ratio. The resulting initial total cell population is $C(0) = G_1(0) + S(0) + G_2(0) = 1$.

#### No CAP Treatment

We first consider the case of no CAP treatment. This represents the natural growth of the cancer cells. For the selected initial stress distribution and the parameters of the step function, the majority of cells, excluding less than 0.3% of the population, have low stress less than $\min\{x_0, x_1, x_2, x_3\} = 0.6$. As such, this case is well approximated by the results presented in section "Special Case: Low-Stress."

Figure 9.3 illustrates the simulation results. As presented in section "Special Case: Low-Stress," the total cell population and the population at each cell cycle grow exponentially with the rate $\lambda$, and consequently, the cell cycle ratio remains fixed.

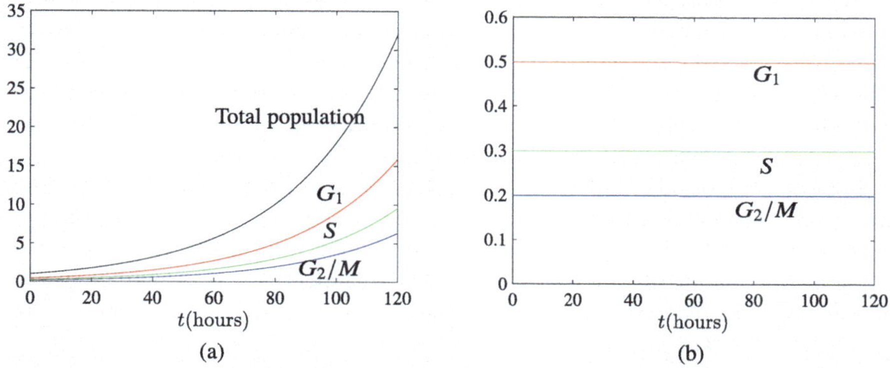


**Fig. 9.3** Simulation results for no CAP treatment. (**a**) Cell population growth. (**b**) Cell cycle ratio

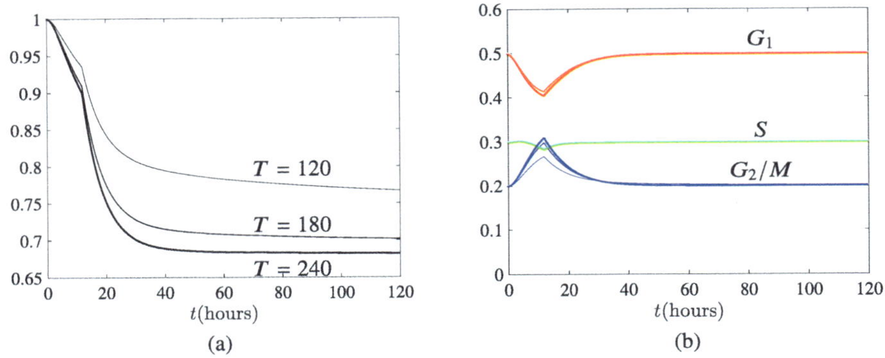


**Fig. 9.4** Simulation results with CAP treatments. (**a**) The ratio of the total cell population to the untreated growth for varying treatment duration. (**b**) Cell cycle ratio ($T = 120$:thin, $T = 180$:medium, $T = 240$:thick): the $G_2/M$ cell cycle arrest occurs when $t = 12$ h where the population of $G_2/M$ exceeds those of $S$

### CAP Treatments

Next, we consider the simulation results with CAP treatments. Three cases are presented for varying treatment duration of 120, 180, and 240 s.

The total cell population relative to the untreated growth is shown at Fig. 9.4a. It is also observed that the longer the treatment duration is, the ratio reduces further. There are two phases in the decrease of the ratio. During the first $T_a = 12$ h, the ratio decreases slightly due to the reduced cell division rate, namely $k_2(x)$ for cells with higher stress. Afterwards, the apoptosis contributes to further decrease.

Figure 9.4b illustrates the cell cycle ratio. The ratio for $G_2/M$ increases until $t = 12$ h, and it exceeds the $S$ phase temporarily, representing the $G_2/M$ cell cycle arrest. After the cells with higher stress are destroyed due to apoptosis, the cell cycle ratio asymptotically converges to the initial value, indicating that the remaining cells with lower stress proliferate in the same fashion presented in the preceding section for the natural growth without treatment.

Figures 9.5, 9.6 and 9.7 show the evolution of the stress distribution for three time segments, when the treatment period is 180 s. In those figures, the gradual increase of the intensity of color represents the time evolution. In Fig. 9.5, it is shown that the stress distribution of $S$ increases due to CAP treatment. The next figure, Fig. 9.6, the cells with the increased stress are transferred from $S$ to $G_2/M$, and they remain at $G_2/M$ as the transition rate from $G_2/M$ to $G_1$, namely $k_2(x)$, is discounted for higher stress level. This corresponds to the $G_2/M$ cell cycle arrest, and it explains the peak of the $G_2/M$ cell cycle ratio in Fig. 9.4b. Finally, in Fig. 9.7, the cells with higher stress that have been accumulated at $G_2/M$ go through apoptosis and are destroyed.

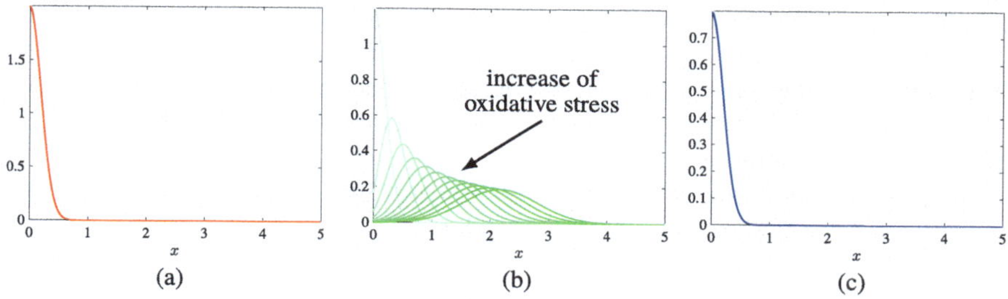


**Fig. 9.5** Evolution of oxidative stress distribution for $0 \le t \le 180$ s: the stress at $S$ is increased due to CAP treatment. (**a**) Population density $g_1$. (**b**) Population density $s$. (**c**) Population density $g_2$

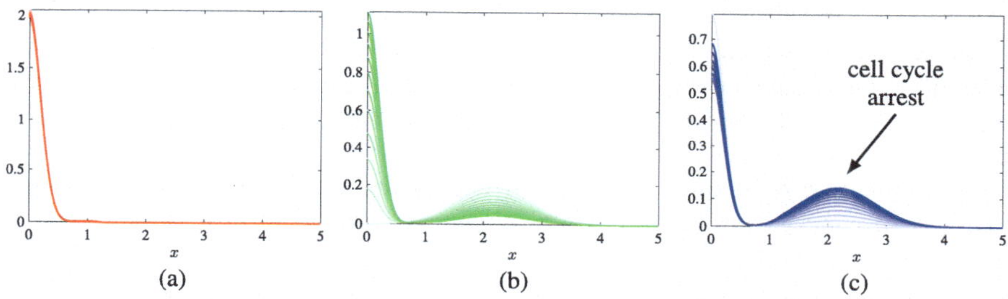


**Fig. 9.6** Evolution of oxidative stress distribution for $180\,\text{s} \le t \le 12\,\text{h}$: the cells with higher stress level are transferred from $S$ to $G_2/M$, and they become accumulated at $G_2/M$, while representing $G_2/M$ cell cycle arrest. (**a**) Population density $g_1$. (**b**) Population density $s$. (**c**) Population density $g_2$

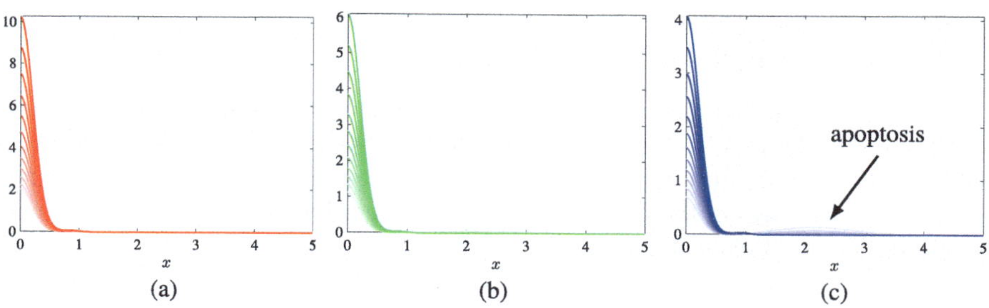


**Fig. 9.7** Evolution of oxidative stress distribution for $12\,\text{h s} \le t \le 72\,\text{h}$: the cells with higher stress level that have been accumulated at $G_2/M$ go through apoptosis. (**a**) Population density $g_1$. (**b**) Population density $s$. (**c**) Population density $g_2$

## 9.2.2 *Empirical Model*

Next, we present another mathematical model of CAP treatments [25]. This is developed to represent the dynamic response of cancer cells under CAP treatment, based on the raw data from in vitro experiments presented in [26]. Cancer cell response to CAP is monitored over the course of 48 h for two types of cancer cell, namely U87 and MDA-MB-231, where the treatment duration is varied from 0 to 180 s, and the plasma discharge voltage is selected from 3.16 and 3.71 kV. The resulting CAP-induced cell death was investigated by RealTime-Glo MT Cell Viability Assay several times.

To generalize the experimental data over arbitrary treatment conditions and time, the following form of growth model is considered:

$$\dot{p} = pF(t, p), \tag{9.14}$$

where $p \in \mathbb{R}$ denotes the population of cancer cell measured in terms of the metabolic activities of cells. To have the consistent value of $p$ for several experiments presented in [26], we normalize the cancer cell viability under CAP treatments with the initial cancer cell viability just before the CAP exposure. Therefore the initial value is $p(0) = 1$ always, and the variable $p$ is unit-less. Next, $F : \mathbb{R} \to \mathbb{R}$ models its net exponential proliferation rate depending on the current viability and the time.

The objective is to find an analytical expression of $F$ that characterizes the viability of cancer cells under CAP treatment as reported in [26], which exhibit the following properties.

- immediately after CAP treatment, cell viability is reduced instantaneously;
- shortly afterwards, from 0 min to 6 h, the cell viability increases rapidly;
- from 6 to 24 h, the cell viability decreases when the treatment duration is sufficiently large;
- from 24 to 48 h, the cell viability approaches its steady state value;
- for the effect of treating duration and voltage, the cell numbers decrease with the increase of the treating duration and voltage.

Based on these common features, we formulate an expression for the net proliferation rate as follows. To represent the instantaneous reduction of the cell viability, the cell viability immediately after the treatment is given by $p(0^+) = p_0$ for $p_0 \in \mathbb{R}$. Afterwards, the cell viability evolves according to (9.14), where the net proliferation rate is chosen as

$$F(t, p) = (c_1 - c_2 t) \exp\left( - c_3^{-t} p^{c_4} \right) - c_5, \tag{9.15}$$

where $c_1, c_2, c_3, c_4, c_5 \in \mathbb{R}$ are parameters to be determined. The above expression is applied to both types of cancer cells, namely U87 and MDA-MB-231, but $c_5$ is set to zero for U87.

Next, the values of the free parameters in (9.15) are determined according to optimal system identification [27]. This is to minimize the discrepancy between the experimental data and (9.14) measured by

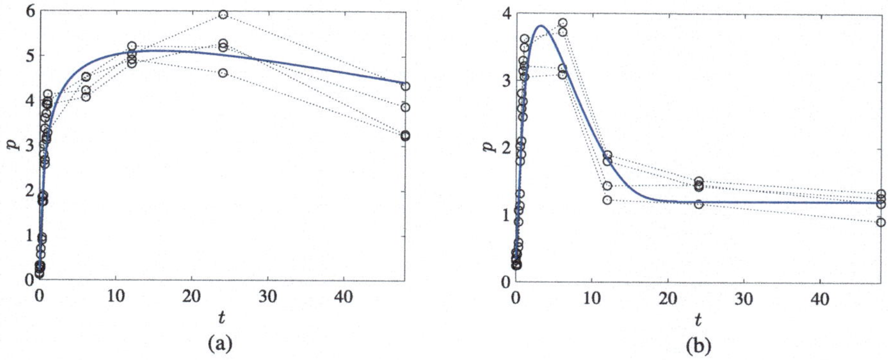


**Fig. 9.8** Normalized cell viability for U87 with the discharge voltage of 3.16 kV (blue: analytical model of (9.14); black circle: experimental data). (**a**) 60 s treatment. (**b**) 180 s treatment

$$J(c) = \sum_{i=1}^{n} \int_0^{48} \| p_{\exp_i}(t) - p(t; c) \|^2 dt, \tag{9.16}$$

where $p_{\exp_i}(t)$ denotes the cell viability at $t$ for the $i$-th experimental data, and $p(t; c)$ corresponds to the value obtained by the mathematical model (9.14) with a given parameter $c = (c_1, c_2, c_3, c_4, c_5) \in \mathbb{R}^5$. The optimal values of the free parameters are selected by

$$c_{\text{opt}} = \arg\min\{J(c)\}. \tag{9.17}$$

This is solved by the nonlinear programming solver, namely `fmincon` in MATLAB for each discharge voltage of $U = 3.16$ and 3.71 kV.

The time evolution of the cell viability predicted by (9.14) is illustrated in Fig. 9.8 against the experimental data of [26] for two selected treatment durations. While the experimental data are noisy, the presented analytical model reflects the overall trend of the data successfully.

The above parameters are optimized for the particular set of the treatment duration and discharge voltages considered in [26]. However, it can be generalized for arbitrary treatment conditions, by assuming that such parameters vary linearly. For example, the cell viability for the treatment duration $\Delta t = 100$ s can be constructed by interpolating the parameters of $\Delta = 90$ and $\Delta t = 180$. Similarly, the effects of the discharge voltage can be generalized as well. They are illustrated in Fig. 9.9.

In Fig. 9.9a, the treatment duration is varied linearly from $\Delta t = 60$ (cyan) to $\Delta t = 180$ (purple) for a fixed discharge voltage $U = 3.16$ kV. Similarly, in Fig. 9.9b, the discharge voltage is varied from $U = 3.16$ kV (cyan) to $U = 3.71$ kV (purple) when $\Delta t = 90$ s. As such, the presented empirical model (9.14) can be utilized to predict the dynamics of cancer cell viability for arbitrary treatment conditions.

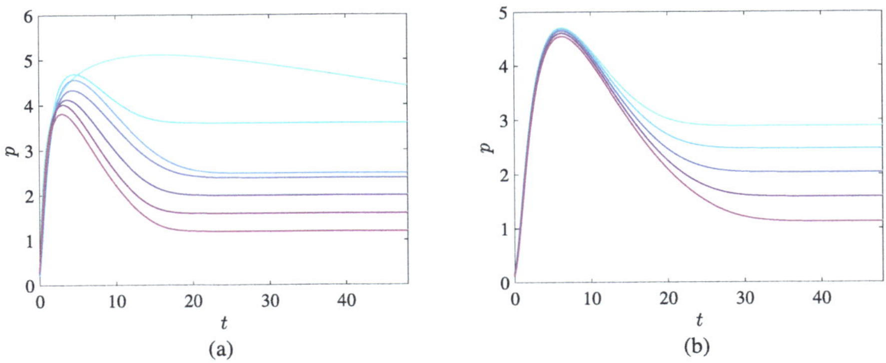


**Fig. 9.9** Normalized cell viability for U87 generalized for arbitrary treatment conditions (blue: smaller values; purple: larger values). (**a**) Treatment duration varying from 60 to 180 s. (**b**) Discharge voltage varying from 3.16 to 3.71 kV

## 9.3 Adaptive Plasma

The mathematical model of the prior section can be utilized to plan a baseline treatment schedule for a given objective. For example, the treatment duration and the plasma discharge voltage can be chosen such that the cancer cell viability is reduced to a desired level after a certain period. However, the effects of CAP on cancer cells vary substantially depending on the various factors, such as the size and the type of the cancer cell under treatment, ambient temperature, and humidity. It is also susceptible to exogenous disturbances not accounted in the mathematical model. Consequently, it cannot be expected that the actual cancer response would follow the one predicted by the model. The key component of adaptive plasma is to address this issue by adjusting the treatment conditions based on in situ diagnostics of the actual response to compensate the discrepancy between the model and the actual response. This section presents three such approaches, namely model predictive control, adaptive learning control, and reinforcement learning, based on the empirical model in Sect. 9.2.2.

### *9.3.1 Model Predictive Control*

A model predictive control (MPC) is a control strategy to convert the solution of open-loop optimal control into a feedback control [28]. The idea is applying the optimal control repeatedly over a certain time period. As each optimization is initiated by the current state vector, the corresponding control input constructed from MPC is a feedback.

We first formulate an optimal control problem as follows. Let the control parameters be the CAP treatment duration $\Delta t$. The objective is to minimize the treatment time, while ensuring that the cancer cell viability is reduced to the desired level. This is to maximize the therapeutic effect of CAP treatments for a prescribed level of cancer growth inhibition. More explicitly, the objective function that is to be minimized is

$$J(\Delta t) = \Delta t^2. \tag{9.18}$$

It is subject to a terminal inequality constraint to reduce the ratio of the terminal cancer viability to the untreated case up to a desired ratio, i.e.,

$$\frac{p(t = 48\,\text{h};\, \Delta t)}{p(t = 48\,\text{h};\, \Delta t = 0)} \leq r_d, \tag{9.19}$$

where $r_d \in \mathbb{R}$ is the desired ratio of the cancer cell viability.

Once the value of $\Delta t$ is given, the above objective function and the inequality constraint can be evaluated by integrating the dynamic model (9.14). As such, the presented optimization can be addressed by any numerical parameter optimization tool. Figure 9.10a illustrates the corresponding results, showing the treatment duration $\Delta t$ required to reduce the relative cancer cell viability to the desired value $r_d$, where $\Delta t$ increases as $r_d$ decreases.

These provide a CAP treatment schedule for a specific level of cancer cell growth inhibition. However, cancer cell response to CAP treatments depends on various intrinsic and extrinsic factors, and the presented mathematical model may not accurately characterize the actual response of the cancer cells under treatments. This may cause that the terminal value of the relative cancer cell viability becomes greater than the desired level, or it may yield unnecessarily intensive CAP treatments.

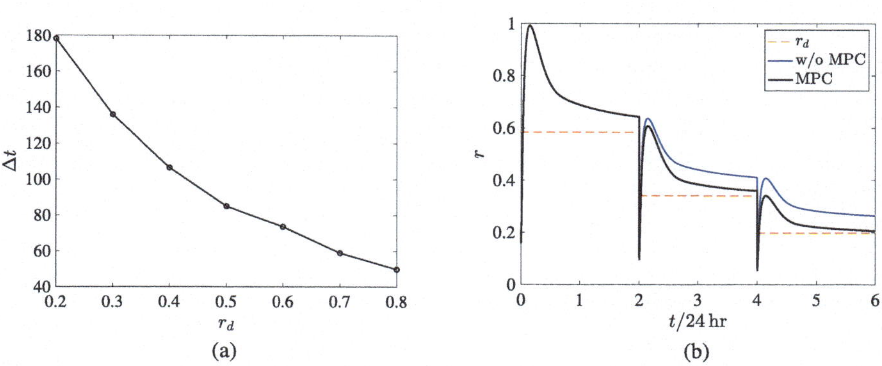


**Fig. 9.10** Simulation results of model predictive control. (**a**) Optimal control to determine the treatment duration in seconds for a given desired ratio. (**b**) Model predictive control

We address this by optimal feedback framework based on model predictive controls (MPC). The objective is to adjust the treatment parameters adaptively based on the actual cell response.

A specific case is considered as below. We assume that a series of three CAP treatments are conducted at the interval of 48 h such that the terminal ratio at the end of 144 h reduces to 0.2. If there is no modeling error, the ratio of the cell viability can be reduced by $(0.2)^{1/3}$ as each treatment. Instead, in the presented MPC, the desired relative cell viability is chosen as

$$r_d = \begin{cases} 0.2^{1/3} & \text{first treatment,} \\ 0.2^{1/3} \times \dfrac{\text{(predicted ratio of cell viability)}}{\text{(actual ratio of cell viability)}} & \text{remaining treatments.} \end{cases} \tag{9.20}$$

The first treatment is scheduled based on the mathematical model. At the end of the first treatment period, the predicted cell viability ratio is compared against the actual value. For the next treatment, the treatment objective is adjusted as in (9.20) to compensate the corresponding discrepancy. For example, this reduces the desired viability of the next treatment further if the actual cell viability at the end of the preceding treatment is greater than its predicted value.

The proposed approach is verified by a numerical simulation, where the preceding mathematical model is considered as the actual cancer response, and the parameters of the mathematical model are altered to represent a mathematical model available to MPC. Therefore, the mathematical model available to the controller is different from the dynamic model representing the actual cancer cell response.

In Table 9.1, the three columns from the second to the fourth correspond to the ideal case when the exact model is available to MPC. In this case, the relative cancer viability reduces exactly by the desired factor $r_d$ at each treatment. The treatment duration for all four treatments is identical to $\Delta t = 75.64$, and the terminal relative viability after four treatments is 0.2 as desired. The next three columns are the results of MPC when the exact model is not available to MPC. The first treatment duration is $\Delta t = 67.5$ for the first treatment, and it is less than the ideal case due to the modeling error. Consequently, the terminal viability ratio 0.64 becomes greater than the desirable value of 0.58. To compensate this, the goal of the second treatment is reduced to 0.47, which results in the viability ratio 0.36 that is slightly greater than the ideal value of 0.34. The final treatment is adjusted similarly so that the actual viability ratio at the end of three treatments achieves the treatment goal of 0.2, just as the ideal case.

**Table 9.1** Modeling predictive control of U87

| | Ideal case | | | MPC | | |
|---|---|---|---|---|---|---|
| Treatment | $r_d$ | $\Delta t$ | Viability ratio | $r_d$ | $\Delta t$ | Viability ratio |
| 1 | 0.58 | 75.64 | 0.58 | 0.58 | 67.5 | 0.64 |
| 2 | 0.58 | 75.64 | 0.34 | 0.47 | 78.8 | 0.36 |
| 3 | 0.58 | 75.64 | 0.20 | 0.50 | 77.0 | 0.20 |

Figure 9.10b illustrates the temporal response of the relative cell viability. The red, dotted lines show the desired relative cell viability at the end of each period. The blue lines are the results of optimization without MPC. Due to the modeling error, the actual cancer cell viability is greater than the desired value. Finally, the black lines correspond to MPC, where the treatment goal is achieved at the end of the three periods. These simulation results suggest that by adjusting CAP treatment conditions adaptive to the actual cancer response, the adverse effects of modeling errors can be mitigated.

### *9.3.2 Adaptive Learning Control*

In the above model predictive control, comparing the actual viability at the end treatment period with the predicted value constitutes the feedback mechanism. As such, the evolution of the actual viability after a CAP treatment is not accounted. Furthermore, the mathematical model itself remains unchanged even after the discrepancy is observed, as the only part that is adjusted is the treatment global of optimization.

It would be more desirable if the adaptive plasma system *learns* the dynamic characteristics of the particular cancer cell under treatment, so that the mathematical model can be refined as the treatment is repeated. This provides more accurate mathematical model that can be used to adjust the prospective treatment plan accordingly.

The conventional adaptive controls deal with unknown parameters in the equations of motion, and as such it is not suitable for the cancer cell dynamics whose uncertainties cannot be represented in a structured form with parametric uncertainties. Next, it is also critical to evaluate the degree of confidence in the learned model, as we cannot rely on untrustworthy information in planning cancer treatment. In other words, the learned model is useful only if we are confident about its validity.

To address these, we propose to utilize Bayesian machine learning, which is a broad field in artificial intelligence to account uncertainties in data-based learning [29]. This is useful as an arbitrary, non-structured model and can be represented and learned in a probabilistic formulation that considers uncertainty distribution explicitly, thereby resolving the aforementioned issues of the conventional adaptive control or deterministic learning.

#### Gaussian Process

In particular, here we adopt the Gaussian process [30] to represent uncertainties in the mathematical model. A Gaussian process is a stochastic process, defined such that any finite number of collection is jointly Gaussian. It is completely described by second-order statistics as follows. Define a mean function $\mathbf{m}(x) : \mathbb{R}^n \to \mathbb{R}$ and a positive-definite covariance function $\mathbf{K}(x, x') : \mathbb{R}^n \times \mathbb{R}^n \to \mathbb{R}$, which is referred to as a kernel function. The corresponding Gaussian process is denoted by

$$g(x) \sim \mathcal{G}(\mathbf{m}(x), \mathbf{K}(x, x')). \tag{9.21}$$

Let $\mathcal{D} = \{(x_i, g_i, \sigma_{g_i})\}_{i \in 1,\ldots N}$ be a set of data, where $g_i \in \mathbb{R}$ is a sample value of $g(x)$ when $x = x_i$, after corrupted by an additive, independent noise. More explicitly,

$$g_i \sim g(x_i) + \epsilon_{g_i}, \tag{9.22}$$

with $\epsilon_{g_i} \sim \mathcal{N}(0, \sigma_{g_i}^2)$.

Define $\mathbf{g}$, $\mathbf{x}$, and $\mathbf{m}(\mathbf{x}) \in \mathbb{R}^N$ be the concatenation of $g_i$, $x_i$, and $\mathbf{m}(x_i)$ for $i \in \{1, \ldots, N\}$, respectively. Also, let the matrix $\mathbf{K}(\mathbf{x}, \mathbf{x}) \in \mathbb{R}^{N \times N}$ be defied such that its $i, j$-th element is $\mathbf{K}(x_i, x_j)$, and let $\Sigma_{\mathbf{g}} = \text{diag}[\sigma_{g_1}^2, \ldots, \sigma_{g_N}^2] \in \mathbb{R}^{N \times N}$. From the definition of the Gaussian process, we have

$$\mathbf{g} \sim \mathcal{N}(\mathbf{m}(\mathbf{g}), \mathbf{K}(\mathbf{x}, \mathbf{x}) + \Sigma_{\mathbf{g}}). \tag{9.23}$$

Let $g_* \in \mathbb{R}$ be a sample value when $x = x_*$. It is jointly Gaussian with $\mathbf{g}$ as

$$\begin{bmatrix} \mathbf{g} \\ g_* \end{bmatrix} = \mathcal{N}\left( \begin{bmatrix} \mathbf{m}(\mathbf{x}) \\ \mathbf{m}(x_*) \end{bmatrix}, \begin{bmatrix} \mathbf{K}(\mathbf{x}, \mathbf{x}) + \Sigma_{\mathbf{g}} & \mathbf{K}(\mathbf{x}, x_*) \\ \mathbf{K}(x_*, \mathbf{x}) & \mathbf{K}(x_*, x_*) \end{bmatrix} \right). \tag{9.24}$$

Therefore, from the conditional distribution of joint Gaussian distributions, the regression equation for $g_*$ is

$$\begin{aligned} g_* | \mathcal{D}, x_* \sim \mathcal{N}(&\mathbf{m}_* + \mathbf{K}_{*\mathbf{x}}(\mathbf{K}_{\mathbf{xx}} + \Sigma_{\mathbf{g}})^{-1}(\mathbf{g} - \mathbf{m}_{\mathbf{x}}), \\ &\mathbf{K}_{**} - \mathbf{K}_{*\mathbf{x}}(\mathbf{K}_{\mathbf{xx}} + \Sigma_{\mathbf{g}})^{-1}\mathbf{K}_{\mathbf{x}*}), \end{aligned} \tag{9.25}$$

where the subscripts for $\mathbf{m}$ and $\mathbf{K}$ denote the input arguments, e.g., $\mathbf{K}_{*\mathbf{x}} = \mathbf{K}(x_*, \mathbf{x}) \in \mathbb{R}^{1 \times N}$.

The desirable feature is that a Gaussian process may represent an arbitrary function explicitly as in (9.25), without need for training or numerical optimization required for common multi-layer neural networks. The uncertainties are represented by Gaussian distributions that are provided by various properties, which can be utilized to simplify the required mathematical analysis.

### Adaptive Learning Control with Gaussian Process

Consider the empirical model (9.14) perturbed by the unknown disturbance or modeling error represented by $\Delta(t, p) \in \mathbb{R}^2 \times \mathbb{R}$:

**Table 9.2** Adaptive learning control of U87

| | Ideal case | | | Adaptive learning control | | |
|---|---|---|---|---|---|---|
| Treatment | $r_d$ | $\Delta t$ | Viability ratio | $r_d$ | $\Delta t$ | Viability ratio |
| 1 | 0.44 | 90 | 0.44 | 0.44 | 75.9 | 0.58 |
| 2 | 0.44 | 90 | 0.20 | 0.37 | 127.6 | 0.20 |

$$\dot{p} = pF(t, p) + \Delta(t, p). \tag{9.26}$$

To simplify the following discussion, suppose the above continuous-time differential equation is discretize over a time sequence $\{t_0, t_1, \ldots, t_N\}$ into

$$p_{k+1} = p_k F_d(t_k, p_k) + \Delta_d(t_k, p_k), \tag{9.27}$$

for $F_d, \Delta_d : \mathbb{R}^2 \times \mathbb{R}$ and the subscript $k$ denotes the value of a variable at $t_k$. Here the first term on the right-hand side $p_k F_d(t_k, p_k)$ corresponds to the mathematical model, and the second term $\Delta_d(t_k, p_k)$ denotes the unknown modeling error or disturbance that may be dependent of the current viability and time. This can be generalized to incorporate other intrinsic and extrinsic factors.

Whenever the viability is measured, the above equation yields a sample data $\{(t_k, p_k), \Delta_d(t_k, p_k)\}$ to be used to represent the unknown disturbance with a Gaussian process described above. The desirable property is that as the treatment is repeated more data become available so that the Gaussian process models the unknown part more accurately, thereby executing the learning process. Once the model is updated, any control strategy can be applied.

The proposed adaptive learning control is applied to the CAP treatment problem formulated in Sect. 9.3.1. The objective is to reduce the relative cancer viability into 0.2. But instead of three consecutive treatment, here we consider two treatments over the interval of 96 h. The first treatment is scheduled based on the optimization with the empirical model. After the first treatment, the cancer viability is measured five times at $t \in \{0.2, 0.4, \ldots, 1\}$ h to produce a set of sample data for a Gaussian process. The second treatment at $t = 48$ is planned based on the learned dynamic model.

The corresponding simulation results are summarized in Table 9.2. The two columns from the second to the fourth are for the ideal case when $\Delta = 0$ without adaptive learning. The desired ratio of each treatment is $r_d = 0.2^{1/2} = 0.44$, which is achieved exactly with $\Delta t = 90$ s. The resulting terminal viability ratio reduces to the desired value exactly. The next three columns are for the proposed adaptive learning control in the presence of non-zero modeling error $\Delta$. Due to the modeling error, the first treatment time $\Delta t = 75.9$ is less than the ideal value of 90. However, the cancer cell response to the first treatment is monitored, and the discrepancy is accounted by the Gaussian process. Based on the learned dynamic model, the second treatment is planned. As the error is properly compensated, the second treatment from the learned model achieves the desired viability ratio exactly.

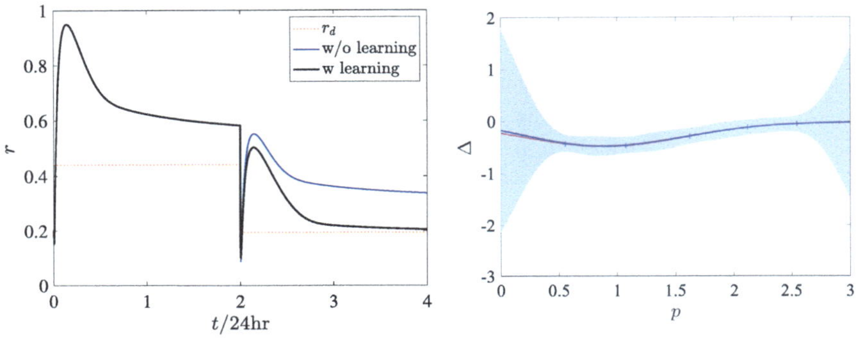


(a) (b)

**Fig. 9.11** Simulation results of adaptive learning control. (**a**) Cell viability ratio. (**b**) Gaussian process learning for modeling error

Figure 9.11a illustrates the evolution of the viability ratio, where the red dotted lines are for the desired treatment goal. The blue lines are for treatments without the Gaussian process learning, and it exhibits a substantial error: the terminal viability ratio is 0.34, which exceeds the desired value by 70%. Finally, the black lines are for the proposed adaptive learning control, which achieves the treatment goal at the end of the second treatment. The next Fig. 9.11b illustrates the modeling error learned by the Gaussian process, where the true values are represented by the red curve, and the learned model is given by the blue curves with $3\sigma$ bounds represented by a shaded area, and the training data denoted by the marks $+$. The learned model accurately represents the true unknown modeling error. More importantly, the $3\sigma$ bounds show that there is less uncertainty over the interval covered by the training data, and outside of the interval is greater uncertainty. As such, it provides the level of confidence in the learned model depending on the given domain.

This information can be utilized in the trade between performance and safety: over the region of smaller uncertainties, we can plan an aggressive plan with confidence; for the region of greater uncertainties, the treatment can be scheduled more conservatively. Accounting uncertainties and gauging confidence are one of the important attributes of the proposed adaptive learning control for adaptive plasma. While this section relies on the Gaussian process, any multi-layer neural network can be utilized to represent the modeling uncertainties. With Bayesian machine learning, it is more suitable to represent more complicated modeling errors and uncertainties depending on various factors.

### *9.3.3 Reinforcement Learning*

Reinforcement learning (RL) deals with optimal strategies for an agent in an environment to take the action to maximize a notion of cumulative reward [18].

The interaction between the agent and the environment is typically modeled by a Markov decision process. As such it is closely related to the dynamic programming in optimal control. The unique feature of RL is that it may not assume the complete knowledge of the dynamics, and the optimal policy does not have to be completely determined in prior. Instead, the optimal policy that may be initialed randomly is revised through each trial after evaluating the possible actions from experience. Therefore, the theory and the practice of RL can be utilized in adaptive plasma, to adjust the treatment plan in an optimal fashion to suppress the growth of the particular cancer cell under treatments, without need for the complete knowledge of the mechanism behind it. However, the successful implementation of RL often requires a large number of trials, which leads to a challenge in cancer treatments.

This section provides a formulation of RL for adaptive plasma. It is based on a partially observable Markov decision process (POMDP) defined by the following components:

- State $s_k$: It represents the current status of the cancer cell. For example, for the empirical model in Sect. 9.2.2, the cancer viability corresponds to the state.
- Action $a_k$: The parameters of CAP that can be adjusted arbitrarily are called action. This may include the treatment duration, plasma discharge voltage, or gas composition.
- Transition probability $P(s_{k+1}|s_k, a_k)$: According to the assumption of the Markov process, the future state is completely determined by the current state and the action taken, without need for the prior history. The transition probability describes the distribution of the state in the next time step, for the given current state and action.
- Reward $R(s_k, a_k)$: This represents the reward by choosing the action $a_k$ at the given state $s_k$. This can be designed to reflect the objectives of the treatment. For example, a reward can be the amount of the reduction of cancer cells.
- Observation $o_k$: The observation corresponds to the characteristics of the cancer that can be measured by a sensor. For example, the level of metabolism is measured in [26]. Recently, the impedance is utilized for real-time diagnostics.
- Observation probability $O(o_k|s_k, a_k)$: this characterizes the type of sensor used. It represents the distribution of the measurement for a given state and action.

In short, an agent (adaptive plasma) in an environment with the state $s_k$ (cancer) chooses an action $a_k$ (CAP), which causes the environment to transfer the state according to the transition probability (cancer dynamics). At the same time, the agent receives a reward (cancer treatment) and an observation (measurement) from the environment. The objective of the agent is to choose an action at each time step that maximizes its expected cumulative reward, or *return* defined as

$$G_k = \mathrm{E}\left[\sum_{l=0}^{\infty} \gamma^l R(s_{k+l+1}, a_{k+l+1})\right], \tag{9.28}$$

where $0 < \gamma < 1$ is a discount factor. The strategy of the agent is called *policy* $\pi(a|s)$, which describes the distribution of the action for a given state at each step. As the state is not directly measured, the belief of the state is determined by its history of actions and observations. For the belief of the current state $b(s_k)$, the next one is updated by

$$b(s_{k+1}) \propto O(o_k|s_k, a_k) \sum_{s_k} P(s_{k+1}|s_k, a_k) b(s_k).$$

The problem of constructing the optimal policy maximizing (9.28) can be addressed by the dynamic programming in optimal control. However, it is subjected to complexities, referred to as *curse of dimensionality*, especially when the system is of higher dimensions, and it requires the complete knowledge of the dynamics.

The heart of RL is avoiding these issues by iteratively updating the policy as described below. For a given policy $\pi$, let $Q^\pi(s_k, a_k)$ be the expected return when an arbitrary action $a_k$ chosen as the $k$-th step, and the prospective actions at the $k+1$-th step and afterwards are chosen from the policy $\pi$, i.e.,

$$Q^\pi(s, a) = \mathrm{E}_\pi[G_k|s_k = s, a_k = a]. \tag{9.29}$$

Once the $Q$ function is computed, the policy can be improved by

$$\pi' = \arg\max\ Q^\pi(s, a), \tag{9.30}$$

which is a *greedy* policy improvement that seeks for the best possible alternative choice from $\pi$ always. Instead of seeking the optimal action at every step greedily, the above policy improvement can be relaxed to explore other possibilities.

However, constructing the $Q$ function from (9.29) is not practical due to the same issues of the dynamic programming: this may take nontrivial computational efforts and it requires the complete knowledge of the dynamics. RL circumvents these by revising the $Q$ function continuously. First, the $Q$ function is initialized by some random values. Let the action $a$ be chosen from the state $s$ so that the state is transferred to $s'$, while resulting in the reward $r$. From this experience, the best guess of the correct value of $Q(s, a)$ is $r + \gamma \max_{a'} Q(s', a')$, which is referred to as the *TD target*. Then the $Q$ function is updated from the current value toward the TD target according to

$$Q(s, a) = Q(s, a) + \alpha\{\underbrace{r + \gamma \max_{a'} Q(s', a')}_{\text{TD target}} - Q(s, a)\},$$

where $\alpha > 0$ corresponds to a learning rate. Once the $Q$ function is updated, the policy can be improved according to (9.30). These process for improving the $Q$ function and the policy is repeated at each time step. It is a learning process as both are improved by an experience represented by the set $(s, a, s', r)$.

To implement this, the $Q$ function can be defined as a lookup-table when there a limited number of discrete states and actions. For continuous state and action, it can be represented by any function approximator. In particular, when the dimension of state and action is large, a deep neural network can be adopted for the $Q$ function, resulting in the so-called deep $Q$-learning [31].

As discussed above, the desirable feature is that it does not require the complete prior knowledge of the cancer dynamics, and the optimal policy representing adaptive plasma is improved as the set of experiences accumulates. The challenge is that it may take a lot of trials until the $Q$ function converges. Also to ensure convergence to the global optimum and to facilitate the process, the agent may need to take an unreasonable action through the course. This might be problematic for cancer treatment, where it is infeasible to repeat numerous CAP treatments, and the treatment should remain within a reasonable bound. This might be mitigated by pre-training the $Q$ function based on the mathematical models formulated in Sect. 9.2. Even with the potential challenges, it is expected that an innovative advance in adaptive plasma can be achieved by utilizing the theory and the practice of RL.

## 9.4 Conclusions

This chapter has presented mathematical models and control strategies for adaptive plasma.

**Mathematical Modeling**

The mathematical models can be used to predict the cancer cell response to CAP for a given treatment conditions, namely the treatment duration and the plasma discharge voltage. The first oxidative DNA stress model is constructed to mathematically formulate the effects of the CAP on the underlying cell cycle. This represents one of the current understandings for the mechanism how CAP eliminates cancer cells, and as such, there is a great potential that this model is further generalized and revised to account various treatment conditions and exogenous factors to reflect cancer cell dynamics accurately.

However, identifying the parameters would require a set of experiments to monitor the evolution of cell cycle population.

On the other hand, the next empirical model is solely based on the observation of cancer cell response, after treating the cancer cell dynamics as a completely unknown system. As such, it does not reflect our comprehension behind the CAP cancer treatment. However, it is relatively simple, and it can be adjusted to make a reasonable prediction consistent with the experimental data.

**Adaptive Plasma**

Next, three control strategies, namely model predictive control, adaptive learning control, and reinforcement learning, are presented. Model predictive control is aligned with the traditional control engineering, and it can ensure optimality under the feedback controls. However, the feedback mechanism is focused on adjusting the desired goal of optimization, and there are no adjustments in the mathematical model and the control algorithm.

In contrast, the adaptive learning control attempts to utilize all of the information available from the experience to refine the model and the control. Furthermore, the proposed stochastic framework evaluates the level of uncertainties throughout the information fusion. As such, the objective of control can be strategically adjusted between higher performance in the region of lower uncertainties and safety in the area of higher uncertainties.

Finally, reinforcement learning can be utilized for the complete model free control. However, a successful implementation of reinforcement learning may require numerous experiments, which can be mitigated by integrating with a mathematical model. As the field of artificial intelligence is rapidly advancing, there is a great potential for reinforcement learning utilized in innovative adaptive plasma.

**Remarks**
All of these mathematical models and control strategy would greatly benefit from real-time in situ diagnostics. As it is nearly impossible to characterize the cancer cell response to CAP from fundamental principles, any of the mathematical model should rely on the experimental data, and the accuracy of such models is limited by the richness and the quality of data. However, it is often that a cell viability and proliferation assay or flow cytometry need to destroy the cell to detect and measure its chemical and physical properties. As such, to measure the time evolution of cancer response to CAP over multiple instances, a set of experiments should be performed in parallel under the identical condition. Any real-time diagnostics can be utilized to generate a variety of valuable data to be adopted for more reliable mathematical models.

Furthermore, the critical component of adaptive plasma is monitoring the cellular response in real-time so that the treatment conditions are adjusted accordingly. The information that can be acquired from real-time diagnostics will play a critical role in the success of adaptive plasma.

# Appendix

## *Step Function*

Here, we construct a smooth step function $\rho(x; x_0, x_1) : \mathbb{R} \to [0, 1]$ such that

$$\rho(x; x_0, x_1) = \begin{cases} 0 & x \leq x_0, \\ 1 & x \geq x_1, \end{cases} \tag{9.31}$$

for $x_0 < x_1$.

We introduce a $C^\infty$ function,

$$f(x) = \begin{cases} e^{-1/t} & t > 0, \\ 0 & t \leq 0. \end{cases}$$

Consider

$$g(x) = \frac{f(x)}{f(x) + f(1-x)},$$

which is a smooth step function from 0 for $x \leq 0$ to 1 for $x \geq 1$. Utilizing this, we can define a function satisfying (9.31) as

$$\rho(x; x_0, x_1) = g\left(\frac{x - x_0}{x_1 - x_0}\right). \tag{9.32}$$

## References


1. G. Fridman, D. Dobrynin, G. Friedman, A. Fridman, Physical and biological mechanisms of plasma interaction with living tissue, in *2008 IEEE 35th International Conference on Plasma Science* (IEEE, Piscataway, 2008), pp. 1–1
2. M. Keidar, I. Beilis, *Plasma Engineering: Applications from Aerospace to Bio and Nanotechnology* (Academic, Cambridge, 2013)
3. N. Knake, K. Niemi, S. Reuter, V. Schulz-von der Gathen, J. Winter, Absolute atomic oxygen density profiles in the discharge core of a microscale atmospheric pressure plasma jet. Appl. Phys. Lett. **93**(13), 131503 (2008)
4. J. Sousa, K. Niemi, L. Cox, Q.T. Algwari, T. Gans, D. O'connell, Cold atmospheric pressure plasma jets as sources of singlet delta oxygen for biomedical applications. J. Appl. Phys. **109**(12), 123302 (2011)
5. G.E. Conway, A. Casey, V. Milosavljevic, Y. Liu, O. Howe, P.J. Cullen, J.F. Curtin, Non-thermal atmospheric plasma induces ROS-independent cell death in U373MG glioma cells and augments the cytotoxicity of temozolomide. Br. J. Cancer **114**(4), 435 (2016)
6. M. Vandamme, E. Robert, S. Lerondel, V. Sarron, D. Ries, S. Dozias, J. Sobilo, D. Gosset, C. Kieda, B. Legrain, et al., ROS implication in a new antitumor strategy based on non-thermal plasma. Int. J. Cancer **130**(9), 2185–2194 (2012)
7. J. Schlegel, J. Köritzer, V. Boxhammer, Plasma in cancer treatment. Clin. Plasma Med. **1**(2), 2–7 (2013)
8. M. Keidar, R. Walk, A. Shashurin, P. Srinivasan, A. Sandler, S. Dasgupta, R. Ravi, R. Guerrero-Preston, B. Trink, Cold plasma selectivity and the possibility of a paradigm shift in cancer therapy. Br. J. Cancer **105**(9), 1295–1301 (2011)
9. J.A. Cook, D. Gius, D.A. Wink, M.C. Krishna, A. Russo, J.B. Mitchell, Oxidative stress, redox, and the tumor microenvironment, in *Seminars in Radiation Oncology*, vol. 14, no. 3 (Elsevier, Amsterdam, 2004), pp. 259–266
10. P.T. Schumacker, Reactive oxygen species in cancer cells: live by the sword, die by the sword. Cancer Cell **10**(3), 175–176 (2006)
11. M. Keidar, A prospectus on innovations in the plasma treatment of cancer. Phys. Plasmas **25**(8), 083504 (2018)

12. J.W. Bradley, J.-S. Oh, O.T. Olabanji, C. Hale, R. Mariani, K. Kontis, Schlieren photography of the outflow from a plasma jet. IEEE Trans. Plasma Sci. **39**(11), 2312–2313 (2011)
13. T. Darny, J.-M. Pouvesle, J. Fontane, L. Joly, S. Dozias, E. Robert, Plasma action on helium flow in cold atmospheric pressure plasma jet experiments. Plasma Sources Sci. Technol. **26**(10), 105001 (2017)
14. B. Klarenaar, O. Guaitella, R. Engeln, A. Sobota, How dielectric, metallic and liquid targets influence the evolution of electron properties in a pulsed he jet measured by Thomson and Raman scattering. Plasma Sources Sci. Technol. **27**(8), 085004 (2018)
15. N. Georgescu, A.R. Lupu, Tumoral and normal cells treatment with high-voltage pulsed cold atmospheric plasma jets. IEEE Trans. Plasma Sci. **38**(8), 1949–1955 (2010)
16. M. Keidar, Therapeutic approaches based on plasmas and nanoparticles. J. Nanomed. Res **3**, 3–5 (2016)
17. M. Keidar, D. Yan, I.I. Beilis, B. Trink, J.H. Sherman, Plasmas for treating cancer: opportunities for adaptive and self-adaptive approaches. Trends Biotechnol. **36**(6), 586–593 (2018)
18. R.S. Sutton, A.G. Barto, *Reinforcement Learning: An Introduction* (MIT Press, Cambridge, 2018)
19. D. Gidon, B. Curtis, J.A. Paulson, D.B. Graves, A. Mesbah, Model-based feedback control of a kHz-excited atmospheric pressure plasma jet. IEEE Trans. Radiat. Plasma Med. Sci. **2**(2), 129–137 (2018)
20. D. Gidon, D.B. Graves, A. Mesbah, Effective dose delivery in atmospheric pressure plasma jets for plasma medicine: a model predictive control approach. Plasma Sources Sci. Technol. **26**(8), 085005 (2017)
21. A. Mesbah, D.B. Graves, Machine learning for modeling, diagnostics, and control of non-equilibrium plasmas. J. Phys. D Appl. Phys. **52**(30), 30LT02 (2019)
22. D.O. Morgan, *The Cell Cycle: Principles of Control* (New Science Press, London, 2007)
23. O. Volotskova, T.S. Hawley, M.A. Stepp, M. Keidar, Targeting the cancer cell cycle by cold atmospheric plasma. Sci. Rep. **2**, 636 (2012)
24. X. Yan, F. Zou, S. Zhao, X. Lu, G. He, Z. Xiong, Q. Xiong, Q. Zhao, P. Deng, J. Huang, et al., On the mechanism of plasma inducing cell apoptosis. IEEE Trans. Plasma Sci. **38**(9), 2451–2457 (2010)
25. Y. Lyu, L. Lin, E. Gjika, T. Lee, M. Keidar, Mathematical modeling and control for cancer treatment with cold atmospheric plasma jet. J. Phys. D Appl. Phys. **52**(18), 185202 (2019)
26. E. Gjika, S. Pal-Ghosh, A. Tang, M. Kirschner, G. Tadvalkar, J. Canady, M.A. Stepp, M. Keidar, Adaptation of operational parameters of cold atmospheric plasma for in vitro treatment of cancer cells. ACS Appl. Mater. Interfaces **10**(11), 9269–9279 (2018)
27. H. Bryson, *Applied Optimal Control: Optimization, Estimation, and Control* (Taylor & Francis, Abingdon-on-Thames, 1975)
28. W. Kwon, S. Han, *Receding Horizon Control: Model Predictive Control for State Models* (Springer, Berlin, 2005)
29. D. Barber, *Bayesian Reasoning and Machine Learning* (Cambridge University Press, Cambridge, 2012)
30. C. Rasmussen, C. Williams, *Gaussian Process for Machine Learning* (MIT Press, Cambridge, 2006)
31. M. Hausknecht, P. Stone, Deep recurrent Q-learning for partially observable MDPS, in *2015 AAAI Fall Symposium Series* (2015)

# Chapter 10
# Fundamental Studies of the Effect of Plasma on Plasmid DNA, Cancer Cells, and Virus

**Xu Yan and XinPei Lu**

**Contents**



**Abstract** This chapter focuses on the interactions of plasma with various biological objects, ranging from the naked plasmid DNA to genes, as well as the impact on cancer cells and viral particles.

## 10.1 Introduction: Scope and Structure

Plasma suitable for biomedical applications is frequently produced in the open air environment. Specifically, $N_2$, $O_2$, and a small amount of $H_2O$ in the air react with electrons, ions and excited particles in the plasma to generate large amounts of reactive nitrogen species (RNS) and reactive oxygen species (ROS), such as NO, OH, O, $O_2$ ($^1\Delta$), $O_2^-$, and $H_2O_2$ [1]. Typically, ROS and RNS exert vital parts in the redox biochemistry within numerous living organisms, which are known to have important biological implications. This chapter focused on the interactions of plasma with various biological objects, ranging from the naked plasmid DNA to

X. Yan
Beijing Neurosurgical Institute, Beijing, People's Republic of China

X. Lu (✉)
School of Electrical and Electronic Engineering, Huazhong University of Science and Technology, Wuhan, People's Republic of China



M. Keidar (ed.), *Plasma Cancer Therapy*, Springer Series on Atomic, Optical, and Plasma Physics 115, https://doi.org/10.1007/978-3-030-49966-2_10

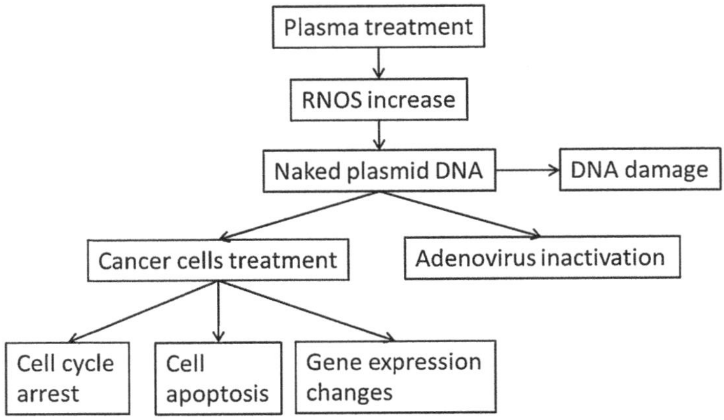


**Fig. 10.1** Research flow chart of this chapter

genes, as well as the impact on cancer cells and viral particles (an envelope of lipids that contain the DNA or RNA surrounding the protein coat) (Fig. 10.1).

## 10.2 Impact of Plasma on Plasmid DNA

Plasmids are the small cellular DNA molecules, which replicate independently and are physically isolated based on the chromosomal DNA [2]. Generally, chromosomes, which are large, possess any necessary genetic data to liver at physiological condition; by contrast, plasmids, which are quite little, only involve the genes necessary for organism at some specific conditions or situations [2]. On the other hand, artificial plasmids have been extensively utilized to be vectors for the sake of molecular cloning, and they can trigger recombinant DNA sequence replication in the hosts.

In this chapter, the influence of atmospheric pressure nonequilibrium plasmas (APNPs) on plasmid pAHC25 was first evaluated. pAHC25 possesses bar and gus, the selectable marker genes controlled by the maize ubiquitin promoter. pAHC25 has been extensively utilized to be the vector in transgene for plants [3].

Figure 10.2 displays the schematic for the experimental setting. As observed, the wire electrode with high voltage (HV), which consisted of one 2-mm-diameter copper wire, was inserted into a 4-cm quartz tube with one end closed. The quartz tube had the outer and inner diameters of 4 and 2 mm, separately. Afterwards, both HV electrode and quartz tube were advanced to the syringe in the hollow barrel. The syringe nozzle and hollow barrel were 1.2 and 6 mm in diameter, separately. Meanwhile, the nozzle was 1 cm away from the tip of HV electrode. The device consumed approximately 2 W of power when the applied voltage was 8 kV and the pulse frequency was 8 kHz. In addition, the following parameters were fixed during

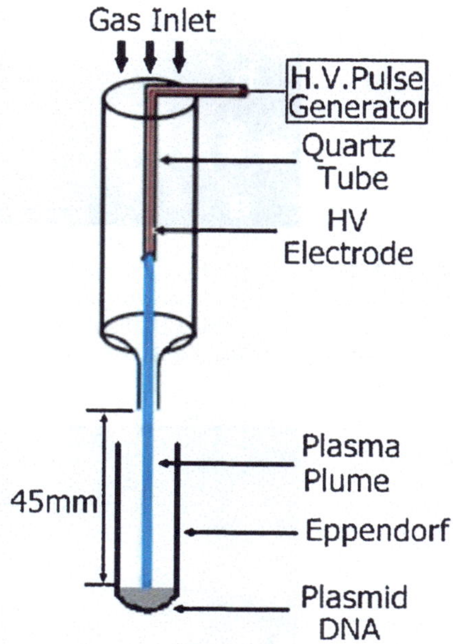


**Fig. 10.2** Schematic diagram of the experimental setup for plasmid treatment [4]

the experiments, namely, applied voltage (V) = 8 kV, pulse width $t_{pw} = 1.0$ μs, pulse frequency = 8 kHz. Thereafter, plasmid DNA was prepared, followed by dilution using distilled water (DW) at the final concentration of 0.1 μg/μL to 15 μL within the Eppendorf (2 ml), which was put directly below the nozzle. Meanwhile, nozzle was always 45 mm away from the plasmid surface. He/$O_2$(1%) was adopted as the working gas at 2 L/min flow rate. Plasmid DNA solution in Eppendorf was then subjected to direct APNPs treatment and analysis through 0.7% agarose gel electrophoresis (AGE) under ethidium bromide (EB) staining.

It was found that APNPs treatment damaged the plasmid DNA (Figs. 10.3 and 10.4) [4], and the degree of plasmid DNA damage was dependent on the APNPs treatment duration, which resulted in the changed conformation of plasmid DNA. The supercoiled torsional energy preserved is released when a single scission is made on the plasmid strand due to APNPs treatment. Moreover, relaxing the supercoiled plasmid facilitates the conversion from tight supercoiled plasmid to the form of open circle. When the other cleavage event occurs near or at the identical site of the opposing strand, then the form of open circle will be transformed into the linear polynucleotide.

Thereafter, the optical emission of plasma plume was examined through a half meter spectroscopy. As suggested by these results, excited He*, $N_2^{+*}$, $N_2^{*}$, and O* dominated the emission spectrum. Besides, UV radiation intensity at 200–280 nm wavelengths was not detected, indicating that UV radiation did not markedly affect plasmid DNA (Fig. 10.5). Therefore, the chemically active species were the most possible factors causing transformation in plasmid conformation.

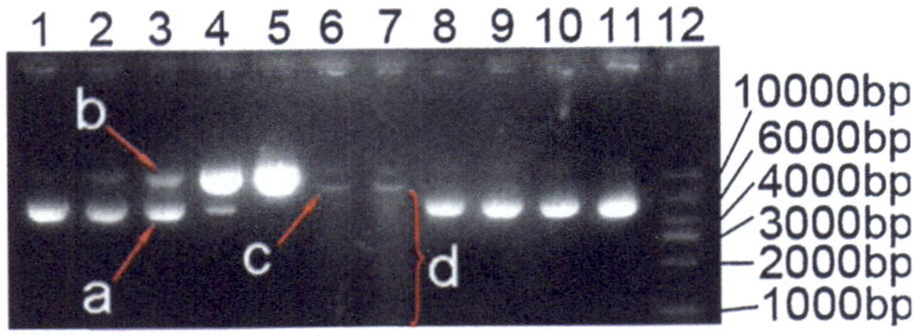


**Fig. 10.3** Impact of APNPs on the plasmid DNA. Lanes 1–7: treatments with plasma for 10 and 30 s, as well as 1, 2, 4, 8, and 16 min, separately. Lanes 8–10: treatments using identical flow rate working gas with plasma off. Lane 11: control with no treatment. Lane 12: DNA marker. a-d: supercoiled, open circular, linearized and fragmentized conformations in succession [4]

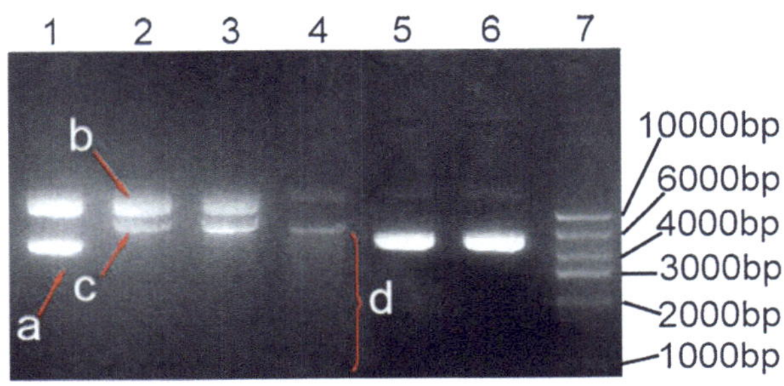


**Fig. 10.4** Lanes 1–4: treatments with plasma for 1, 5, 6, and 7 min, separately. Lane 5: identical flow rate working gas treatments with 7 min of plasma off. Lane 6: control with no treatment. Lane 7: DNA marker. a-d: supercoiled, open circular, linearized and fragmentized conformations in succession [4]

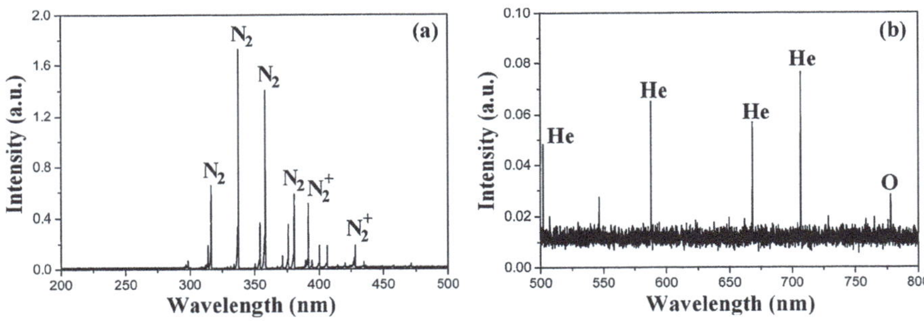


**Fig. 10.5** Typical plasma plume emission spectra of $He/O_2$ (1%) (total flow rate, 2 L/min) (**a**) 200–500 nm; (**b**) 500–800 nm [4]

Taken together, this study provided an important implication that, the plasma characteristics that induced DNA damage might also be adopted to induce cancer cell death, since DNA damage was also a feature of apoptotic (a form of cell death) cells [5]. Based on this hypothesis, the following sections were mainly about the impact of plasma on cancer cell death according to the production of ROS and RNS by the plasma.

## 10.3 Effect of Plasma on Cancer Cells

### *10.3.1 Reactive Species Generated by Plasma Treatment*

Hepatocellular carcinoma (HCC) has been recognized to be the frequently seen cancer, which is associated with a particularly high prevalence in China [6]. The following part investigated the effect of plasma against cancer, together with the underlying mechanisms within the HepG2 cell line.

In brief, the HepG2 cell line was cultured into the 96-well plates, then the culture medium was replaced with 80 μL phosphate-buffered saline (PBS), and plasma treatments were carried out for different durations (5–640 s). The results showed that, relative to non-treated control groups, the viable cell percentage was rapidly reduced with the increase in plasma exposure duration (Fig. 10.6).

ROS and RNS exert vital parts in the inactivation of cancer cells. The plasma-induced changes in ROS and NO species were evaluated using some biological methods [7]. In brief, the plasma-treated cells were lysed to detect the NO content using Griess reagent. As suggested by our results, both intracellular and extracellular NO contents were markedly increased after plasma treatment relative to those in untreated control group. To detect the intracellular ROS levels, cells in each group were stained with DCFH-DA, a specific ROS intracellular probe. DCFH-DA,

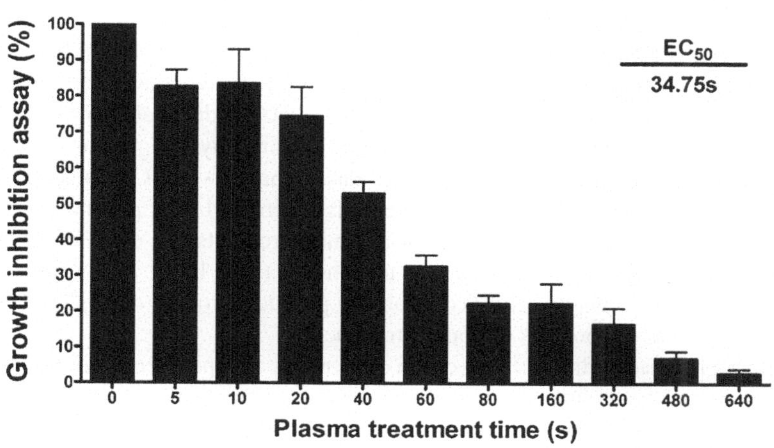


**Fig. 10.6** Anti-proliferation effect of plasma on HepG2 cell viabilities. The viability of cells was measured through MTT assay, and the results were presented in the form of mean ± standard deviation (SD) from 3 independent assays [7]

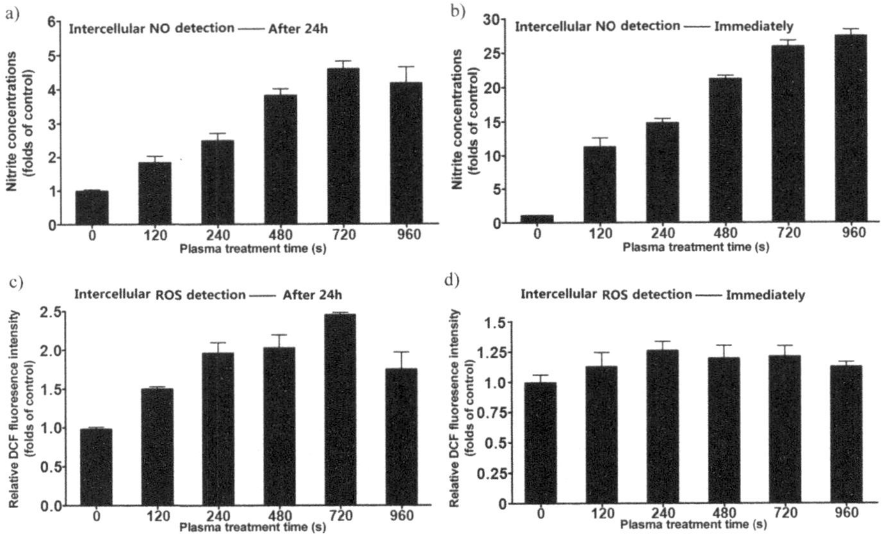


**Fig. 10.7** Variations of ROS and NO contents within HepG2 cells immediately and at 24 h after treatment with plasma. (**a**) Plasma effect on NO concentrations within HepG2 cells following culture for 24 h; (**b**) Plasma effect on ROS concentrations within HepG2 cells cultured for 24 h; (**c**) Plasma treatment effect on the NO contents within HepG2 cells immediately following treatment with plasma; (**d**) Plasma effect on ROS contents within the HepG2 cell line immediately following treatment with plasma. All data were presented in the manner of mean ± SD from 3 independent assays [7]

the extensively investigated compound for detecting and quantifying ROS levels in cells, is subjected to deacetylation by esterases for the formation of nonfluorescent DCFH, and it is also transported in and out of the cell membrane. Thereafter, DCFH will be trapped within cells, followed by the ROS-mediated conversion into DCF, the greatly fluorescent material, while the resultant DCF is detectable and calculated on the intensity of fluorescence. DCF fluorescence intensity was measured through fluorospectrophotometry in this study. Similar to changes in NO contents, it was found that the intracellular ROS levels were also increased after plasma treatment. Interestingly, different from extracellular NO measurements, the extracellular DCF fluorescence intensity remained largely unchanged (Fig. 10.7). According to the above findings, ROS inside cells were probably produced in cells during 24 h of cell culture, but not in the process of plasma treatment.

Increased intracellular ROS levels in turn result in lipid peroxidation (LPO), which can thereby induce extensive cell damage. Specifically, ROS attacks the unsaturated fatty acids (UFAs) contained within the membrane lipids at their side chains, as a result, lipid hydroperoxides will come into being and are accumulated into the cell membranes. This will destroy the normal membrane functions and even induce the collapse of membrane. In turn, this results in extensive loss of selective permeability and dramatic leaks, finally causing cell necrosis or apoptosis.

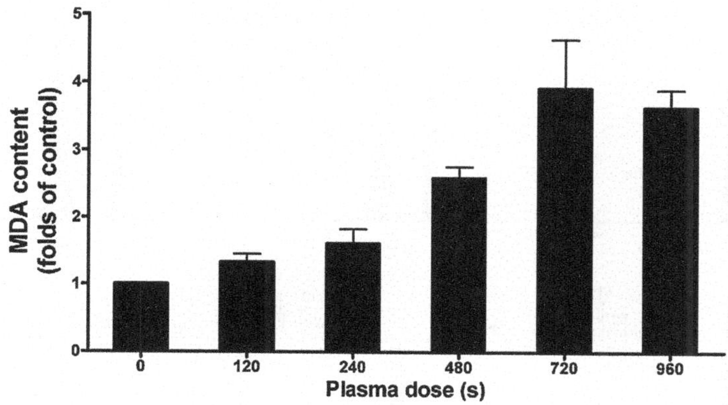


**Fig. 10.8** MDA contents in plasma-treated HepG2 cells. All data were presented in the form of mean ± SD from 3 independent assays [7]

Therefore, MDA content, a widely accepted and reliable indicator of LPO level, was detected to examine the plasma effect on LPO in cells. As suggested, plasma treatment dramatically increased MDA contents depending on its dose; specifically, the MDA contents were increased by about four times in HepG2 cells after plasma treatment for 720 s (Fig. 10.8).

### *10.3.2 Biologic Targets of Plasma in Cancer Cell Death*

#### Plasma Effect on the Arrest of Cell Cycle

Afterwards, the biologic mechanism regarding the plasma-induced cell death was explored [8]. The cell cycle is often constituted by a closely modulated series of events, so as to guarantee the ready replication, faithful copy, and appropriate segregation of genomic material to the daughter cells. Additionally, genomic DNA integrity was observed continuously, while the DNA repair was adjusted based on cell cycle through those G2/M, intra-S, and G1/S checkpoints [9]. Besides, flow cytometry was carried out for detecting variations within the cell cycle of plasma-treated HepG2 cells through analyzing the cellular DNA content, which was stained by propidium iodide (PI). Flow cytometry allows to identify cell distribution in different cell cycle phases, and 4 different phases are detected among the proliferating cells, including M (mitosis), G2, S (DNA synthesis phase), and G1 phases. On this account, the DNA content at G2 phase is twice as high as that at G1 phase, but DNA contents at M and G2 phases are the same, so it is difficult to discriminate them according to the heterogeneities in DNA concentrations. Moreover, M phase is generally quite short, which is thereby frequently referred to as the G2/M phase biologically. Therefore, the one-parameter histogram can present

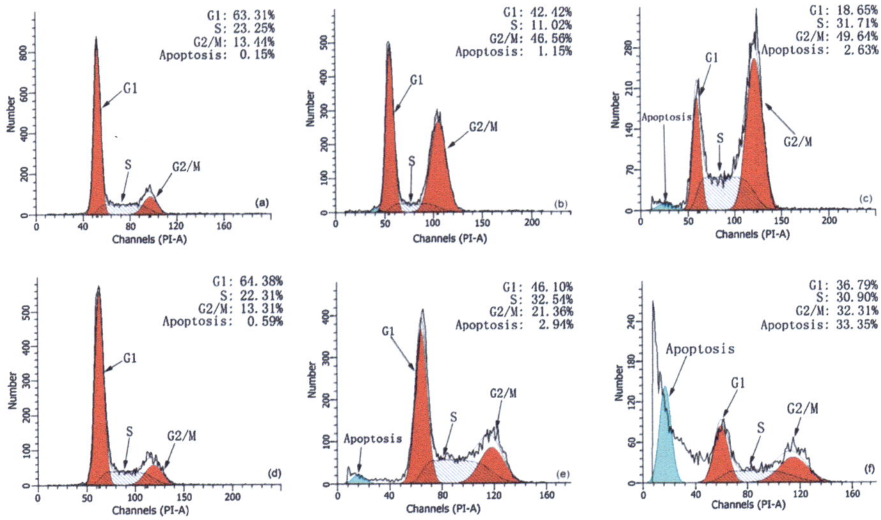


**Fig. 10.9** Histograms of cellular DNA contents displaying the distribution of cell cycle and HepG2 cell apoptosis after treatments with plasma. All data were derived based on 1 out of 3 similar independent assays. (**a–c**) Plasma treatments in cells for 0, 240, and 480 s, separately, followed by 24 h of culture. (**d–f**) Plasma treatments in cells for 0, 240, and 480 s, separately, followed by 48 h of culture [7]

cell content distribution; namely, it exhibits the number of cells containing a given quantity of DNA. In this histogram, the cell concentration is indexed to each class or channel, and presented on x-axis; while cell number indexed to one certain channel would be deemed simple count or channel content, and presented on y-axis. Then, those cells with nearly the same quantity of cell contents (like DNA) can generate one peak. In the representative DNA histogram, one peak stands for G1 phase, while the other (channel value at twofold) stands for G2/M phase. It was found in this study that, plasma treatment causes G2/M arrest in the cell cycle, which was related to the elevated apoptotic cell percentage (Figs. 10.9 and 10.10).

Thereafter, the following research focused on those genes related to apoptosis and the cell cycle. Briefly, those genes that took part in the G2/M phase arrest in the cell cycle and apoptosis were examined through reverse transcription-polymerase chain reaction (RT-PCR). Besides, genes within cells with different states were subjected to selective transcription to mRNA and later translation to proteins for exerting the diverse biological effects. First of all, the RNA strand was separated for reverse transcription to complementary DNA (cDNA) through enzyme RT. Later, that as-prepared cDNA is utilized to be the PCR amplification template for one gene using its related primers. Then, those PCR products will be examined, images are taken, and DNA fragments are separated through agarose gel electrophoresis (AGE) to various lengths. Moreover, those bands displayed on AGE image indicate those

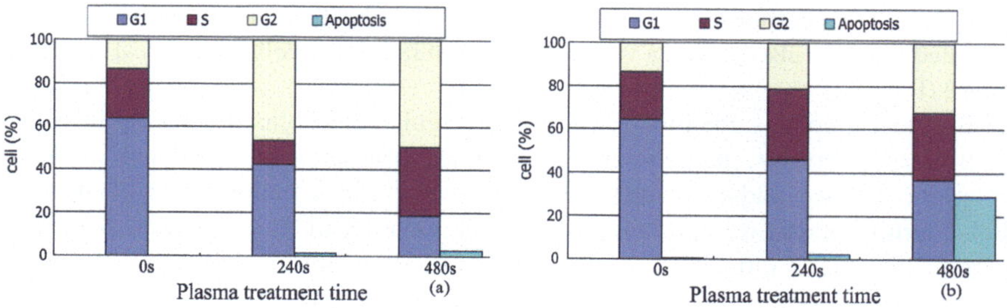


**Fig. 10.10** Distribution of cell cycle within plasma-treated HepG2 cells detected by FACS after incubation for various durations. Cells were divided into cycling and apoptotic cells, while cycling cells were adopted for analyzing the cell cycle. Cell number distributed at every cell cycle phase was presented. (**a**) 24 h of culture after plasma treatment. (**b**) 48 h of culture after plasma treatment [7]

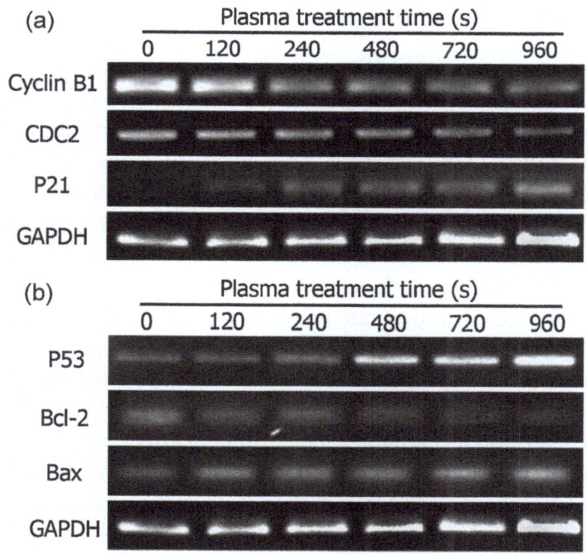


**Fig. 10.11** Plasma effect on (**a**) genes related to cell cycle and (**b**) those related to apoptosis at mRNA levels within HepG2 cells. Plasma treatments in cells for 0, 120, 240, 480, 720, and 960 s, followed by culture for 48 h before they were collected. Thereafter, total RNA was collected, followed by transcription into cDNA for PCR with indicated primers. Afterwards, products obtained from the reaction were separated through 1% AGE, and EB staining was conducted for visualization. GAPDH served as internal reference [8]

PCR-amplified DNA fragments possessing certain lengths, whereas band brightness reflects PCR product quantity.

Cyclin B1/Cdc2 kinase complex, a primary regulatory element, governs G2-M phase transition [10]. As found based on semi-quantitative RT-PCR in this study, Cdc2 and cyclin genes had been down-regulated at transcription level, which suggested that plasma potentially exerted some impact on that Cdc2/cyclin B1 complex (Fig. 10.11). Based on such results, plasma-induced G2 phase arrest in the cell cycle, and delayed cells from entering the M phase through preventing the Cdc2 and cyclin B1 mRNA from forming that Cdc2/cyclin B1 complex [8].

P21 has been recognized to be the vital regulator for the p53-induced cell cycle arrest responding to DNA injury [11]. As suggested in this study, treatment with plasma enhanced transcription levels of p53 and the Cdk inhibitor p21, which revealed that p21 also played a vital role in maintaining the cell cycle arrest at G2/M phase (Fig. 10.11).

Besides, apoptosis, the biological process regulated genetically, mainly involves 2 primary pathways, namely, the mitochondria apoptosome-mediated intrinsic apoptotic pathway and the death receptor-induced extrinsic pathway [12]. Typically, Bcl-2 family plays an important part in the regulation of mitochondrial pathway; specifically, those anti- and pro-apoptotic proteins in Bcl-2 family can initiate or terminate the apoptosis, since heterodimers are formed across those proteins [13]. It has been demonstrated that Bax is increased in plasma-treated HepG2 cells, whereas Bcl-2 is concomitantly decreased, and such results indicate that, plasma induces apoptosis by changing the ratio of Bax/Bcl-2 to apoptosis (Fig. 10.11).

## Effect of Plasma on Apoptosis of Cells

Furthermore, the biologic mechanisms by which plasma-induced cell apoptosis were further investigated [14]. First of all, the HepG2 cell apoptosis induced by plasma treatment was identified through PI/Annexin V-FITC staining as well as Hoechst 33342 staining, and observed using flow cytometry (Fig. 10.12a) and fluorescence microscopy (Fig. 10.12b), respectively. Annexin-V, the protein that binds to phospholipid, displays great affinity for phosphatidylserine (PS), which is a vital cellular membrane component. Under normal conditions, the PS molecules are localized on the inner cell membrane surface, but in apoptotic cells, they are translocated onto outer membrane surface, and annexin-V easily binds to the PS molecules on that surface. Therefore, the apoptotic cells are rapidly discriminated using this approach. Meanwhile, cells that are suffering from apoptosis are frequently associated with typical changes in morphology, such as DNA degradation to nucleosomal fragments and chromatin condensation. As a result, the morphological changes in nuclei were monitored by Hoechst 33342 staining, a dye specific to nucleic acid, in this study. The results revealed that, compared with untreated control group, plasma treatment dramatically increased the apoptotic cell rate (including the early and late stages of apoptosis), accompanying with cell chromatin condensation and nuclear fragmentation.

## Oxidative and Nitrative Stress Induced by Plasma Treatment

The RNS and ROS contents in cells were examined through DCFH-DA staining and the Griess method, respectively. Consistent with previous results, plasma treatment was demonstrated to boost RNS and ROS contents in cells in a dose-dependent manner (Fig. 10.13).

The elevated NOS production in cells is frequently relative to nitric oxide synthase (NOS) activation or inducible NOS (iNOS) up-regulation [15]. Therefore, the expression and activity of NOS were examined in this study using the NOS assay

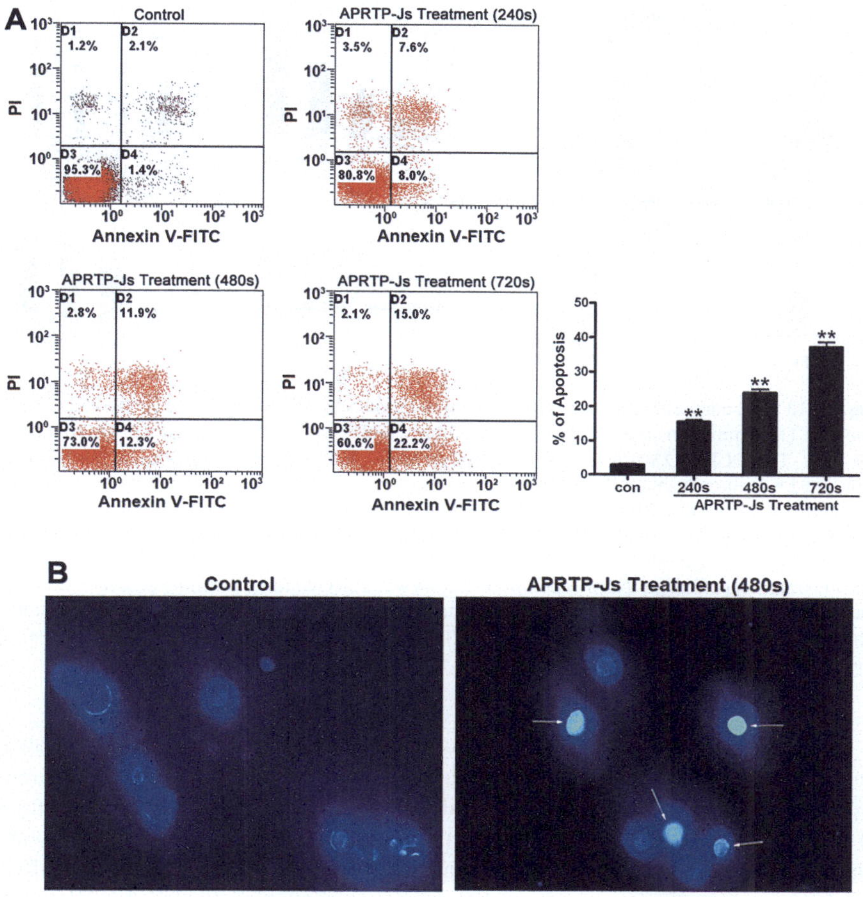


**Fig. 10.12** Treatments with plasma resulted in HepG2 cell apoptosis. Plasma treatments in cells for 240, 480 and 720 s, respectively, followed by 24 h of culture. (**a**) Cells were stained by PI and annexin V-FITC and PI and analyzed by flow cytometry. Annexin V-FITC positive while PI staining negative cells were at early apoptotic stage; PI and annexin V-FITC staining positive cells were at late of apoptotic stage; and PI and annexin V-FITC staining negative cells were alive, in which apoptosis was not detectable. Histogram was plotted to show the apoptotic cell percentage (positive annexin V-FITC staining). (**b**) Fluorescence microscopy was performed to observe the Hoechst 33342 staining for apoptosis. White arrows indicate representative apoptotic nuclei [14]

kit. It was found that, the enzymatic activities of both constitutive NOS (cNOS) and total cellular iNOS (endothelial NOS (eNOS) and neuronal NOS (nNOS) were both included) were enhanced in the plasma dose-dependent manner compared with control group. Meanwhile, iNOS expression was also verified through Western blotting, and the results suggested that, the iNOS protein expression was up-regulated as the plasma treatment duration increased (Fig. 10.13).

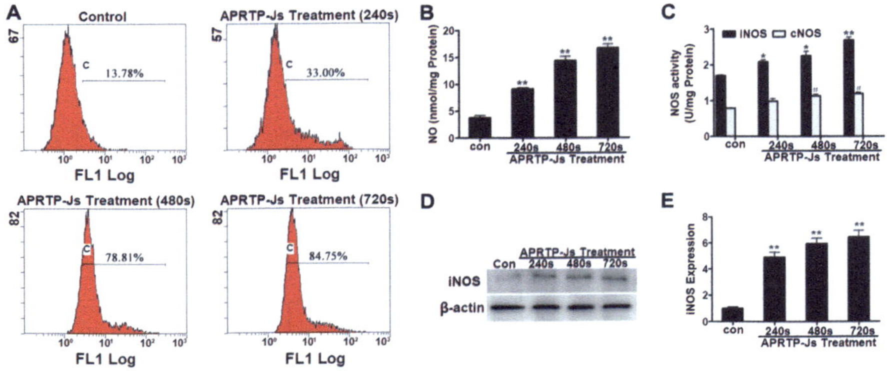


**Fig. 10.13** Plasma treatment enhanced the accumulation of RNS and ROS within the HepG2 cell line. (**a**) Determination of ROS production within HepG2 cells following APRTP-Js treatment for 24 h. The DCFH-DA probe was used to stain cells, then, the flow cytometer was utilized to measure DCF fluorescence. (**b**) APRTP-Js treatment effect on NO contents within HepG2 cells cultured for 24 h. (**c**) APRTP-Js treatment effect on the activities of NOS within HepG2 cells cultured for 24 h. Statistical significance was examined relative to related control (**$P < 0.01$ and *$P < 0.05$ compared with iNOS control, and # $P < 0.05$ compared with cNOS control). (**d**) Plasma treatment effect on iNOS expression. β-actin was detected for confirming equivalent protein loading. (**e**) Relevant densitometry for iNOS. The values were standardized based on β-actin signal. The data were expressed as means ± SEM for 3 independent assays. Asterisk represented statistical significance relative to control (**$p < 0.01$ and *$p < 0.05$) [14]

On the other hand, RNS and ROS accumulation indicates that there are nitrative and oxidative stresses in cells. On the other hand, the suitable balance between anti-oxidation and oxidation in cells is crucial for cells to resist oxidative/nitrative stress [16]. Therefore, HepG2 cell oxidation resistance was assessed at 24 h following plasma treatments. Briefly, cells were harvested for lysis with the cell lysis buffer; later, antioxidant activity assays would be carried out using commercial GSH, SOD, and catalase activity detecting kits. The results demonstrated that, plasma treatment remarkably reduced the GSH and SOD levels, as well as catalase activity in cells, which indicated that the antioxidant capacity of cells was markedly declined, and that balance between cellular anti-oxidation and oxidation had been destructed (Fig. 10.14).

Proteins may potentially serve as the major targets for nitrative and oxidative injuries, which is ascribed to the abundances within cells. RNS mainly play the biological roles by nitrating various target proteins possessing the tyrosine residues. Therefore, the cell tyrosine nitration was determined through Western blotting in this study. As suggested by our results, the nitrated protein levels, which were low under normal conditions, were largely up-regulated within HepG2 cells under plasma treatment, which revealed that plasma exposure in HepG2 cells led to increased protein tyrosine nitration depending on their doses (Fig. 10.15a).

Carbonyl groups are introduced into proteins due to the oxidative cleavage in peptide backbone; as a result, carbonyl formation is used to be a vital factor

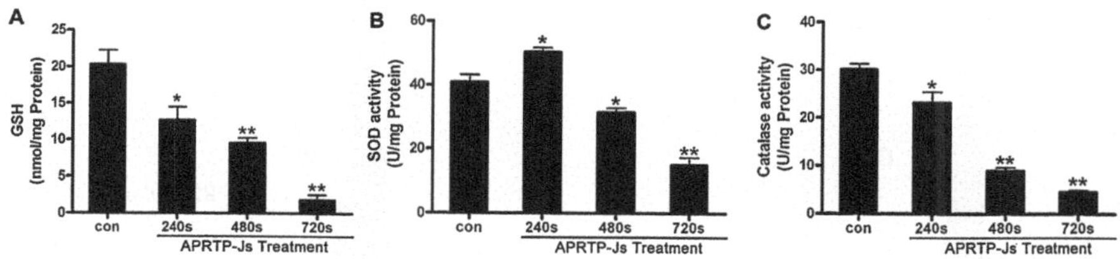


**Fig. 10.14** Treatment with plasma impaired the antioxidative system within HepG2 cells. Thereafter, the cells were subjected to APRTP-Js treatment for 240, 480, and 720 s or non-treatment, separately. Then, cells were collected after 24 h, the cell lysis buffer was used for lysis, and supernatants were collected from samples through centrifugation to carry out antioxidant activity assays. (**a**) APRTP-Js treatment effect on the contents of total GSH within HepG2 cells. (**b**) APRTP-Js treatment effect on activity of SOD within the HepG2 cells. (**c**) APRTP-Js treatment effect on activity of catalase within the HepG2 cells. The data were expressed as means ± SEM of 3 independent assays. Asterisk represented statistical significance relative to control (**$p < 0.01$ and *$p < 0.05$) [14]

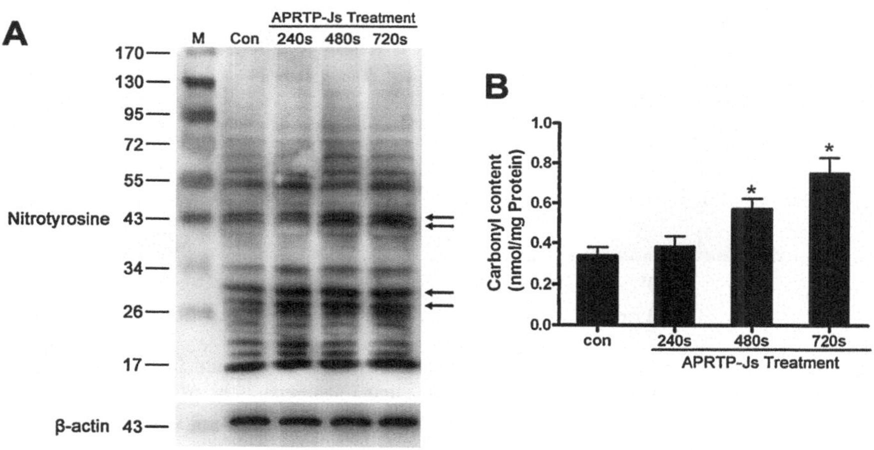


**Fig. 10.15** Nitrative and oxidative injuries within HepG2 cells induced by plasma treatment. (**a**) Total nitration of proteins. Protein total nitration states were examined through Western blotting using the nitrotyrosine antibody in cells in the presence or absence of APRTP-Js treatment. Each arrows denoted a deepened nitrotyrosine epitope. (**b**) Carbonyl level statuses of proteins. Carbonyl level of protein was detected by the commercially available assay kit. The data were expressed as means ± SEM of 3 independent assays. Asterisk represented the statistical significance relative to control (**$p < 0.01$ and *$p < 0.05$) [14]

predicting oxidative damage of proteins [17]. In addition, carbonyl formation is generally accepted to induce changes in structure and biological function loss in protein. On this account, this study further assessed plasma effect on the oxidative damage in proteins by detecting the formation of carbonyls within cells using the aprotein carbonyl assay kit. This study indicated that, relative to control group, protein carbonyl levels had been markedly elevated within the groups subjected

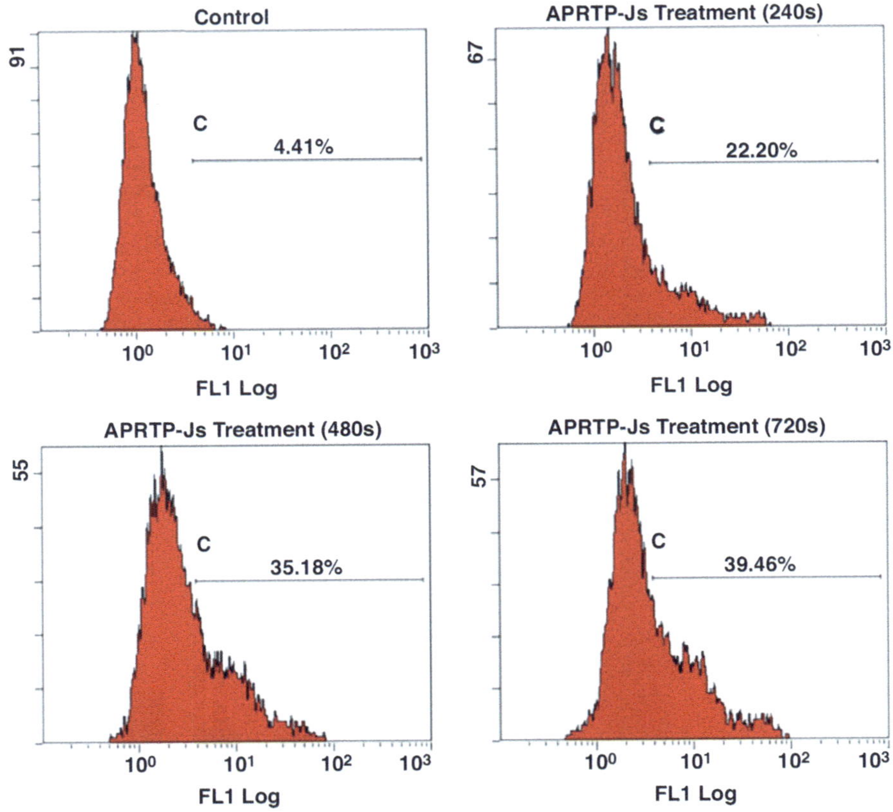


**Fig. 10.16** Treatment with plasma destroyed the homeostasis of calcium in cells. Cells were subjected to APRTP-Js treatment or not for 240, 480, and 720 s, separately. Afterwards, cells were cultured for 24 h continuously, and loaded using the Fluo-3/AM calcium probe, and free Ca2+ level in cells was detected through the flow cytometry (FACScan) [14]

to plasma treatment, suggesting that cells under plasma treatments undergoing oxidative damage in cells, which aggravated the cell damage (Fig. 10.15b).

## Calcium Perturbation and ER Stress Induced by Plasma Treatment

Calcium homeostasis plays an essential role in numerous cell activities, including signal transduction, as well as protein transport, processing, and folding. Additionally, calcium induces apoptosis responding to various stimuli [18]. The effect of plasma exposure on homeostasis of calcium within the HepG2 cells was determined by the Fura-3/AM fluorescence approach. Our results revealed that, the free Ca2+ was accumulated in cells, while the homeostasis of calcium was destroyed within the plasma-treated HepG2 cells (Fig. 10.16).

Ca2+ exerts its pro-apoptotic effect via the regulation by various Ca2+ sensitive molecules, which are located within different cell organelles, such as endoplasmic reticulum (ER). Calcium homeostasis perturbation is found to result in ER stress [18], which serves as the adaptive mechanism for cell survival under their exposed challenging conditions [19]. Nonetheless, the above signal transduction pathways account for the pro-death program to eliminate the non-surviving cells when those challenging conditions have not been mitigated. Typically, the ER stress signaling downstream is primarily transduced via CHOP and GRP78, and these two factors can boost apoptosis induced by ER stress. Caspase12, which is the ER stress-induced apoptosis executioner, is located in ER and cleaved in the case of apoptosis induced by ER stress [20, 21]. In this study, plasma effects on pro-caspase12, CHOP, and GRP78 expression were detected through Western blotting. After exposure to plasma for different durations, the relative CHOP and GRP78 expression was dramatically increased depending on dose, compared with that in control group. Conforming to the above observations, caspase12 was triggered following plasma treatments, while its precursor was evidently decreased (Fig. 10.17). These changes in protein expression within cells under plasma treatment suggested that ER stress occurred.

### Effect of Antioxidant Combined with Plasma Treatment on Cancer Cells

To obtain direct evidence that supports a causative link between overproduction of ROS/RNS and HepG2 cells apoptosis, a comparative study using antioxidant was carried out. Before treatment with plasma, HepG2 cells had been subjected to pretreatment with no or with N-Acetylcysteine (NAC), the scavenger of free radicals. After treatment with plasma, HepG2 cells were subjected to 24 h continuous culture before collected, and PI/annexin V-FITC staining was performed to determine the apoptotic activity. As suggested by Fig. 10.18, pretreatment with NAC entirely mitigated cell apoptosis, relative to plasma treatment at the identical dose without NAC. Besides, NAC made no difference to HepG2 cells by itself. The above findings potently indicated that, ROS/RNS overproduction accounted for the leading mechanism mediating HepG2 cell apoptosis induced by plasma.

Subsequently, the following work investigated the effect of plasma combined with NAC on HepG2 Cells. In brief, the HepG2 cells were subjected to 1 h of NAC pretreatment at various contents, and then 15 s of treatment with plasma. Afterwards, cells were cultivated for another 24 h before the following experiments. As suggested by Fig. 10.19, HepG2 cell proliferation was promoted for NAC combined with plasma treatment group, while cells without NAC treatment underwent death or apoptosis following plasma treatment.

Further investigation indicated that NAC combined with plasma treatment accelerated the G1-S phase transition in HepG2 cells (Fig. 10.20); besides, cyclinD1 (Fig. 10.21), a proliferation indicator, was markedly increased for the plasma combined with NAC group. Such findings provided a rationale for developing the approach to promote proliferation, which also provided more insights into possible applications

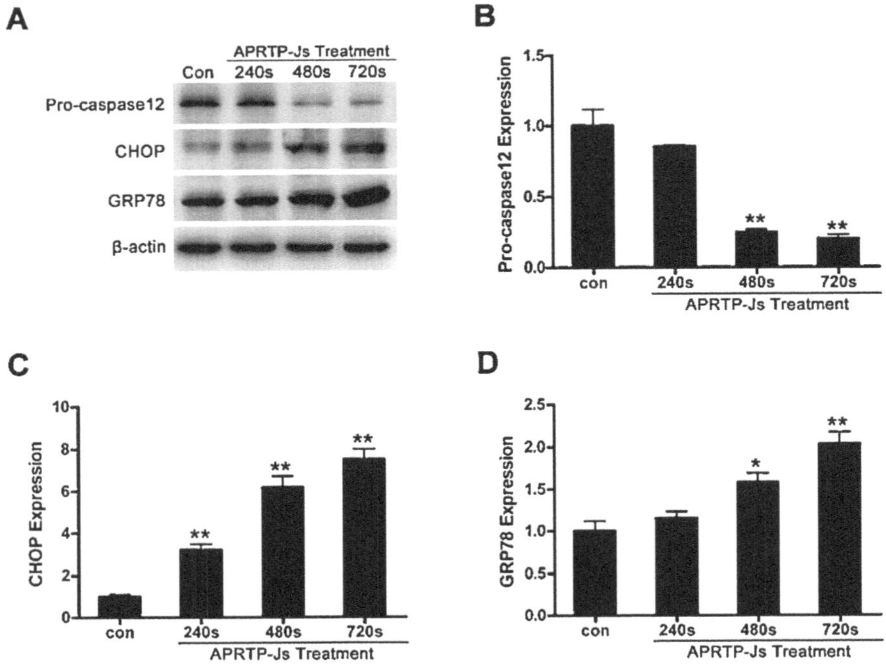


**Fig. 10.17** Treatment with plasma-induced ER stress within the HepG2 cells. (**a**) Typical autoradiographs for procaspase12, CHOP, and GRP78. β-actin was detected for confirming equivalent loading of protein. (**b–d**) Related densitometry for CHOP, GRP78, and pro-caspase12, separately. All values were standardized based on β-actin signal. Asterisk represented the statistical significance relative to control (**$p < 0.01$ and *$p < 0.05$) [14]

of cold plasma in enhancing the incorporation of transplanted tosses; for instance, the large-scale in vitro stem cell culture for transplantation of stem cells.

## 10.4 Influence of Plasma on Adenovirus

The following section investigated the inactivation effect of plasma on adenovirus (AdV) [23]. AdV is the frequently seen pathogens resistant to treatment in enterogastritic, ophthalmic, and respiratory diseases, which is identified to be the primary obstacle towards the potentially revolutionary next-generation stem cell treatments [24–26]. However, the fatal systemic infections with AdV, which impact the whole body and is confirmed through the adenoviremia positiveness biologically, are associated with the risk of as high as 50% in the process of hematopoietic stem cell transplantation, and it may be even increased in the recipients of cord blood transplant [27]. In this study, the AdV infectivity was

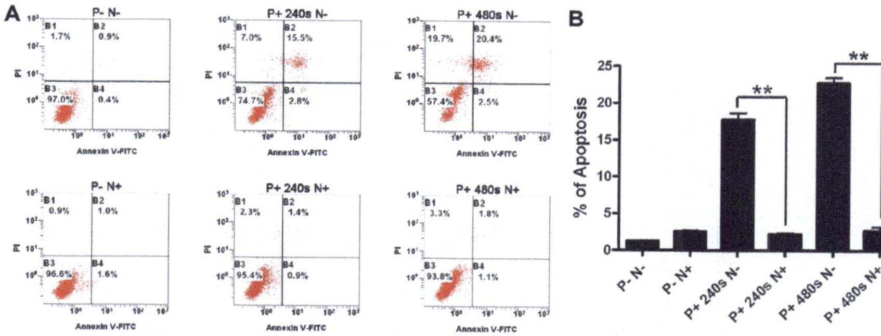


**Fig. 10.18** NAC mitigated apoptosis resulted from treatment with APRTP-Js. (**a**) HepG2 cells were subjected to 1 h incubation using 10 mM NAC (N+) or with the culture medium (N-), and plasma treatment was carried out at a specific dose. Afterwards, cells were collected for annexin V-FITC/PI staining to measure apoptosis following plasma treatment for 24 h. Later, the samples were detected through flow cytometry. P- N-: control group under no treatment; P+ 240 s N-: cells treated with plasma for 240 s alone; P+ 480 s N-: cells treated with plasma for 480 s alone; P-N+: cells pretreated with NAC alone; P+ 240 s N+: cells pretreated with NAC and then treated with plasma for 240 s; P+ 480 s N+: cells pretreated with NAC and then treated with plasma for 480 s. (**b**) Flow cytometry was carried out to determine the apoptotic cell percentage [22]

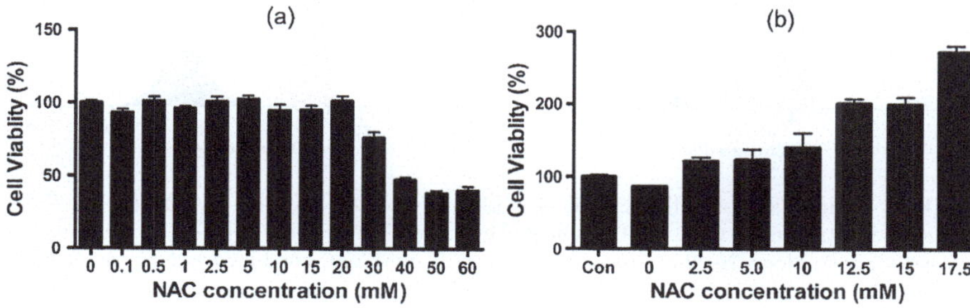


**Fig. 10.19** Relative HepG2 cell proliferation detected according to the MTT experiment. y-axis represented proliferation rate compared with the control. Means and SEM were displayed ($n = 3$). (**a**) NAC effect on HepG2 cell proliferation. (**b**) NAC (at various contents) combined with plasma treatment (15 s) on HepG2 cell proliferation at a low original content. The data were presented in the form of mean ± SEM of 3 independent assays [22]

monitored based on the expression of AdV genes, together with the related green fluorescent protein emissions from the host HEK 293A cells. Specifically, HEK 293A cells were treated with plasma and cell lesions in each well are observed under the fluorescence microscope. Our results showed that, the plasma generated via a plasma needle device successfully inactivated the adenovirus functions. Moreover, as suggested by the survival curves obtained based on green fluorescence in those infected HEK 293A cells, over 90% AdV virions were efficiently inactivated after plasma treatment for a short time (8 min) (Fig. 10.22). Besides, optical emission spectroscopy for the He+O2 plasma was also investigated for examining the candidate effects of gas temperature and reactive species on inactivating pathogen.

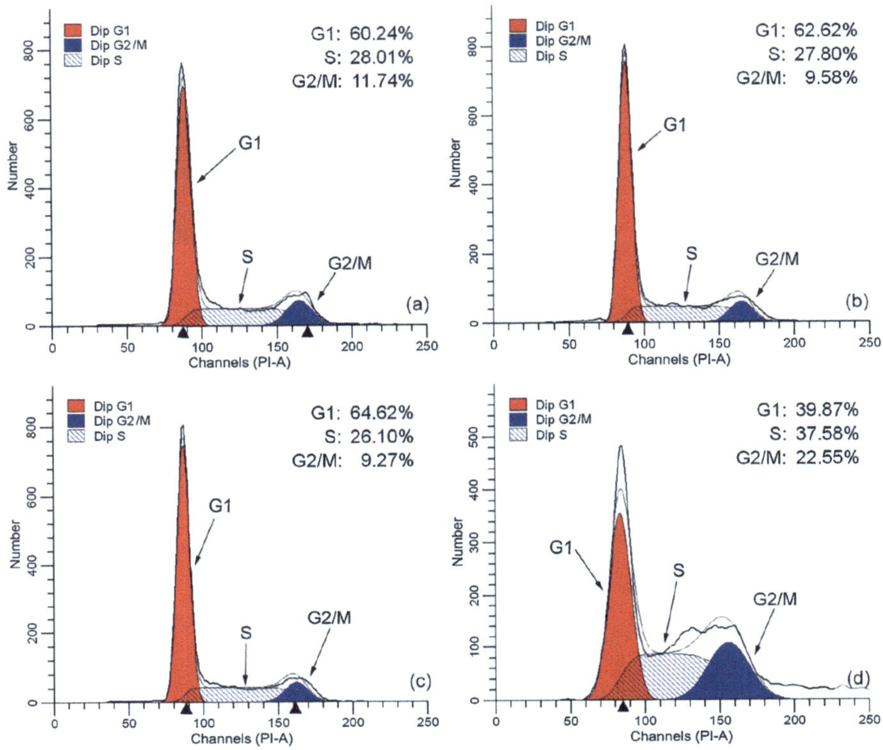


**Fig. 10.20** Distribution of cell cycle in HepG2 cells detected by flow cytometry. (**a**) Cells subjected to no treatment (Con). (**b**) Cells treated by 15-mM NAC alone (N+P−). (**c**) Cells treated by plasma for 60 s alone (N−P+). (**d**) Cells treated by 15-mM NAC combined with plasma for 60 s (N+P+). The collected cells were fixated and dyed by PI, and flow cytometry was performed to analyze the DNA contents. Results from 1 of the 3 independent assays were shown. x- and y-axes denoted the DNA content and the number of cells, separately. All phases were computed by cell ModFit software. Cell percentages at G2/M, S, and G1 phases were displayed [22]

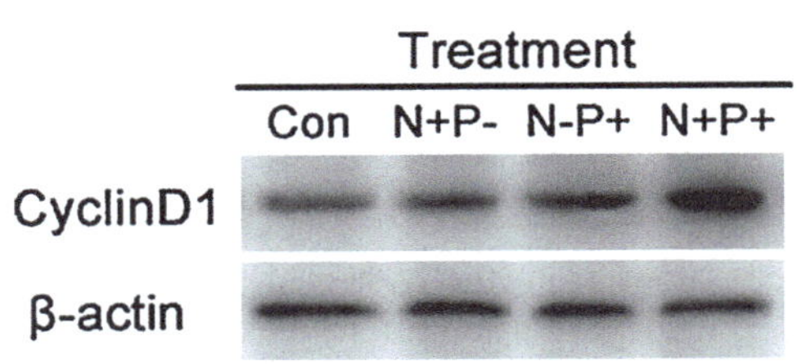


**Fig. 10.21** Expression of cyclinD1 in HepG2 cells among various treatments. Cells with no treatment (Con). Cells subjected to 15-mM NAC treatment alone (N+P−). Cells subjected to plasma treatment for 60 s alone (N−P+). Cells subjected to 15-mM NAC combined with plasma treatment for 60 s (N+P+). Total proteins were collected based on those collected HepG2 cells under various treatments, so as to detect the expression of cyclinD1 through Western blotting. Protein abundances were assessed according to corresponding band intensity, with β-actin as an internal reference [22]

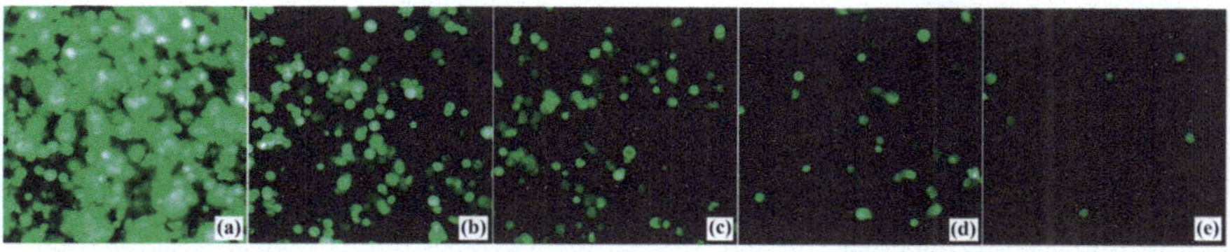


**Fig. 10.22** Fluorescence images for those HEK 293A cells infected with AdV following treatment with plasma needle. Panels (**a**)–(**e**) showed fluorescence images of those cells infected with AdV following plasma treatments for 0, 2, 4, 8, and 16 min, separately. All images were taken on day 10 following treatment, which corresponded to $10^{-3}$ dilution of those infected cells [23]

To search for the suitable therapeutic protocols, APPs exert parallel effect on various viral pathogens, which thus remarkably broaden our treatment options while improving the pre-emptive anti-viral efficacy.

## References

1. D.B. Graves, The emerging role of reactive oxygen and nitrogen species in redox biology and some implications for plasma applications to medicine and biology. J. Phys. D. Appl. Phys. **45**(26), 263001 (2012)
2. C. Smillie, M.P. Garcillán-Barcia, M.V. Francia, E.P. Rocha, F. de la Cruz, Mobility of plasmids. Microbiol Mol Biol Rev **74**(3), 434–452 (2010)
3. Q. Yao, L. Cong, G. He, J. Chang, K. Li, G. Yang, Optimization of wheat co-transformation procedure with gene cassettes resulted in an improvement in transformation frequency. Mol. Biol. Rep. **34**(1), 61–67 (2007)
4. X. Yan, F. Zou, X.P. Lu, G. He, M.J. Shi, Q. Xiong, X. Gao, Z. Xiong, Y. Li, F.Y. Ma, Effect of the atmospheric pressure nonequilibrium plasmas on the conformational changes of plasmid DNA. Appl. Phys. Lett. **95**(8), 083702 (2009)
5. C.J. Norbury, B. Zhivotovsky, DNA damage-induced apoptosis. Oncogene **23**(16), 2797 (2004)
6. R.X. Zhu, W.-K. Seto, C.-L. Lai, M.-F. Yuen, Epidemiology of hepatocellular carcinoma in the Asia-Pacific region. Gut Liver **10**(3), 332 (2016)
7. X. Yan, Z. Xiong, F. Zou, S. Zhao, X. Lu, G. Yang, G. He, K. Ostrikov, Plasma-induced death of HepG2 cancer cells:intracellular effects of reactive species. Plasma Process. Polym. **9**(1), 59–66 (2012)
8. X. Yan, F. Zou, S. Zhao, X.P. Lu, On the mechanism of plasma inducing cell apoptosis. IEEE Trans. Plasma Sci. **38**(9), 2451–2457 (2010)
9. M.S. Chaurushiya, M.D. Weitzman, Viral manipulation of DNA repair and cell cycle checkpoints. DNA Repair **8**(9), 1166–1176 (2009)
10. R. Visconti, R. Della Monica, D. Grieco, Cell cycle checkpoint in cancer: a therapeutically targetable double-edged sword. J. Exp. Clin. Cancer Res. **35**(1), 153 (2016)
11. J. Chen, The cell-cycle arrest and apoptotic functions of p53 in tumor initiation and progression. Cold Spring Harb. Perspect. Med. **6**(3), a026104 (2016)
12. S. Fulda, Targeting apoptosis for anticancer therapy. Semin. Cancer Biol **31**, 84 (2015)
13. A.N. Hata, J.A. Engelman, A.C. Faber, The BCL2 family: key mediators of the apoptotic response to targeted anticancer therapeutics. Cancer Discov. **5**(5), 475–487 (2015)

14. S. Zhao, Z. Xiong, X. Mao, D. Meng, Q. Lei, Y. Li, P. Deng, M. Chen, M. Tu, X. Lu, Atmospheric pressure room temperature plasma jets facilitate oxidative and nitrative stress and lead to endoplasmic reticulum stress dependent apoptosis in HepG2 cells. PLoS One **8**(8), e73665 (2013)
15. F. Vannini, K. Kashfi, N. Nath, The dual role of iNOS in cancer. Redox Biol. **6**, 334–343 (2015)
16. E.B. Kurutas, The importance of antioxidants which play the role in cellular response against oxidative/nitrosative stress: current state. Nutr. J. **15**(1), 71 (2015)
17. I. Dalle-Donne, R. Rossi, D. Giustarini, A. Milzani, R. Colombo, Protein carbonyl groups as biomarkers of oxidative stress. Clin Chim Acta **329**(1-2), 23–38 (2003)
18. S. Marchi, S. Patergnani, S. Missiroli, G. Morciano, A. Rimessi, M.R. Wieckowski, C. Giorgi, P. Pinton, Mitochondrial and endoplasmic reticulum calcium homeostasis and cell death. Cell Calcium **69**, 62–72 (2018)
19. C. Kim, B. Kim, Anti-cancer natural products and their bioactive compounds inducing ER stress-mediated apoptosis: a review. Nutrients **10**(8), 1021 (2018)
20. Y. Lin, M. Jiang, W. Chen, T. Zhao, Y. Wei, Cancer and ER stress: mutual crosstalk between autophagy, oxidative stress and inflammatory response. Biomed. Pharmacother. **118**, 109249 (2019)
21. S.G. de la Cadena, L. Massieu, Caspases and their role in inflammation and ischemic neuronal death. Focus on caspase-12. Apoptosis **21**(7), 763–777 (2016)
22. S. Zhao, Z. Xiong, X. Mao, X. Lu, G. He, F. Han, G. Yang, Combined effect of N-acetylcysteine (NAC) and plasma on proliferation of HepG2 cells. IEEE Trans. Plasma Sci. **40**(9), 2179–2184 (2012)
23. Z. Xiong, X. Lu, Y. Cao, Q. Ning, K. Ostrikov, Y. Lu, X. Zhou, J. Liu, Room-temperature, atmospheric plasma needle reduces adenovirus gene expression in HEK 293A host cells. Appl. Phys. Lett. **99**(25), 253703 (2011)
24. C. Soria, F.E. Estermann, K.C. Espantman, C.C. O'Shea, Heterochromatin silencing of p53 target genes by a small viral protein. Nature **466**, 1076 (2010)
25. N.R. Douglas, S. Reissmann, J. Zhang, B. Chen, J. Jakana, R. Kumar, W. Chiu, J. Frydman, Dual action of ATP hydrolysis couples lid closure to substrate release into the group II chaperonin chamber. Cell **144**(2), 240–252 (2011)
26. H. Jeulin, A. Salmon, P. Bordigoni, V. Venard, Diagnostic value of quantitative PCR for adenovirus detection in stool samples as compared with antigen detection and cell culture in haematopoietic stem cell transplant recipients. Clin. Microbiol. Infect. **17**(11), 1674–1680 (2011)
27. M. Robin, S. Marque-Juillet, C. Scieux, R.P. de Latour, C. Ferry, V. Rocha, J.-M. Molina, A. Bergeron, A. Devergie, E. Gluckman, P. Ribaud, G. Socié, Disseminated adenovirus infections after allogeneic hematopoietic stem cell transplantation: incidence, risk factors and outcome. Haematologica **92**(9), 1254–1257 (2007)

# Chapter 11
# Current Understanding of Mechanisms in Plasma Cancer Therapy and Recent Advances in Technology

**Dayun Yan, Li Lin, Eda Gjika, Carles Corbella, Alisa Malyavko, Isak I. Beilis, Jonathan H. Sherman, and Michael Keidar**

## Contents



**Abstract** This chapter outlines some recent advances in plasma application for cancer therapy including cancer cell sensitization by plasma, recently discovered effect of activation and development of new plasma sources.

---

D. Yan
The George Washington University, School of Engineering and Applied Science, Washington, DC, USA

L. Lin · E. Gjika · C. Corbella · A. Malyavko
The George Washington University, Washington, DC, USA

I. I. Beilis
Tel Aviv University, Tel Aviv, Israel

J. H. Sherman
The George Washington University, Department of Neurological Surgery, Washington, DC, USA

West Virginia University, Martinsburg, WV

M. Keidar (✉)
Mechanical and Aerospace Engineering, School of Engineering and Applied Science, The George Washington University, Washington, DC, USA
e-mail: keidar@gwu.edu

 
M. Keidar (ed.), *Plasma Cancer Therapy*, Springer Series on Atomic, Optical, and Plasma Physics 115, https://doi.org/10.1007/978-3-030-49966-2_11

## 11.1 The CAP-Based Activation on Cancer Cells

### *11.1.1 The Definition of the Cap-Based Activation*

The traditional understanding of the biological effect of CAP treatment on cells is focused on the effect of reactive species on cells cultured in vitro [1]. The effect of CAP treatment on cells in vitro is not different from the effect of treatment with cytotoxic chemicals, particularly long-lived reactive species such as $H_2O_2$, $NO_2^-$, $NO_3^-$, and $ONOO^-$. The pattern of cell death initiated by CAP treatment appears to be similar to the cell death triggered by the cytotoxicity of the reactive species. However, to draw a more complete picture of how cold atmospheric plasma can play an important role in the field of plasma medicine, the impact of CAP on mammalian cells needs to be further investigated.

One big puzzle in plasma medicine is understanding the specific roles of the long-lived and short-lived reactive species in the interaction between CAP and various types of cells. The current understanding is that reactive species are toxic to cancer cells [2]; however, the exact mechanisms and effects of long-lived versus short-lived reactive species are not known. The comparison between direct CAP treatment and indirect CAP treatment provides many clues to understanding the specific roles of the reactive species. To understand the direct CAP treatment, three factors need to be considered: the long-lived reactive species, the short-lived reactive species, and the physical CAP factors. Indirect CAP treatment solely involves the long-lived reactive species by observing the effects that CAP treatment media has on cells that have not been exposed to CAP directly. This comparison has been limited to a few cases, but the results of such experiments have shown that direct CAP treatment has a much stronger anti-cancer effect on cancer cells than indirect CAP treatment under the same experimental conditions [3, 4]. These experiments also showed that quick removal of the media containing reactive species after direct CAP treatment can nearly counteract the cytotoxicity of CAP on cancer cells and various other cell types [3]. Therefore, the cytotoxicity of the CAP treatment still relies on the reactive species originating from the CAP treatment, like the previous studies found. To understand the whole picture of the cytotoxicity of the reactive species exerted on cancer cells, a new concept needs to be proposed.

Based on these results, we have gravitated toward the following rationale. We propose that cancer cells directly treated by CAP enter an activation state (Fig. 11.1). When the cancer cells enter this unique activation state, they become sensitive to the CAP-generated reactive species, particularly the long-lived reactive species $H_2O_2$ and $NO_2^-$. However, the activation state will only occur after a certain time period of exposure to direct CAP treatment. If the cancer cells are treated for too short of a time period, they may not enter an activated state and will not be sensitive to reactive species. If the cancer cells are solely exposed to reactive species such as $H_2O_2$ or $NO_2^-$ without prior direct CAP treatment, the cells do not enter an activation state and are not as sensitive to the toxicity of these species. This concept

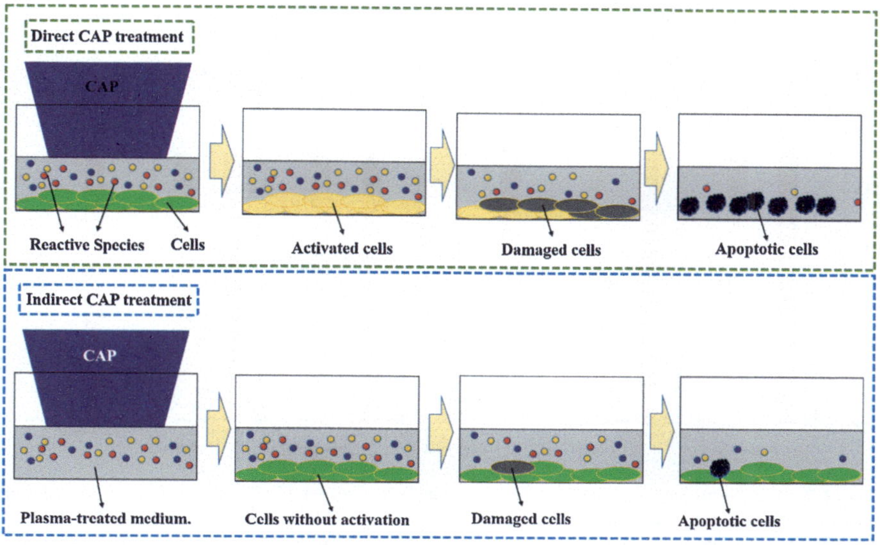


**Fig. 11.1** A schematic illustration of direct and indirect CAP treatment. The concentration of the CAP-originated reactive species is assumed to be the same in the two cases. The direct CAP treatment forces the cancer cells into an activation state where they are more sensitive to reactive species such as $H_2O_2$ and $NO_2^-$. Indirect CAP treatment results in a significantly weaker cytotoxicity response to reactive species, implying that the cells do not enter an activation state with indirect treatment. The activation of the cancer cells may be a basic feature of CAP cancer treatment (Reproduced with permission from D. Yan, et al., Scientific Reports, 8.1, 15418 (2018))

of an activation state has drastically changed the general understanding of CAP-based cancer treatment.

Based on this rationale, we have a new understanding of two important roles CAP plays in anti-cancer treatment. The first role of CAP is providing abundant reactive species in the extracellular environment, such as the medium bathing the cells. When the concentration of reactive species reaches a threshold, the cancer cells will enter the phase of cell death, mainly through the process of apoptosis [1]. Different types of cancer cells have varying levels of sensitivity to the cytotoxicity of reactive species, particularly reactive oxygen species (ROS). Therefore, the length of CAP treatment to accumulate an adequate amount of reactive species to be at a fatal level for the cells can vary from seconds to several minutes depending on the cell line. The second role of CAP is having the cells enter an activation state after direct CAP treatment. As stated earlier, cells in the activated state are less resistant to reactive species and other chemical factors generated from CAP. The threshold of the cells to enter apoptosis in the activated state is also lower resulting in a higher rate of cell death compared to non-activated cells.

We recently demonstrated this theory of activation of cancer cells after direct CAP treatment through a series of experiments. The initial experiment focused on CAP treatment of pancreatic adenocarcinoma cells (PA-TU-8988 T). As shown in

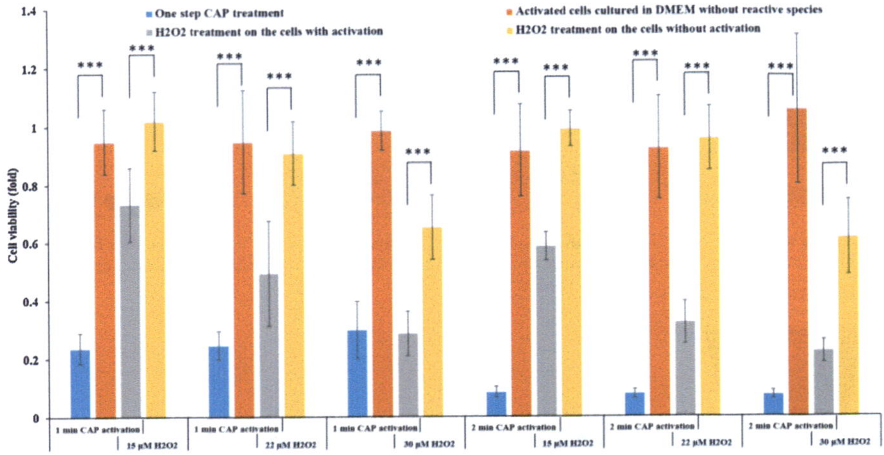


**Fig. 11.2** A demonstration of the activation state after direct CAP treatment and its role in reducing cell resistance to cytotoxicity. The "1 or 2 min CAP activation" refers to direct CAP treatment being performed on the cancer cells for 1 min or 2 min. The "15, 22, or 30 μM $H_2O_2$" refers to DMEM with that specific concentration of $H_2O_2$ which was used to affect the growth of cancer cells with or without activation (Reproduced with permission from D. Yan, et al., Scientific Reports, 8.1, 15418 (2018))

Fig. 11.2 cells were exposed to either 0 min, 1 min, or 2 min of direct CAP treatment followed by no change in media, change in media, or addition of external chemicals. It was found that when cells are not directly treated with CAP, and thus do not enter an activation state, addition of 15 μM or 22 μM $H_2O_2$ has very little impact on growth inhibition of pancreatic cancer cells. However, cancer cells activated by CAP after 1 min of direct CAP treatment experience approximately a 30% decrease in cell viability with 15 μM $H_2O_2$ and approximately a 50% decrease in cell viability with 22 μM $H_2O_2$ treatment. A direct CAP treatment of 2 min resulted in an even further decrease in cell viability under the 15 μM and 22 μM $H_2O_2$ treatment conditions. We also performed an experiment to observe cell viability after the treatment of activated and non-activated pancreatic cancer cells with reactive nitrogen species (RNS) such as $NO_2^-$ and $NO_3^-$ (Fig. 11.3a). Generally, $NO_2^-$ is regarded as being a safe chemical for cancer cells even at concentrations as high as 900 μM [3]. However, after activation of cancer cells following direct CAP treatment, a concentration of 50 μM $NO_2^-$ was able to cause noticeable growth inhibition of the pancreatic cancer cells. No significant decrease in cell viability was seen among activated and non-activated cells following treatment with $NO_3^-$.

We also tested the cytotoxicity of CAP stimulated medium (PSM), produced by indirect CAP treatment, on activated pancreatic cancer cells following 1 or 2 min of direct CAP treatment (Fig. 11.2) as well as the cytotoxicity of PSM on non-activated cancer cells (Fig. 11.3b). Plasma activated media contains nearly all of the long-lived reactive species produced by direct CAP treatment but does not

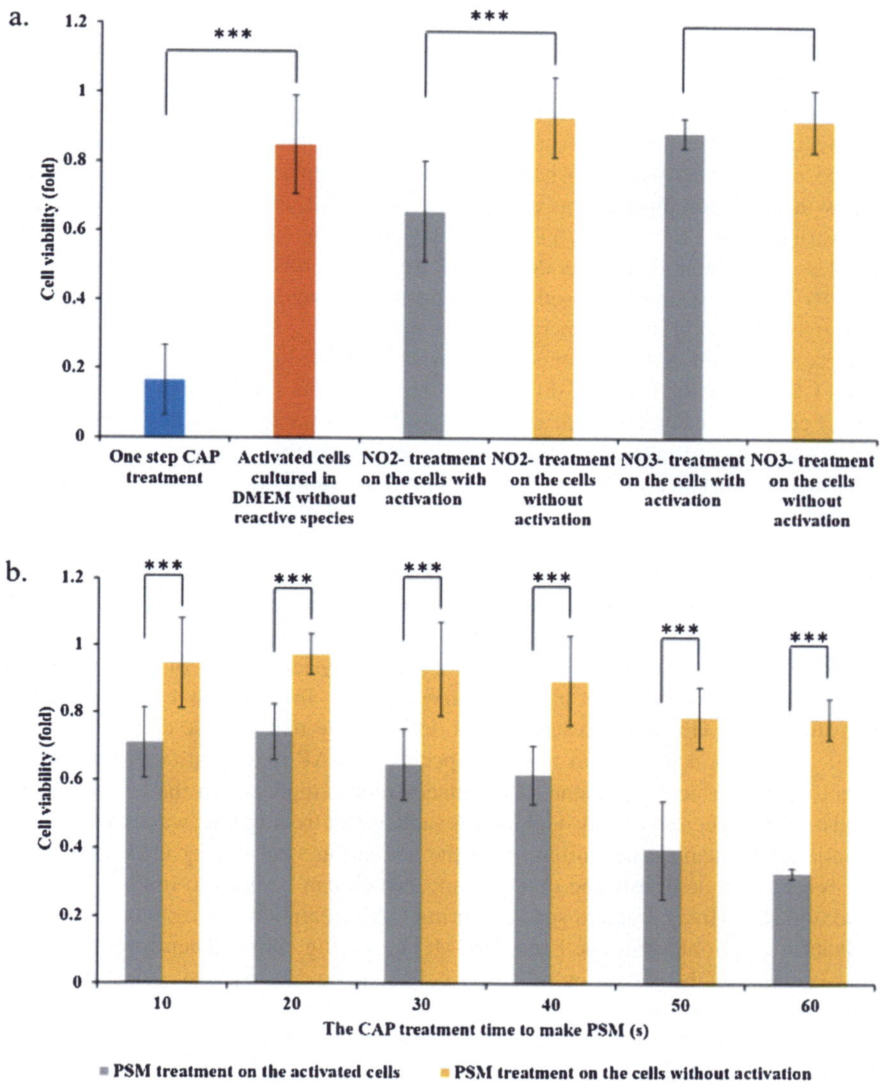


**Fig. 11.3** The activated cells are sensitive to other reactive species. (**a**) The cytotoxicity of $NO_2^-$ (50 μM) and $NO_3^-$ (50 μM) on activated and non-activated cancer cells. (**b**) The cytotoxicity of PSM on activated and non-activated cancer cells. PSM was made by treating DMEM in a 12-well plate by the CAP jet (Reproduced with permission from D. Yan, et al., Scientific Reports, 8.1, 15418 (2018))

activate cancer cells when applied to cells not previously treated by a direct CAP method. Our experiments showed that a higher percentage of cell death resulted when PSM was applied to activated cancer cells compared to non-activated cells.

These results led us to conclude that direct CAP treatment of cancer cells will result in greater cytotoxicity and a greater decrease in cell viability compared to indirect CAP treatment.

In addition, cell activation has two basic features: a quick sensitization and slow desensitization. We further investigated the treatment time point at which activation can be observed as well as when cells lose their sensitivity to reactive species. Following the same protocol as explained in Fig. 11.2, direct CAP treatment time was varied from 1 to 8 s (Fig. 11.4a) and 10–60 s (Fig. 11.4b). Cell activation was noted at approximately 2 s after direct CAP treatment and became more pronounced after 10–20 s of treatment. Based on these results, a longer treatment time results in a stronger activation of the pancreatic cancer cells.

Desensitization of the cancer cells was found to be a slow process as illustrated in Fig. 11.5. Following 1 min of direct CAP treatment, the PA-TU- 8988T pancreatic cancer cells gradually lost their sensitivity to $H_2O_2$ over the span of five hours. At the five-hour post-treatment mark there was no observable activation of the cancer cells. It is important to mention that our previous studies demonstrated that CAP-originated ROS $H_2O_2$ will only exist in the extracellular environment for roughly three hours [3]. The slow desensitization process naturally guarantees the maximum cytotoxicity of the reactive species on CAP treated cancer cells. This process may also exist in vivo which can give this treatment method a large advantage in clinical application.

It is still unknown what the exact chemical and physical factors of CAP treatment play a role as well as the particular mechanisms involved in establishing the activation state. Our unpublished results demonstrate that the use of just $H_2O_2$ or $NO_2^-$, the typical long-lived reactive species of CAP treatment, will not trigger cells to enter an activation state. This places more emphasis on the role of short-lived reactive species and physical factors such as electromagnetic waves, ultraviolet radiation, and heat on the initiation of the activation state during CAP treatment. Our recent data demonstrated that cysteine, which can be used to inhibit the long-lived and short-lived reactive species during CAP treatment [5], did not inhibit the activation of glioblastoma cell line U87MG cells (Fig. 11.6). Therefore, based on this observation, the activation of cancer cells may be due to physical factors rather than short-lived reactive species. In addition, activation by a nano sound pulsed generator has also been demonstrated recently which could provide one mechanism by which the activation state is reached [6].

The activation state of cancer cells after direct CAP treatment can be regulated by various operation parameters of the CAP source used for the treatment. We recently devised a roadmap of controlling cell activation by modulating various paraments of the CAP jet such as the flow rate of the carrying gas (Helium), the discharge voltage, and the discharge frequency. Each one of these parameters has an impact on the activation of the pancreatic adenocarcinoma cells; however, altering the discharge voltage was found to have the most significant impact on the activation state of the cancer cells. Interestingly, the maximum level of cell activation was reached at an intermediate discharge voltage. A 0D chemical simulation revealed that under a particular discharge voltage, the maximum activation level is attained when the

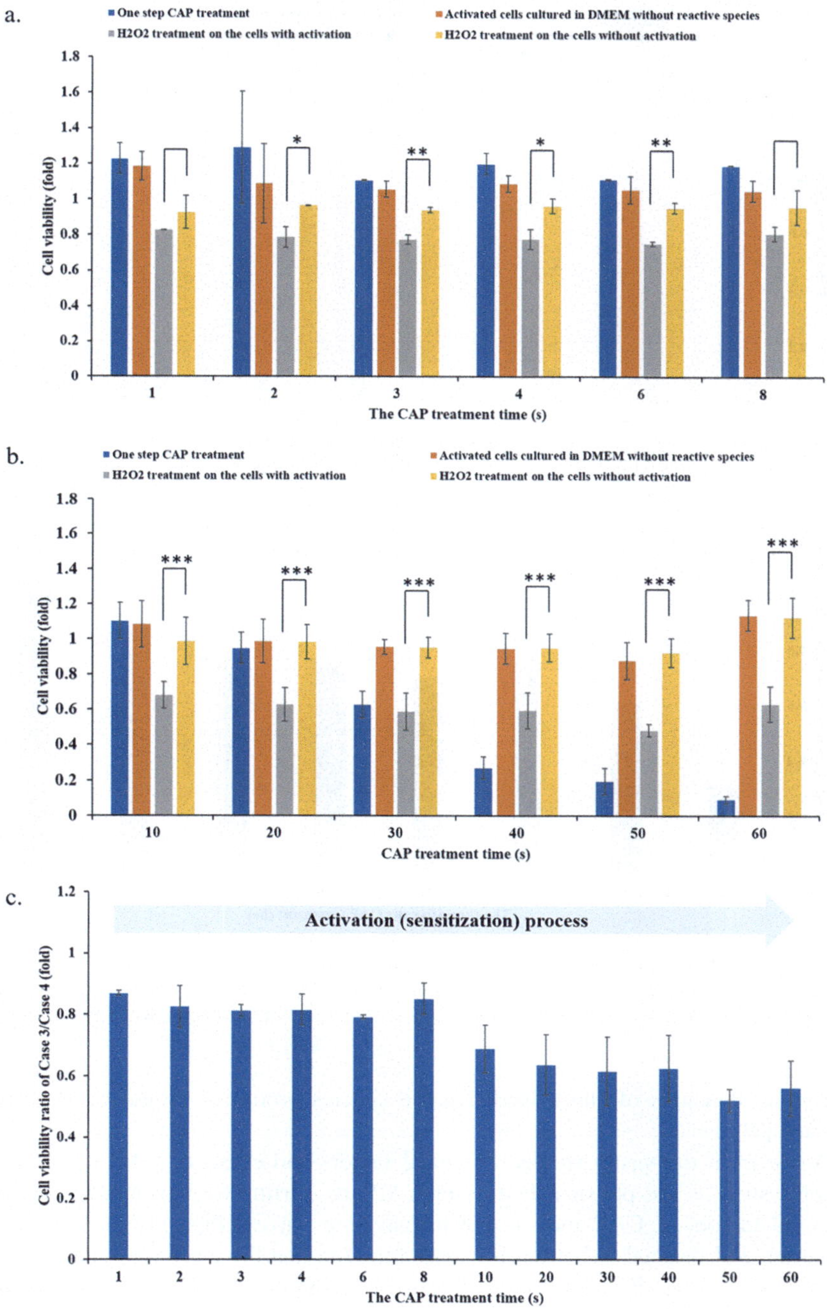


**Fig. 11.4** The activation pattern of CAP treated cancer cells. (**a**) 1 s to 8 s of direct CAP treatment. (**b**) 10–60 s of direct CAP treatment. (**c**) a full picture of the activation process (Reproduced with permission from D. Yan, et al., Scientific Reports, 8.1, 15418 (2018))

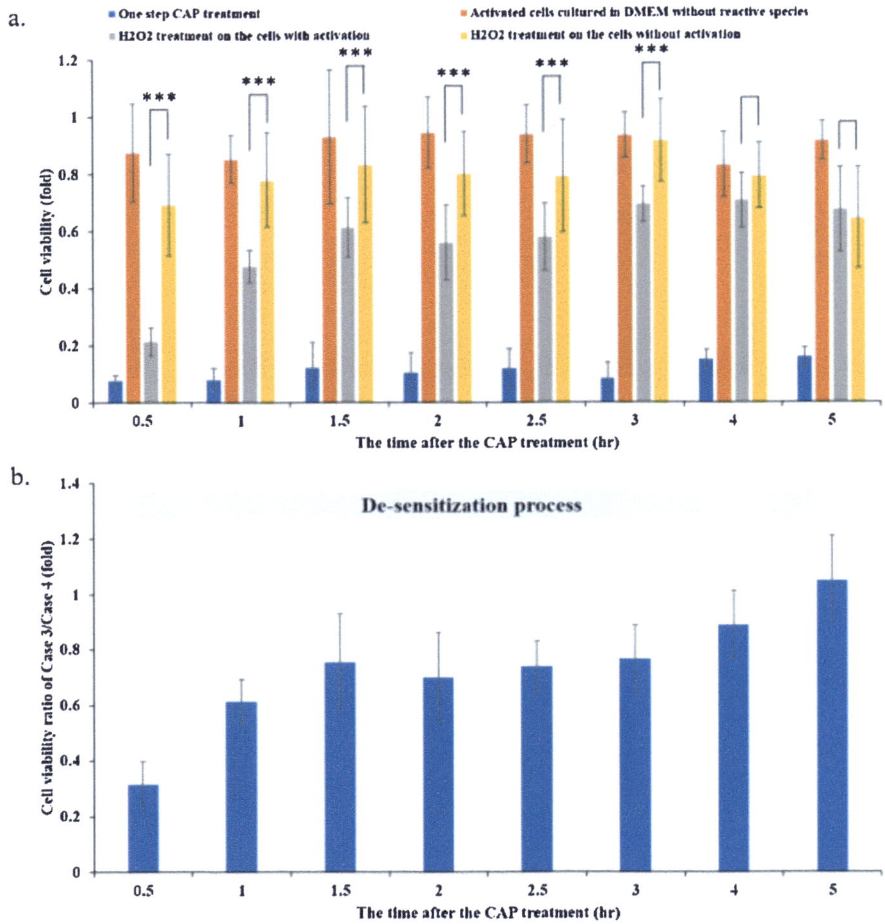


**Fig. 11.5** The desensitization of CAP-activated cancer cells is a slow process. (**a**) The evolution of the activation state after direct CAP treatment. (**b**) The calculated cell viability ratio of Case 3/Case 4 (Reproduced with permission from D. Yan, et al., Scientific Reports, 8.1, 15418 (2018))

maximum densities of short-lived reactive species produced by the CAP jet are reached [7].

Aside from the vastly studied chemical factors and effect of CAP, other components such as the physical factors of CAP are starting to gain more attention. Physical factors of CAP include electromagnetic waves (Fig. 11.6c), ultraviolet radiation, and thermal radiation. It is hypothesized that the electormagnetic waves emmitted during CAP treatment may play a role in inhibiting cell growth, decreasing cell viability, and sensitizing cells to addition treatment such as chemotherapy medications. It has been previously shown that cancer cells express a greater number of aquaporins on their cell membranes, which are believed to play a role in the effect

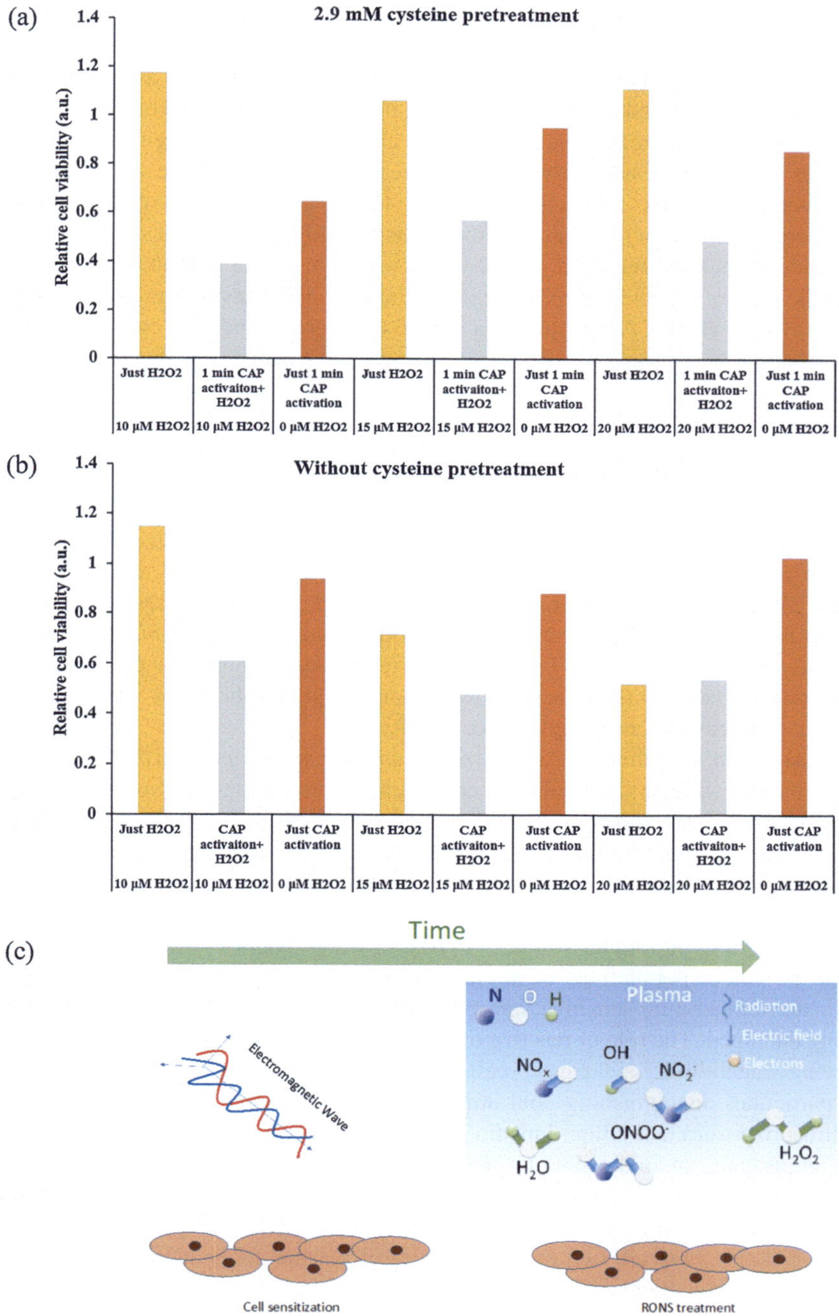


**Fig. 11.6** Cysteine does not inhibit the activation of glioblastoma U87MG cells after direct CAP treatment. (**a**) With the cysteine pretreatment. (**b**) Without the cysteine pretreatment. The activation can be visualized by comparing the gray bar with the yellow bar in each case. (**c**) Physical factors of CAP, such as electromagnetic waves, effecting the cancer cells versus the impact of reactive oxygen and nitrogen species on cancer cells

that CAP has on the cells [8]. The effect of electric fields on aquaporin permeability has been studied and there is evidence that permeability increases under certain field conditions due to changes in important amino acid residues such as histidine and arginine [9, 10]. Based on this information, it is theorized that the electromagnetic emission of CAP increases the permeability of aquaporins by changing the states of various amino acid residues resulting in increased influx of reactive species into the activated cancer cells. This is an explanation for the physical triggered activation.

There are many factors that go into producing and regulating the activation state of cancer cells after treatment with CAP. As we delve deeper into understanding the specific factors involved, we can begin to theorize and construct mechanisms and pathways by which CAP treatment is able to transform aggressive cancer cells into cells that are weaker and less resistant to the current treatment methods.

## 11.2 Morphing Plasma Sources

Plasma medical treatments offer the key advantages of being selective, instantaneous, and tunable in contrast with other techniques (laser, chemicals). According to the 2019 Roadmap of plasma tools for medicine and hygiene, different challenges in the field of biomedicine have been identified: *plasma decontamination, wound healing, and onco-therapy* [11]. All of them require developing new plasma sources that could be adapted to surfaces showing non-trivial morphologies, like asperities, sharp edges, cavities, or complex accessibility.

Plasma can be defined as an ionized gas that responds collectively to external perturbations, like applied voltages and boundary conditions. A gas composed of free electrons, ions, and neutral species (atoms, molecules, and radicals) constitutes a very rich environment for reactive processes. Plasmas are also natural sources of photons in the UV-visible-IR region of the spectrum. Most of plasma applications in Industry are based on the so-called low-temperature plasmas. Such discharges are characterized by a hot electron gas (1–10 eV) which is not in thermal equilibrium with the colder populations of atomic and molecular species ($\approx$0.02 eV). Thus, delicate, temperature-sensitive surfaces may be exposed to cold plasmas without being damaged. The role of reactive oxygen and nitrogen species (RONS) in plasma healing capabilities and cancer cell activation is currently under investigation. In particular, distinguishing cold atmospheric plasma treatments with direct and indirect exposure using liquid media is of special interest [12]. These items justify the importance of understanding and controlling plasma chemistry and plasma-surface interactions. Plasmas tested in liquids are also important to assess operation in real tissues in which biological fluids are involved [13].

In general, there are two basic principles of plasma application in the medical field: (1) *Treatment of surfaces, materials, or devices* to improve their properties for subsequent medical applications, such as implants. In this case, living tissues are not exposed to plasma. (2) *Plasma application with direct interaction with living tissue* for therapeutic purposes (lethal/non-lethal plasma effects). A criterion to classify plasma sources can be based on their specificity.

### *11.2.1 Basic Plasma Sources for Biomedical Applications*

Two elementary approaches in plasma sources for biomedicine have been considered so far: compact plasma jet source with all integrated components (APPJ concept) and floating-electrode DBD plasma source (FE-DBD concept) (Fig. 11.7). The latter consists, basically, in a DBD arrangement, where the very target sample plays the role of electrode, being the plasma volume well adapted to the sample surface [16]. However, its performance is limited due to the strong sensitivity to surface parameters, like surface topography, skin moisture, or skin pH. On the other hand, APPJ devices do not require any counter-electrode and permit full control over particle fluxes yet with very compact circuitry [17]. However, an accurate prediction of plasma-surface interactions in APPJ treatments is challenging due to the streaming gas fluid problem coupled with the inherently complex dynamics of atmospheric plasmas [18].

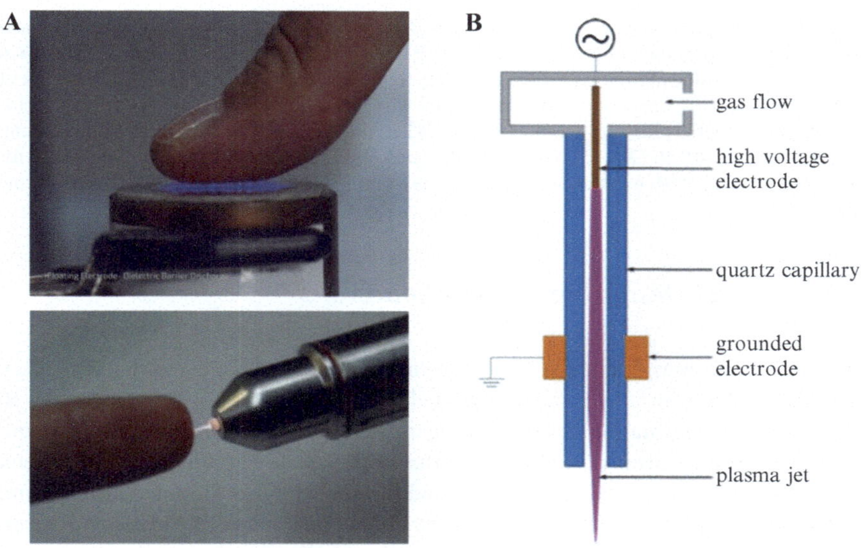


**Fig. 11.7** (**a**) Examples of cold atmospheric plasmas in contact with human skin. Top: volume dielectric barrier discharge (DBD) (Image courtesy of Prof. A. Fridman, Drexel University). Bottom: atmospheric pressure plasma jet (APPJ) [14] (Reproduced from M. Laroussi, Plasma 1, 47–60 (2018). Published under CC BY license). (**b**) Example of working principle and electrode structure of an APPJ [15] (Reproduced from S. Van Vrekhem et al., Scientific Reports 8, 4720 (2018). Published under CC BY license). Link to CC BY license: http://creativecommons.org/licenses/by/4.0/

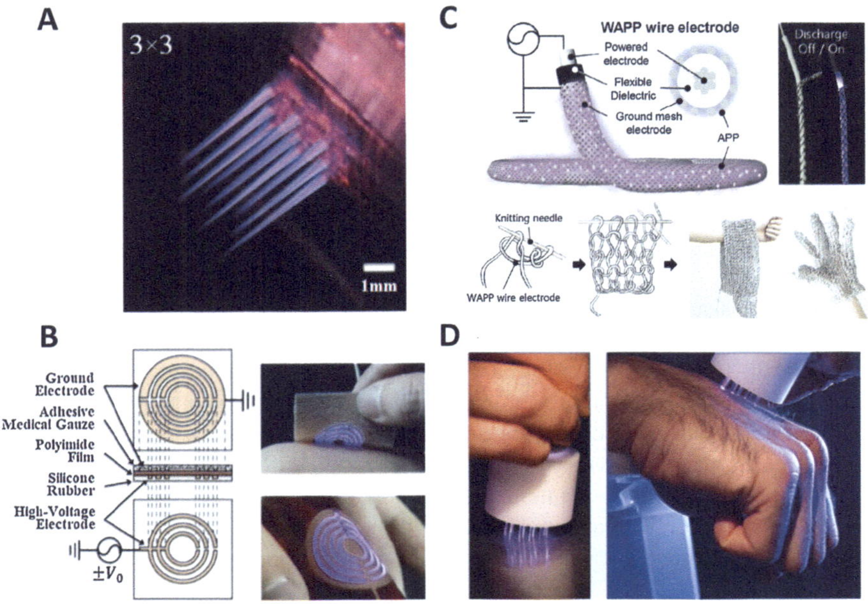


**Fig. 11.8** Recent prototypes of extended plasma sources: (**a**) 3 × 3 array of microjets operated with He [19] (reproduced with permission from P.P. Sun et al., IEEE Trans. Plasma Sci. 40, 2946 (2012)). (**b**) Thin-film plasma pad discharge [20] (reproduced with permission from J. Kim et al., Appl. Sci. 7, 1308 (2017)). (**c**) Sketches of the wire electrode structure and its utilization for wearable atmospheric pressure plasma fabrics [21] (reproduced with permission from H. Jung et al., Sci. Rep. 7, 40746 (2017)). (**d**) One of the latest generations of multijet plasma source for cancer treatments [22] (Reproduced with permission from T. Maho et al., Clinical Plasma Medicine 9, 2 (2018))

## *11.2.2 Novel Designs of Plasma Sources*

APPJs are very convenient for local plasma treatment but are restricted to very small areas (~1 mm$^2$). To overcome such technical issue, it is convenient to design extended plasma sources that mimic the substrate topography—the *morphing* concept. Therefore, arrays of APPJ and other sophisticated architectures have been investigated to increase the surface area and homogeneity of the interacting beam, like the multijet plasma source presented by Maho et al. (Fig. 11.8) [22]. However, moving parts are required so that plasma effects become uniform over the treated area. Johnson et al. designed a novel APPJ device, in which the volume occupied by plasma jets has been extended by coupling a primary He APPJ beam with a number of additional He streams orthogonal to the former beam [23]. In another example, the circuitry has been reduced by implementing a piezoelectric transformer in an effort to miniaturize APPJ devices [24]. The system is then operated at the resonant frequency of the solid state transformer at relatively low input voltages (10 V).

Some prototypes consist of flexible DBD devices generating plasmas confined within their internal electrode structure. The main advantage relies on their ambient air operation without requiring the addition of an external gas. A relevant example is constituted by the paper-based antimicrobial plasma sanitizers manufactured by Xie et al., which basically consist of alternate stacks of metallic and dielectric layers deposited on a porous cellulose substrate [25]. The performance of such sources based on porous fibers was tested by treating colonies of S. cerevisiae and E. coli. Also, to improve surface conformity, the referred planar devices can be stretched and unfolded into 3D geometries in the so-called Kirigami-like plasma generators. In another case, wearable plasma pads for antibacterial dermatologic treatments have been prepared with an attached miniaturized power supply that can be conveniently modulated [20]. Finally, flexible and lightweight large-area fabrics assembling co-axial wire electrodes have been "knitted" with the aim of skin decontamination of chemical warfare agents (mustard, soman, and nerve agents) [21].

In conclusion, the scope of applications in plasma medicine is being definitely expanded by upgrading the conventional plasma sources limited to localized treatment area into flexible, large-area plasma sources adaptable to complex topographies thanks to very specific device architectures.

## 11.3 Plasma and Temozolomide (TMZ) in the Treatment of Glioblastoma

### *11.3.1 Resistance to Chemotherapy and the Advantages of CAP Treatment*

Cold atmospheric plasma has consistently exhibited a positive anti-cancer activity as a stand-alone therapy [26–28] and it was proposed for combination therapy with temozolomide in the treatment of glioblastoma [29]. The current standard of care for GBM includes maximum debulking surgery, radiation therapy, and treatment with the alkylating agent temozolomide also referred to as Temodar. The response to TMZ treatment is often associated with $O^6$-methylguanine-DNA methyltransferase (MGMT) expression. MGMT is a repair enzyme that protects cells against DNA damage induced by alkylating agents such as TMZ. Overexpression of the MGMT gene results in resistance to TMZ [30, 31]. While treatment of glioblastoma with TMZ may prolong survival up to 2 years in best case scenarios when the MGMT gene is silent, glioblastoma still remains incurable. The incurability of GBM has been attributed to its profound therapy resistance [32–35]. Major challenges that must be addressed for successful treatment of infiltrative tumors, such as glioblastomas, include: penetration to the tumor site, specificity to cancerous tissue only, and resistance to drug treatment. Conventional therapies are unable to overcome some of these challenges as they primarily rely on vascular delivery of drugs. Several elements, including abnormal vasculature and elevated interstitial fluid pressure, can act to prevent proper drug diffusion leading to poor penetration effect. Unlike drugs, the effect of CAP is not diffusion-dependent because the CAP jet is very focal and can be positioned to treat particular regions in a tumor [36]. Additionally, the CAP technology has shown specificity towards several types of cancers [37, 38] including glioblastoma [36]; however, the specific parameters for glioblastoma treatment are yet to be elucidated. CAP's unique composition (ultraviolet rays, electromagnetic fields, and RONS) [26, 39] can provide both genotoxic and phototoxic effects. This dual effect can offer a distinctive advantage over radiation therapy which relies primarily on the effect of reactive species. This claim was also supported by a recent study which revealed that plasma could significantly enhance treatment response to radiation therapy [40].

### 11.3.2 Sensitization of Glioblastoma with CAP

At present, CAP-produced reactive oxygen species (ROS) are one of the main thoroughly investigated chemical components linked to the therapeutic effects of this technology [2, 41]. However, it remains questionable that plasma-generated ROS can reach the nucleus and damage DNA directly. This concept of the ROS-dependent CAP mechanism of action is starting to shift. In fact, it has been reported that DNA damage characterized by the phosphorylation of γ-H2AX occurs as a consequence of oxidative stress and apoptosis and it is not caused by the plasma-generated ROS [42]. Therefore, it is important to recall that CAP can also induce physical effects via ultraviolet rays and electromagnetic fields, both potential mechanisms of action that may better explain the impact of CAP technology in cancer cell death.

Cancer cells respond to DNA damage with cell cycle arrest and entry into apoptosis or other cell death mechanism. In previous studies with glioma cells, treatment with CAP-only led to impairment of DNA and cell cycle arrest which resulted in apoptosis [36, 43]. Our recent investigation utilizing γ-H2AX (pSer139), an indicator of DNA damage, confirmed the positive role of CAP in the treatment with TMZ. Figure 11.9a shows representative images of cells stained with γ-H2AX clearly exhibiting a steady increase in γ-H2AX expression levels following treatment with CAP, TMZ, and CAP-TMZ combined. The study ultimately showed that CAP induced a greater degree of DNA damage when combined with TMZ (Fig. 11.9b). DNA damage is central to the cellular response to chemotherapy and is a good indicator of cell cycle arrest in G2/M and apoptosis in glioblastoma [43].

An additional study also revealed that CAP treatment can restore the responsiveness of resistant glioma cells towards TMZ therapy. Cells with highly expressed MGMT gene responded to CAP treatment with cell cycle arrest. The study indicated that CAP can successfully sensitize cells to TMZ treatment. Thus, CAP effectiveness seems not necessarily limited to cells with silent MGMT gene [44]. Collectively, both studies support the notion that CAP can significantly improve response to chemotherapy by sensitizing cells to TMZ treatment. CAP is

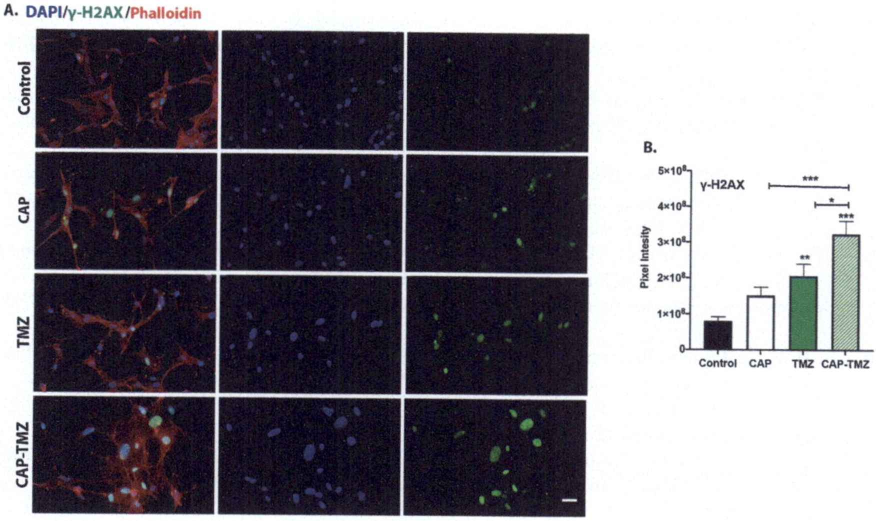


**Fig. 11.9** Influence of CAP on TMZ induced DNA damage. (**a**) The immunofluorescence images of glioblastoma U87MG cells antibodies against γ-H2AX for the following variables: untreated control, 180 s CAP (1 treatment), 50 μM TMZ (3 treatments), and combined CAP-TMZ treatments on day 6. (**b**) The data represents the level of phosphorylated H2AX in each treatment group and is expressed as the mean pixel intensity per cell. These values are calculated by dividing the sum intensity of each image by the total number of cells in the image. A total of 24 fields were analyzed per condition. Error bars indicate the standard error of mean, and the asterisk indicates statistical significance to untreated control conditions unless otherwise noted. Scale bar: 14 μm

a promising option to be utilized in future investigations for sensitizing cells to TMZ treatment. Ultimately, the question still remains the depth of penetration of combined CAP-TMZ will be effective for patients with glioblastoma.

## References

1. M. Keidar, A prospectus on innovations in the plasma treatment of cancer. Phys. Plasmas **25**(8), 083504 (2018)
2. D.B. Graves, The emerging role of reactive oxygen and nitrogen species in redox biology and some implications for plasma applications to medicine and biology. J. Phys. D. Appl. Phys. **45**(26), 263001 (2012)
3. D. Yan, H. Cui, W. Zhu, N. Nourmohammadi, J. Milberg, L.G. Zhang, J.H. Sherman, M. Keidar, The specific vulnerabilities of cancer cells to the cold atmospheric plasma-stimulated solutions. Sci. Rep. **7**(1), 1–12 (2017)
4. F. Saadati, H. Mahdikia, H.A. Abbaszadeh, M.A. Abdollahifar, M.S. Khoramgah, B. Shokri, Comparison of direct and indirect cold atmospheric-pressure plasma methods in the B 16 F 10 melanoma cancer cells treatment. Sci. Rep. **8**(1), 1–15 (2018)
5. D. Yan, A. Talbot, N. Nourmohammadi, X. Cheng, J. Canady, J. Sherman, M. Keidar, Principles of using cold atmospheric plasma stimulated media for cancer treatment. Sci. Rep. **5**, 18339 (2015)

6. W. Xu, D. Yan, J. Sun, J. Chen, X. Yao, J.H. Sherman, M. Keidar, The activation of cancer cells by a nanosecond-pulsed magnetic field generator. J. Phys. D. Appl. Phys. **53**(12), 125401 (2020)
7. D. Yan, L. Lin, W. Xu, N. Nourmohammadi, J.H. Sherman, M. Keidar, Cold plasma-based control of the activation of pancreatic adenocarcinoma cells. J. Phys. D. Appl. Phys. **52**(44), 445202 (2019)
8. D. Yan, A. Talbot, N. Nourmohammadi, J.H. Sherman, X. Cheng, M. Keidar, Toward understanding the selective anticancer capacity of cold atmospheric plasma—a model based on aquaporins. Biointerphases **10**(4), 040801 (2015)
9. M. Bernardi, P. Marracino, M.R. Ghaani, M. Liberti, F. Del Signore, C.J. Burnham, J.A. Gárate, F. Apollonio, N.J. English, Human aquaporin 4 gating dynamics under axially oriented electric-field impulses: a non-equilibrium molecular-dynamics study. J. Chem. Phys. **149**(24), 245102 (2018)
10. J.S. Hub, C. Aponte-Santamaría, H. Grubmüller, B.L. de Groot, Voltage-regulated water flux through aquaporin channels in silico. Biophys. J. **99**(12), L97–L99 (2010)
11. S. Bekeschus, P. Favia, E. Robert, T. von Woedtke, White paper on plasma for medicine and hygiene: future in plasma health sciences. Plasma Process. Polym. **16**(1), 1800033 (2019)
12. D. Yan, W. Xu, X. Yao, L. Lin, J.H. Sherman, M. Keidar, The cell activation phenomena in the cold atmospheric plasma cancer treatment. Sci. Rep. **8**(1), 1–10 (2018)
13. A. Khlyustova, C. Labay, Z. Machala, M.P. Ginebra, C. Canal, Important parameters in plasma jets for the production of RONS in liquids for plasma medicine: a brief review. Front. Chem. Sci. Eng. **13**(2), 238–252 (2019)
14. K.D. Weltmann, T. Von Woedtke, Plasma medicine—current state of research and medical application. Plasma Phys. Controlled Fusion **59**(1), 014031 (2016)
15. J.L. Walsh, F. Iza, N.B. Janson, V.J. Law, M.G. Kong, Three distinct modes in a cold atmospheric pressure plasma jet. J. Phys. D. Appl. Phys. **43**(7), 075201 (2010)
16. G. Fridman, G. Friedman, A. Gutsol, A.B. Shekhter, V.N. Vasilets, A. Fridman, Applied plasma medicine. Plasma Process. Polym. **5**(6), 503–533 (2008)
17. J. Golda, J. Held, B. Redeker, M. Konkowski, P. Beijer, A. Sobota, G. Kroesen, N.S.J. Braithwaite, S. Reuter, M.M. Turner, T. Gans, Concepts and characteristics of the 'COST reference microplasma jet. J. Phys. D. Appl. Phys. **49**(8), 084003 (2016)
18. L. Martinez, A. Dhruv, L. Lin, E. Balaras, M. Keidar, Interaction between a helium atmospheric plasma jet and targets and dynamics of the interface. Plasma Sources Sci. Technol. **28**(11), 115002 (2019)
19. P.P. Sun, J.H. Cho, C.H. Park, S.J. Park, J.G. Eden, Close-packed arrays of plasma jets emanating from microchannels in a transparent polymer. IEEE Trans. Plasma Sci. **40**(11), 2946–2950 (2012)
20. J. Kim, K.H. Choi, Y. Kim, B.J. Park, G. Cho, Wearable plasma pads for biomedical applications. Appl. Sci. **7**(12), 1308 (2017)
21. H. Jung, J.A. Seo, S. Choi, Wearable atmospheric pressure plasma fabrics produced by knitting flexible wire electrodes for the decontamination of chemical warfare agents. Sci. Rep. **7**(1), 1–9 (2017)
22. J. Canady, B. Trink, J. Sherman, M. Keidar, T. Maho, X. Damany, S. Dozias, J. Pouvesle, E. Robert, Gwu-Usmi plasma medicine research program atmospheric pressure multijet plasma sources for cancer treatments air dielectric barrier discharge plasma source for in vitro cancer studies. Clin. Plasma Med. **9**, 9–10 (2018)
23. M.J. Johnson, D.R. Boris, T.B. Petrova, S.G. Walton, Extending the volume of atmospheric pressure plasma jets through the use of additional helium gas streams. Plasma Sources Sci. Technol. **29**(1), 015006 (2020)
24. M.J. Johnson, D.R. Boris, T.B. Petrova, S.G. Walton, Characterization of a compact, low-cost atmospheric-pressure plasma jet driven by a piezoelectric transformer. IEEE Trans. Plasma Sci. **47**(1), 434–444 (2018)
25. J. Xie, Q. Chen, P. Suresh, S. Roy, J.F. White, A.D. Mazzeo, Paper-based plasma sanitizers. Proc. Natl. Acad. Sci. **114**(20), 5119–5124 (2017)

26. M. Keidar, A. Shashurin, O. Volotskova, M. Ann Stepp, P. Srinivasan, A. Sandler, B. Trink, Cold atmospheric plasma in cancer therapy. Phys. Plasmas **20**(5), 057101 (2013)
27. M. Keidar, Plasma for cancer treatment. Plasma Sources Sci. Technol. **24**(3), 033001 (2015)
28. M. Keidar, D. Yan, I.I. Beilis, B. Trink, J.H. Sherman, Plasmas for treating cancer: opportunities for adaptive and self-adaptive approaches. Trends Biotechnol. **36**(6), 586–593 (2018)
29. M. Vandamme, E. Robert, S. Dozias, J. Sobilo, S. Lerondel, A. Le Pape, J.M. Pouvesle, Response of human glioma U87 xenografted on mice to non thermal plasma treatment. Plasma Med. **1**(1), 27 (2011)
30. A.M. Barciszewska, D. Gurda, P. Głodowicz, S. Nowak, M.Z. Naskręt-Barciszewska, A new epigenetic mechanism of temozolomide action in glioma cells. PLoS One **10**(8), e0136669 (2015)
31. J. Zhang, M. FG Stevens, T. D Bradshaw, Temozolomide: mechanisms of action, repair and resistance. Curr. Mol. Pharmacol. **5**(1), 102–114 (2012)
32. H.S. Friedman, T. Kerby, H. Calvert, Temozolomide and treatment of malignant glioma. Clin. Cancer Res. **6**(7), 2585–2597 (2000)
33. S.Y. Lee, Temozolomide resistance in glioblastoma multiforme. Genes Dis. **3**(3), 198–210 (2016)
34. C.P. Haar, P. Hebbar, G.C. Wallace, A. Das, W.A. Vandergrift, J.A. Smith, P. Giglio, S.J. Patel, S.K. Ray, N.L. Banik, Drug resistance in glioblastoma: a mini review. Neurochem. Res. **37**(6), 1192–1200 (2012)
35. S. Jiapaer, T. Furuta, S. Tanaka, T. Kitabayashi, M. Nakada, Potential strategies overcoming the temozolomide resistance for glioblastoma. Neurol. Med. Chir. **58**(10), 405 (2018)
36. A. Siu, O. Volotskova, X. Cheng, S.S. Khalsa, K. Bian, F. Murad, M. Keidar, J.H. Sherman, Differential effects of cold atmospheric plasma in the treatment of malignant glioma. PLoS One **10**(6), e0126313 (2015)
37. S. Iseki, K. Nakamura, M. Hayashi, H. Tanaka, H. Kondo, H. Kajiyama, H. Kano, F. Kikkawa, M. Hori, Selective killing of ovarian cancer cells through induction of apoptosis by nonequilibrium atmospheric pressure plasma. Appl. Phys. Lett. **100**(11), 113702 (2012)
38. M. Keidar, R. Walk, A. Shashurin, P. Srinivasan, A. Sandler, S. Dasgupta, R. Ravi, R. Guerrero-Preston, B. Trink, Cold plasma selectivity and the possibility of a paradigm shift in cancer therapy. Br. J. Cancer **105**(9), 1295–1301 (2011)
39. A. Dubuc, P. Monsarrat, F. Virard, N. Merbahi, J.P. Sarrette, S. Laurencin-Dalicieux, S. Cousty, Use of cold-atmospheric plasma in oncology: a concise systematic review. Thera. Adv. Med. Oncol. **10**, 1758835918786475 (2018)
40. L. Lin, L. Wang, Y. Liu, C. Xu, Y. Tu, J. Zhou, Non-thermal plasma inhibits tumor growth and proliferation and enhances the sensitivity to radiation in vitro and in vivo. Oncol. Rep. **40**(6), 3405–3415 (2018)
41. D. Yan, J.H. Sherman, M. Keidar, Cold atmospheric plasma, a novel promising anti-cancer treatment modality. Oncotarget **8**(9), 15977 (2017)
42. S. Bekeschus, C.S. Schütz, F. Nießner, K. Wende, K.D. Weltmann, N. Gelbrich, T. von Woedtke, A. Schmidt, M.B. Stope, Elevated H2AX phosphorylation observed with kINPen plasma treatment is not caused by ROS-mediated DNA damage but is the consequence of apoptosis. Oxidative Med. Cell. Longev. **2019**, 1 (2019)
43. M. Akter, A. Jangra, S.A. Choi, E.H. Choi, I. Han, Non-thermal atmospheric pressure bio-compatible plasma stimulates apoptosis via p38/MAPK mechanism in U87 malignant glioblastoma. Cancers **12**(1), 245 (2020)
44. J. Köritzer, V. Boxhammer, A. Schäfer, T. Shimizu, T.G. Klämpfl, Y.F. Li, C. Welz, S. Schwenk-Zieger, G.E. Morfill, J.L. Zimmermann, J. Schlegel, Restoration of sensitivity in chemo—resistant glioma cells by cold atmospheric plasma. PLoS One **8**(5), e64498 (2013)

# Chapter 12
# Clinical Applications of Cold Atmospheric Plasma

**Neil D. Almeida, Kenneth Sack, and Jonathan H. Sherman**

**Contents**



**Abstract** The aim of this review chapter is to give a concise overview of clinical applications of plasma energy and future prospects for this promising modality.

## 12.1 Plasma Medicine: A Novel Application of Plasma Energy

Plasma medicine is a dynamic emerging field that combines physics and biomedicine. The understanding that plasma can be used with medical devices and instrumentation has been well established, but only recently has plasma energy been translated directly into biomedical applications. Physicians and life scientists are beginning to expand the use of physical plasma and for the first time are gaining a sense of how plasma interacts with living tissue [1]. Currently, there are three types of plasma sources, which are utilized in a multitude of applications in clinical medicine including discharges, plasma jets, and corona discharges [2]. Recently,

N. D. Almeida · K. Sack
The George Washington University, Washington, DC, USA

J. H. Sherman (✉)
The George Washington University, Department of Neurological Surgery, Washington, DC, USA

West Virginia University, Martinsburg, WV
e-mail: jsherman@mfa.gwu.edu



M. Keidar (ed.), *Plasma Cancer Therapy*, Springer Series on Atomic, Optical, and Plasma Physics 115, https://doi.org/10.1007/978-3-030-49966-2_12

there have been numerous studies that have expanded the use of these plasma applications to a multitude of disciplines within clinical medicine. The aim of this review chapter is to give a concise overview of clinical applications of plasma energy and future prospects for this promising modality.

## 12.2 Plasma Energy Enhancing Coagulation and Hemostasis

Plasma energy has not only expanded the boundaries of surgical specialties but also shown great promise in coagulation, hemostasis, and tissue devitalization. Argon plasma coagulation (APC) is a technique that utilizes argon gas, which is ignited into a plasma to conduct current directly to the grounded tissue resulting in coagulation necrosis [3].

The thermal energy that is produced is capable of denaturing proteins and evaporates water (Chua, Santacruz, & Gildea, 2011), resulting in the destruction of tissue and ultimately coagulation [4]. There are an increasing number of manufacturers that are employing APC as a coagulation agent in surgical interventions for indications such as hemostasis and tissue devitalization. Procedures involving treatment of ulcers, blood vessels, and tumors are especially at risk for being complicated by hemorrhage. APC has also been shown to be an effective hemostatic agent in procedures involving breast surgery and urologic surgery [5]. In addition, APC has been utilized to shrink tumors and obstructive tissues through the effects of coagulation and desiccation. Numerous studies have illustrated the utilization of APC to devitalize tissues, specifically in stomach cancer, airway obstructions, and bladder tumors [5]. Furthermore, nonthermal plasma can be used as an agent to cauterize the blood and achieve coagulation without damaging surrounding tissue [6].

Plasma coagulation was initially applied in the setting of endoscopy within the digestive tract [7]. Recent studies have also shown the utilization of plasma coagulation as an effective treatment option of early gastric cancer with intramucosal invasion [8]. APC can be directly used as an agent in hemostasis, and in these settings can directly cauterize superficial surfaces.

The mechanism of argon plasma coagulation resulting in tissue necrosis and coagulation allows for direct application within the field of pulmonology. Often, vascular tumors located within the airway can be complicated by bleeding. Plasma is an effective treatment option and has shown multiple advantages over laser including the ability to coagulate the lesion in view of the bronchoscope and the capability to debulk tissue [4, 9, 10].

It is evident that the ability of cold plasma to coagulate blood efficiently and sterilize tissue allows it to be a very useful modality for surgical subspecialties. Specifically, numerous applications of cold plasma are plausible with the potential to control bleeding within organs and during endoscopic procedures.

## 12.3 Regenerative Medicine: Nonthermal Plasma and Cellular Proliferation

The interaction of nonthermal plasma and living cells is a novel field of study. Nonthermal plasma has been applied in a variety of disciplines as a medical treatment and its applications within biomedical engineering are now increasing. Recent studies have been especially interested in investigating the ability of nonthermal plasma to increase the proliferation of vascular endothelial cells [11–13]. Kalghatgi et al. results illustrated the effects of nonthermal plasma on the vasculature [14]. Specifically, their results indicated that cellular proliferation following the application of nonthermal plasma is tied to the production of reactive oxygen species. These preliminary data are highly promising and show that nonthermal plasma treatment has enormous potential in angiogenesis.

Furthermore, recent studies by Liu et al. showed that the application of plasma to murine fibroblast cells was able to induce a significant increase in proliferation [15]. Interestingly, their results showed that with an increased treatment duration of nonthermal plasma, there was increased production of reaction oxygen species during this period. The mechanism by which nonthermal plasma induces proliferation of fibroblasts has not yet been elucidated and is currently under investigation.

Another field of study with the potential to be advanced through the application of nonthermal plasma involves synthetic implants. Subcutaneous synthetic biologic implants are utilized in a multitude of surgical subspecialties to reconstruct patient anatomy following abdominal, breast, or facial surgery procedures [16]. Currently, increased risk of infection or rejection of the foreign synthetic material of the implant contributes to patient morbidity and mortality. It has been postulated by Griffin et al. that the demise of the subcutaneous synthetic implant within the patient is due to suboptimal angiogenesis and integration with the nearby surrounding tissue. Their results indicated that after 3 months, plasma surface modification utilizing argon is a cost-effective modality that greatly enhances the integration of the synthetic implant with the surrounding tissue via the augmentation of angiogenesis.

Further studies by this research team also illustrated the efficacy of argon plasma surface modification in conjunction with adipose-derived stem cells to augment vascularization and tissue formation [17]. This combination is especially promising and could potentially increase the survival of tissue implants. These findings could have a profound impact on regenerative medicine, tissue integration, and angiogenesis. Future research aims to apply argon surface modification to various clinically approved biomedical substances to optimize function. It is especially intriguing that the physical chemistry principles of plasma surface modification highlighted in this study can be translated to a multitude of other surfaces in biomedicine, with an enormous potential to advance applications within surgery that require extended use of a subcutaneous synthetic implant.

## 12.4 Plasma: A Substitute for Antibacterial Treatment?

A multitude of methods has been developed over the past decades to sterilize contaminated media including autoclaving, incineration, and radiation [18]. Current antimicrobial research is focused on developing a novel methodology to improve sterilization. Irradiation by an electron beam was first investigated in 1996 by Laroussi et al [18]. Their initial results indicated that discharge plasma can be generated at atmospheric pressure and exposure for several minutes can eliminate microorganisms that are thriving in the exposed medium. Since their initial findings, numerous other groups have investigated utilizing single plasma modalities in order to eliminate microorganisms [19, 20]. As mentioned previously, these plasma agents lead to the production of molecular species including OH and $NO_X$, which are increasingly reactive agents and play a major role in the sterilization process [20]. Recent studies have also utilized low-temperature physical plasma as a means to counter pathogen and biofilm resistance. Ermolaeva et al. utilized nonthermal plasma to eliminate pathogenic bacteria from biofilms and wound surfaces [21]. This modality demonstrated particular efficacy against *Pseudomonas aeruginosa* and *Staphylococcus aureus,* which are commonly implicated in biofilms and complicate chronic wound infections. Two minutes of cold atmospheric plasma treatment have shown efficacy against bacteria including *Escherichia coli*, group A Streptococcus, Methicillin-resistant *Staphylococcus aureus* (MRSA), and *Pseudomonas aeruginosa* [22]. In addition, studies have illustrated the effect of CAP on various fungal strains and a significant reduction in bacteria [23]. These findings pave the way for future applications of cold atmospheric plasma, which could have a greater role in addressing wound infections and decontamination procedures.

At the same time, despite the advances in using single plasma agents in the sterilization process, there are numerous obstacles that this modality faces and have prevented the technique from being widely utilized. The key issue is that the current modalities utilized in disinfection are simply much more cost-effective [19]. This has relegated the use of plasma agents to fields such as endoscopy where they serve a unique niche in the decontamination process. Hot plasmas are also used to disinfect the surgical equipment [21]. It is evident that plasma has enormous potential particularly against multiple-antibiotic resistant and pathogenic bacteria. In addition, plasma has the advantage of being painless and minimizing contamination of the environment [21]. Furthermore, optimizing plasma for clinical applications of decontamination and sterilization will continue to require the collaborative efforts of physicists, biologists, and healthcare professionals in order to surmount the current obstacles of time and cost-effectiveness and expand the use of plasma energy.

## 12.5 Dermatological Applications of Cold Atmospheric Plasma

The understanding that cold atmospheric plasma has applications as an antimicrobial agent and microbial inactivation has led to applications within dermatology. The human skin is a dynamic surface and is often complicated by infectious microorganisms and disease processes, which can complicate wound healing [1, 24]. Atopic dermatitis, the most common form of eczema, is an inflammatory skin disease that affects one-fifth of all individuals at some point in their life [25]. A case study in 2008 showed a significant reduction in itch after daily application of CAP treatment [26]. Another study by Isbary et al. illustrated that the application of cold atmospheric plasma can lead to a significant pain reduction in patients with a chronic postoperative ear infection [27]. The preliminary data of these studies highlight that cold atmospheric plasma has a beneficial impact in atopic eczema, and within specific applications can also relieve itch and pain.

In addition, Helmke et al. studied the effect of cold atmospheric plasma on pH of the hydrolipid film of the skin [28]. Researchers utilized a DBD plasma source to treat lipid films with CAP and observed a significant decrease in pH values. The decreased pH would lead to decreased growth of pathogens and optimize wound healing [28]. Another recent research team showed the efficacy of CAP on acne scar treatment [29]. Ten patients received a single treatment with CAP and 30% experienced an improvement in their acne scar. Given the established antimicrobial effect of CAP, treatment with plasma could have enormous potential in standard of acne care in the future [24].

Actinic keratosis is an erythematous, scaly skin lesion, which portends an increased risk of squamous cell carcinoma and other skin malignancies [30]. Wirtz et al. results showed that cold atmospheric plasma has a positive impact on the healing of actinic keratosis [31]. Specifically, CAP treatment administered twice a week had a favorable outcome in all seven patients in the study [32]. These preliminary results indicate that CAP has enormous potential in the field of dermato-oncology as research continues to further elucidate the beneficial effects of cold atmospheric plasma on precancerous lesions [24].

Another major focus of the application of plasma sources has shifted to addressing wound healing and treatment of skin disease. It is evident that CAP, and the generation of plasma, involves the production of multiple molecular species. Multiple attributes of CAP can promote wound healing including UV radiation and the production of ROS, generation of nitric oxide, and angiogenesis, which is a byproduct of electric current [24]. A recent study illustrated that CAP treatment led to increased angiogenesis, resulting in accelerated in vivo wound healing and increased cellular proliferation [33]. Until recently, CAP has been largely characterized by in vitro microbiology; as in vivo studies rapidly expand such as the aforementioned study, we will be able to gain a better sense of the therapeutic potential of CAP.

Plasma medicine is gradually being adapted into clinical applications and incorporated into medical devices. It is evident that for a multitude of skin diseases, application of CAP is already a standard of care [24]. There are an increasing number of studies that are investigating patient outcomes following delivery of CAP through such devices.

One such study utilizing devices administering CAP enrolled 50 patients to study the positive impact of CAP on ulcers. Twenty-five individuals received treatment with cold atmospheric plasma delivered with a bioplasma jet device [34]. Their results showed a positive impact of CAP on healing of pressure ulcers. Another study by Ulrich et al. investigated the impact of CAP on the healing of chronic leg ulcers. Their results showed that delivery of CAP via a plasma source kINPen MED had a similar reduction in bacterial load compared to octenidine, an antiseptic, treatment [35].

Furthermore, cold atmospheric plasma is currently involved in three clinical trials within Germany investigating ulcer treatment [36]. In 2013, two sources of CAP got a CE marking as a medical device [1].

## 12.6 Cold Atmospheric Plasma in Oncology

CAP has both physical and chemical properties, which lend itself being selective toward cancer cells [37]. As treatments move toward targeted therapies with decreased systemic effects, CAP emerges as a promising modality. Two clinical trials have investigated the palliative effects of CAP on head and neck cancers [37]. As previously mentioned, CAP is an appropriate modality for this pathology due to its capability of treating ulcerations, applications within wound healing, and ability to induce coagulation necrosis.

Plasma is an ionized gas that is conventionally generated under high-temperature conditions. The advent of cold plasmas led to the discovery that CAP is capable of selectively killing cancer cells in both in vitro and in vivo experiments. Utilization of CAP allows for the application of another state of matter for tumor ablation and suggests a possible shift in the current paradigm of cancer treatment moving forward [14]. Metelmann et al. investigated the efficacy of cold plasma treatment for head and neck cancer. It is well established that CAP has antitumor effects on a multitude of cell lines and models. Metelmann et al. were the first to demonstrate the ability of CAP to not just impact the tumor surface but lead to lasting partial remission in the setting of inoperable head and neck cancer patients. Researchers investigated six patients who had squamous cell carcinoma of the oropharynx and were suffering from open infected ulcerations who underwent treatment with a jet plasma source. Their results indicated that following treatment with CAP patients had a reduction in odor and pain levels as well as an increase in social function. Furthermore, two patients experienced a significant response to the treatment and experienced tumor reduction, specifically partial remission for 9 months.

Metelmann et al.'s clinical trial highlights the clinical relevance and potential CAP therapy could have in cancer treatments going forward [38].

In one clinical trial described in Ref. [38], 5 of the patients died between 1 month and 12 months of CAP-treatment. One patient was still under control and was receiving treatment (by the time of publication) for more than 9 months (as shown in Fig. 12.1). It was concluded that result described in Fig. 12.1 corresponds to expectations: With regard to the very limited literature available, the median survival time of patients with advanced cancer of oropharynx just under palliative treatment amounts to 7.5 months.

Researchers also went one step further to investigate the use of CAP as a treatment modality in palliative cancer. As discussed above, individuals with head and neck cancer can often experience superinfection of the chronic wound. Seebauer et al. applied CAP therapy directly to a superinfected necrotic tumor. Results showed that CAP treatment led to a decrease in wound inflammation, odor, and colonization due to bacteria. Interestingly, the tumor had been reduced to nearly one-quarter of the original size prior to application of CAP. Furthermore, histological examination investigating cytology highlighted an increased number

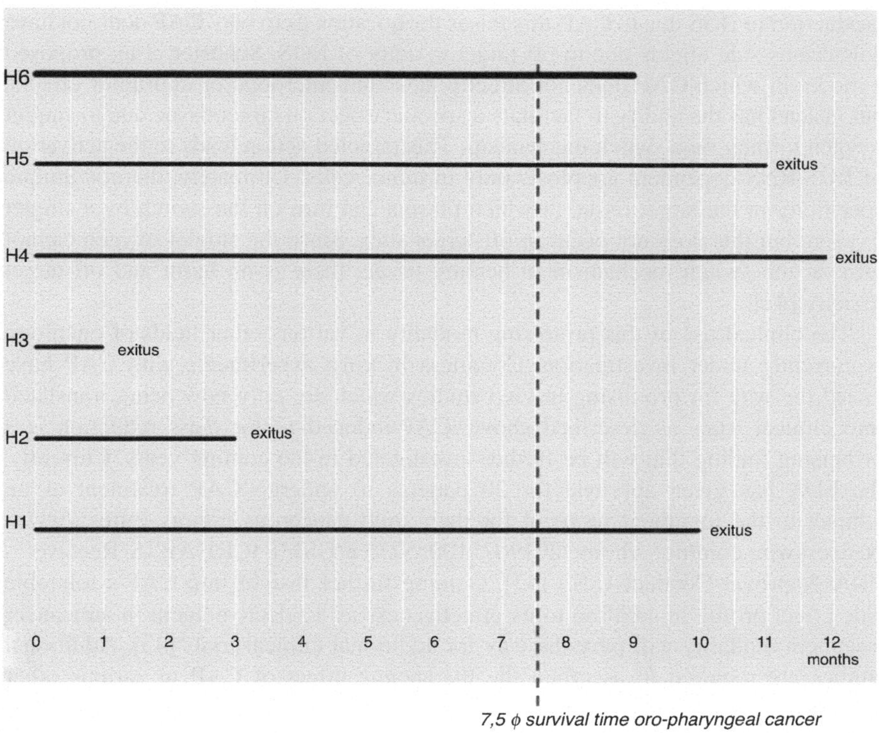


**Fig. 12.1** Survival time, taken from Ref. [38] with permissions

of apoptotic tumor cells and augmented immune system effects. Ultimately, these results highlight that CAP is a valuable treatment modality in palliative cancer care that targeted multiple facets in addition to the known antimicrobial effects [39].

The molecular underpinnings that allow cold atmospheric plasma to have an impact on cancer cells are currently unknown. It is postulated that reactive oxygen species, ionized particles, and charges contribute greatly to the mechanism of plasma driven cell death [40].

Additional research teams have investigated the utility of CAP treatment in advanced head and neck cancers and ulcerations which were contaminated. Schuster et al. were interested in determining the properties of CAP that make it highly effective in palliative cancer treatment with minimal side effects. Ulcerations in oral carcinoma can often become infected and these patients are known to have poor outcomes [41]. This study entailed a retrospective analysis of 20 patients who underwent CAP treatment for decontamination and specifically reviewed the various side effects experienced in these patients due to treatment with CAP. The authors reported that there were no, mild, or moderate unwanted outcomes due to the application of CAP. Furthermore, the researchers postulated as to the etiology of the biochemical basis, which leads CAP to not have the severe side effects as observed. Given the abundant data, and our prior discussion involving the abundant production of ROS due to CAP, this raises the question as to why CAP does not have deleterious side effects due to off-target toxicity of ROS. Schuster et al. proposed a model in which CAP does not directly have anti-microbial or antitumor effects, but instead has the ability to facilitate a specific effect on target tissue due to singlet oxygen turning on a switch mechanism. This targeted action leads to the activation of ROS/RNS-dependent apoptosis only in tumor cells. Ultimately, there is unique specificity in the target tissue in which plasma can turn on the switch by a singlet oxygen, but this does not occur in off-target sites. Since the singlet oxygen cannot turn on the switch mechanism in healthy tissue, there is no harm and off-target toxicity [41].

The clinical use of this promising modality in various other fields of oncology is currently under investigation. Countless in vitro experiments with CAP have paved the way for promising in vivo studies which are only now being translated into clinical trials as described above. CAP-induced tumor mass reduction is a promising finding that will be further investigated in the coming years. Currently, the FDA has given approval for 20 patients to undergo CAP treatment as an adjunct to the treatment protocol for their solid cancerous tumors (https://www.businesswire.com/news/home/20190731005521/en/USMI-JCRI-ABTS-Receive-FDA-Approval-Conduct-U.S.) [42]. Gaining further insight into CAP's tolerable side effect profile in addition to its effectiveness as a selective agent in anticancer treatment modality will pave the way for additional clinical trials [43]. Additional studies are required to ascertain the therapeutic effect of CAP in various other cancers.

## References


1. T. von Woedtke, H.-R. Metelmann, K.-D. Weltmann, Clinical plasma medicine: state and perspectives of in vivo application of cold atmospheric plasma. Contrib. Plasma Physics **54**, 104–117 (2014)
2. T. von Woedtke, S. Reuter, K. Masur, K.-D. Weltmann, Plasmas for medicine. Phys. Rep. **530**, 291–320 (2013)
3. A. Chua, J.F. Santacruz, T.R. Gildea, Chap. 29. Pulmonary complications of cancer therapy and central airway obstruction, in *Supportive Oncology*, (Saunders, Philadelphia, 2011), pp. 309–325
4. F.D. Sheski, D.J. Feller-Kopman, G. Finlay, Bronchoscopic argon plasma coagulation in the management of airway disease in adults, 1
5. M. Zenker, Argon plasma coagulation. GMS Krankenhaushygiene Interdisziplinär. **3**, Doc15 (2008)
6. S. Kalghatgi, Applications of non thermal atmospheric pressure plasma in medicine, in *Plasma Assisted Decontamination of Biological and Chemical Agents*, ed. by S. Güçeri, A. Fridman, K. Gibson, C. Haas, (Springer, Dordrecht, 2008)
7. G. Farin, K.E. Grund, Technology of argon plasma coagulation with particular regard to endoscopic applications. Endosc. Surg. Allied Technol. **2**, 71–77 (1994)
8. T. Sagawa, T. Takayama, T. Oku, T. Hayashi, H. Ota, T. Okamoto, et al., Argon plasma coagulation for successful treatment of early gastric cancer with intramucosal invasion. Gut **52**, 334–339 (2003)
9. C. Crosta, L. Spaggiari, A. De Stefano, G. Fiori, D. Ravizza, U. Pastorino, Endoscopic argon plasma coagulation for palliative treatment of malignant airway obstructions: early results in 47 cases. Lung Cancer **33**, 75–80 (2001)
10. G. Reichle, L. Freitag, H.J. Kullmann, R. Prenzel, H.N. Macha, G. Farin, Argon plasma coagulation in bronchology: a new method—alternative or complementary? Pneumologie **54**, 508–516 (2000)
11. K.P. Arjunan, G. Friedman, A. Fridman, A.M. Clyne, Non-thermal dielectric barrier discharge plasma induces angiogenesis through reactive oxygen species. J. R. Soc. Interface **9**, 147–157 (2012)
12. S. Kalghatgi, G. Friedman, A. Fridman, A.M. Clyne, Endothelial cell proliferation is enhanced by low dose non-thermal plasma through fibroblast growth factor-2 release. Ann. Biomed. Eng. **38**, 748–757 (2010)
13. C. Kobayashi, T. Komachi, T. Kishimoto, T. Hirata, A. Mori, Effect of neoangiogenesis using micro-spot atmospheric pressure plasma, 1044, 2012
14. M. Keidar, R. Walk, A. Shashurin, P. Srinivasan, A. Sandler, S. Dasgupta, et al., Cold plasma selectivity and the possibility of a paradigm shift in cancer therapy. Br. J. Cancer **105**, 1295–1301 (2011)
15. J. Liu, G. Xu, X. Shi, G. Zhang, Low temperature plasma promoting fibroblast proliferation by activating the NF-κB pathway and increasing cyclinD1 expression. Sci. Rep. **7**, 11698 (2017)
16. M. Griffin, R. Palgrave, V.G. Baldovino-Medrano, P. Butler, M.D. Kalaskar, Argon plasma improves the tissue integration and angiogenesis of subcutaneous implants by modifying surface chemistry and topography. Int. J. Nanomedicine **18**, 6123–6141 (2018)
17. M.F. Griffin, N. Naderi, D.M. Kalaskar, A.M. Seifalian, P.E. Butler, Argon plasma surface modification promotes the therapeutic angiogenesis and tissue formation of tissue-engineered scaffolds in vivo by adipose-derived stem cells. Stem Cell Res Ther **10**, 110 (2019)
18. M. Laroussi, Sterilization of contaminated matter with an atmospheric pressure plasma. IEEE Tran. Plasma Sci. **24**, 1188–1191 (1996)
19. J. Ehlbeck, U. Schnabel, M. Polak, J. Winter, T. von Woedtke, R. Brandenburg, et al., Low temperature atmospheric pressure plasma sources for microbial decontamination. J. Phys. D **44**, 013002 (2010)

20. M. Laroussi, F. Leipold, Evaluation of the roles of reactive species, heat, and UV radiation in the inactivation of bacterial cells by air plasmas at atmospheric pressure. Int. J. Mass Spectrom. **233**, 81–86 (2004)
21. S.A. Ermolaeva, A.F. Varfolomeev, M.Y. Chernukha, D.S. Yurov, M.M. Vasiliev, A.A. Kaminskaya, et al., Bactericidal effects of non-thermal argon plasma in vitro, in biofilms and in the animal model of infected wounds. J. Med. Microbiol. **60**, 75–83 (2011)
22. G. Daeschlein, S. Scholz, A. Arnold, S. von Podewils, H. Haase, S. Emmert, et al., In vitro susceptibility of important skin and wound pathogens against low temperature atmospheric pressure plasma jet (APPJ) and dielectric barrier discharge plasma (DBD). Plasma Process. Polym. **9**, 380–389 (2012)
23. G. Daeschlein, S. Scholz, T. von Woedtke, M. Niggemeier, E. Kindel, R. Brandenburg, et al., In vitro killing of clinical fungal strains by low-temperature atmospheric-pressure plasma jet. IEEE Trans. Plasma Sci. **39**, 815–821 (2011)
24. T. Bernhardt, M.L. Semmler, M. Schäfer, S. Bekeschus, S. Emmert, L. Boeckmann, Plasma medicine: applications of cold atmospheric pressure plasma in dermatology. Oxidative Med. Cell. Longev. **2019**, 10 (2019)
25. S.F. Thomsen, Atopic dermatitis: natural history, diagnosis, and treatment. ISRN Allergy **2014**, 354250 (2014)
26. N. Mertens, A. Goppold, S. Emmert, W. Vioel, Dielectric barrier discharge plasma-A powerful tool for medical applications. In 20th International Conference of the Society for Medical Innovation and Technology (SMIT) (2008)
27. G. Isbary, T. Shimizu, J.L. Zimmermann, H.M. Thomas, G.E. Morfill, W. Stolz, Cold atmospheric plasma for local infection control and subsequent pain reduction in a patient with chronic post-operative ear infection. New Microbes New Infect. **1**, 41–43 (2013)
28. A. Helmke, D. Hoffmeiste, N. Mertens, S. Emmert, J. Schuette, W. Vioel, The acidification of lipid film surfaces by non-thermal DBD at atmospheric pressure in air. New J. Phys. **11**, 115025 (2009)
29. C. Chutsirimongkol, D. Boonyawan, N. Polnikorn, W. Techawatthanawisan, T. Kundilokchai, Non-thermal plasma for acne and aesthetic skin improvement. Plasma Med. **4**, 79–88 (2014)
30. A. Dodds, A. Chia, S. Shumack, Actinic keratosis: rationale and management. Dermatol. Ther. **4**, 11–31 (2014)
31. M. Wirtz, I. Stoffels, J. Dissemond, D. Schadendorf, A. Roesch, Actinic keratoses treated with cold atmospheric plasma. J. Eur. Acad. Dermatol. Venereol. **32**, e37–e39 (2018)
32. R.H. Rosen, A.K. Gupta, S.K. Tyring, Dual mechanism of action of ingenol mebutate gel for topical treatment of actinic keratoses: rapid lesion necrosis followed by lesion-specific immune response. J. Am. Acad. Dermatol. **66**, 486–493 (2012)
33. C. Duchesne, S. Banzet, J. Lataillade, A. Rousseau, N. Frescaline, Cold atmospheric plasma modulates endothelial nitric oxide synthase signalling and enhances burn wound neovascularisation. J. Pathol. **249**, 368–380 (2019)
34. A. Chuangsuwanich, T. Assadamongkol, D. Boonyawan, The healing effect of low-temperature atmospheric-pressure plasma in pressure ulcer: a randomized controlled trial. Int J Low Extrem Wounds **15**, 313–319 (2016)
35. C. Ulrich, F. Kluschke, A. Patzelt, S. Vandersee, V.A. Czaika, H. Richter, et al., Clinical use of cold atmospheric pressure argon plasma in chronic leg ulcers: a pilot study. J. Wound Care **24**, 196 (2015)., 198-200, 202
36. F. Brehmer, H.A. Haenssle, G. Daeschlein, R. Ahmed, S. Pfeiffer, A. Görlitz, et al., Alleviation of chronic venous leg ulcers with a hand-held dielectric barrier discharge plasma generator (PlasmaDerm(®) VU-2010): results of a monocentric, two-armed, open, prospective, randomized and controlled trial (NCT01415622). J. Eur. Acad. Dermatol. Venereol. **29**, 148–155 (2015)
37. A. Dubuc, P. Monsarrat, F. Virard, N. Merbahi, J. Sarrette, S. Laurencin-Dalicieux, et al., Use of cold-atmospheric plasma in oncology: a concise systematic review. Thera. Adv. Med. Oncol. **10**, 175883591878647 (2018)

38. Metelmann H-, Seebauer C, Miller V, Fridman A, Bauer G, Graves DB, et al: Clinical experience with cold plasma in the treatment of locally advanced head and neck cancer. Clin. Plasma Med. 9:6-13, 2018
39. C. Seebauer, S. Kindler, T. von Woedtke, H. Metelmann, Physical plasma in palliative cancer care: introduction and perspectives. New Horiz. Clin. Case Rep. **1**, 28 (2017)
40. A. Lin, N. Chernets, J. Han, Y. Alicea, D. Dobrynin, G. Fridman, et al., Non-equilibrium dielectric barrier discharge treatment of mesenchymal stem cells: charges and reactive oxygen species play the major role in cell death. Plasma Process. Polym. **12**, 1117–1127 (2015)
41. M. Schuster, R. Rutkowski, A. Hauschild, R. Shojaei, T. von Woedtke, A. Rana, et al., Side effects in cold plasma treatment of advanced oral cancer—clinical data and biological interpretation. Clin. Plasma Med. **10**, 9–15 (2018)
42. J. Canady, B. Trink, J.H. Sherman, M. Keidar, GWU-USMI plasma medicine research program, 5th international workshop on plasma for cancer treatment. Clin. Plasma Med. **9**, 3 (2018)
43. H.-R. Metelmann, T. von Woedtke, K.-D. Weltmann, *Comprehensive Clinical Plasma Medicine: Cold Physical Plasma for Medical Application* (Springer, Cham, 2018), pp. 185–195

# Chapter 13
# Outlook

**Michael Keidar and Jonathan H. Sherman**

## Contents



**Abstract** This chapter presents the outlook for the field of cold atmospheric plasma applications in cancer therapy.

Over the last decade, CAP is shown to demonstrate a strong potential modality in cancer therapy. The key driver involves CAP's ability to differentiate between cancer and normal cells in vitro and the demonstrated potential to reduce tumor size in vivo. Multiple plasma sources and various diagnostic tools have been developed to monitor reactive oxygen and nitrogen species in plasma, gas, and liquid phases.

As an evolving research field, the mechanism by which CAP affects cancer has been further elucidated. This includes such observations as the CAP modulated immune response, the cancer cell instantaneous generation of unusually high μM-levels of $H_2O_2$ during the direct CAP treatment on the cells cultured in vitro, and CAP as a sensitizer of chemotherapy. The latter potentially supports the idea of CAP as a new modality in conjunction with chemotherapy and radiation. In addition, evidence that CAP activation leads to cell sensitivity to plasma activated media allows us to formulate possible mechanism of plasma medicine; namely, fast activation of cells by direct plasma action followed by RONS interaction with cells including transport across the cell membrane. Recent data has also demonstrated that the slow deactivation process is an enabler for the strong effect of RONS on cells.

M. Keidar (✉)
Mechanical and Aerospace Engineering, School of Engineering and Applied Science, The George Washington University, Washington, DC, USA
e-mail: keidar@gwu.edu

J. H. Sherman
Department of Neurological Surgery, The George Washington University, Washington, DC, USA

West Virginia University, Martinsburg, WV



M. Keidar (ed.), *Plasma Cancer Therapy*, Springer Series on Atomic, Optical, and Plasma Physics 115, https://doi.org/10.1007/978-3-030-49966-2_13

The unique nature of plasma can be profoundly utilized in an adaptive therapeutic platform. Some aspects of such a system include plasma chemistry modulation, discharge mode control, and a model predictive control-based feedback system. This biomedical approach based on adaptive cold atmospheric plasma could potentially revolutionize therapy by introducing personalized treatment. Consequently, the medical treatment could be tailored to meet peculiarities associated with the patient and the tumor's genetic makeup.

An even more intriguing potential use of CAP can be associated with the plasma self-organization phenomena. As a dynamic system, plasma independent of its type or composition tends to evolve towards a state of equilibrium or an attractor. Such tendencies lead to reducing the uncertainty about the system's state, and therefore to reduction of the system's statistical entropy. Self-organization in plasmas can be described as a process of spontaneous transition from a homogeneous stable state to a regular pattern in a spatially extended system or transition between different patterns.

To this end, some recent observations of CAP self-organization near normal and cancer cells provide evidence of the potential for self-adaptive plasma cancer therapy [1]. The protocol of the double-cell (cancer and normal cells) test procedure and key result are shown in Fig. 13.1. For the setup, a plastic cell barrier was designed to separate the two cell colonies during the incubation. One normal cell line and another cancer cell line were seeded in the two wells of the barrier. After the incubation, the barrier was removed creating a small gap between the two cell colonies. This gap represents the boundary between the two cell colonies.

The observed SOP is a luminous CAPJ-target contact region. The brightness distribution of SOP indicates where the CAPJ propagates. Figure 13.1b shows a side-view example of the CAPJ directed at the boundary between normal brain cell (E6/E7) and cancer brain cell (U87MG) colonies. The contact spot appeared as a bright ellipse. For comparison, the contact spot brightness distribution for the pure DMEM case is symmetric as shown in Fig. 13.1c. In contrast, the contacted spot was inclined to the right (positive $x$ side) where the U87MG colony was seeded, while E6/E7 colony located at the negative $x$ side. The boundary of these two cell colonies was located at $x = 0$ mm. This result might also explain the stronger killing effect of U87MG when compared with E6/E7/hTERT with a metal plate beneath the dish, which had been reported earlier.

In light of recent observations of the possible electromagnetic nature of the plasma effect, it is clear that the plasma therapeutic effect is complicated and involves various chemical and physical pathways [2]. Thus, systematic study of mechanisms of plasma action on cancer and normal cells is paramount for further progress. To this end, it is also important to understand the limitations of plasma's potential in cancer therapy including understanding which types of cancers are more or less affected/resistant to CAP.

In order to effectively take the CAP technology to the next level of cancer treatment (i.e., standard of care treatment), a strong campaign must ensue to surpass the barriers to reach a true proof of clinical efficacy. With the various techniques by which CAP can be delivered, we must first determine the ideal delivery modality at

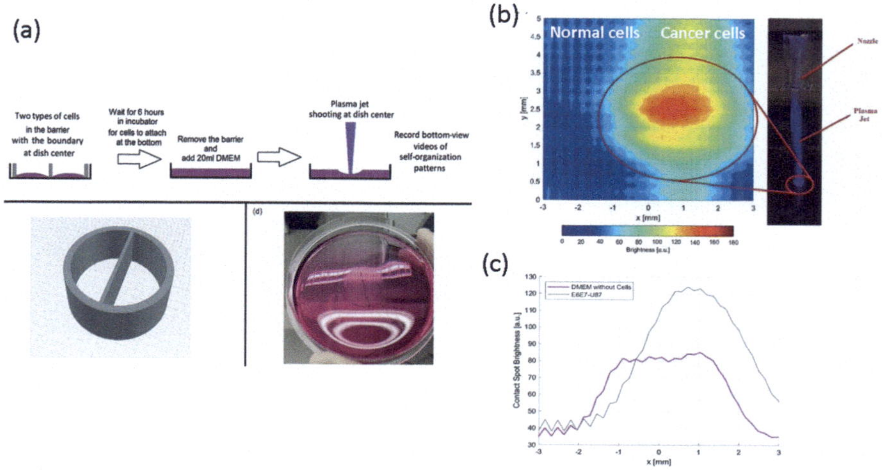


**Fig. 13.1** In vitro cell culture for observing the self-organization patterns of CAPJ. (**a**) The cell preparation procedure of double-cell test. The 3D model of the cell barrier. The E6/E7/hTERT cell colony (left) and U87MG cell colony (right) in 20 mL of DMEM medium. (**b**) A side-view example of normal cells (E6/E7) the negative-$x$ region and cancer cells (U87MG) at positive-$x$ region with a grounded copper plate beneath the dish. (**c**) The brightness summary of the side-view self-organization patterns that taking average brightness from $y = 0.15$ mm to $y = 0.35$ mm. One can see plasma shift towards the cancer cells. Li et al., Ref. [1]

the bench top. The next major hurdle involves incorporating CAP in the treatment of various tumor subtypes in a large animal model. This data can then be utilized to justify human clinical trials. Traditionally these trials serve to validate clinical equipoise with existing treatments. With effective pre-clinical work, we look to show how CAP can go beyond this low validation bar. We look to show true efficacy such that treatment with CAP improves patient survival by synergistically increasing the efficacy of radiation and chemotherapy. With persistence and dedication to defining the mechanism by which CAP selectively affects cancer cells, this goal can certainly be realized.

## References


1. L. Lin, D. Yan, E. Gjika, J.H. Sherman, M. Keidar, Atmospheric plasma meets cell: plasma tailoring by living cells. ACS Appl. Mater. Interfaces **11**(34), 30621–30630 (2019)
2. Dayun Yan, Qihui, Wang, Manish Adhikari, Alisa Malyavko, Li Lin Denis Zolotukhin, Xiaoliang Yao, Megan Kirschner, Jonathan Sherman, and Michael Keidar, A Physically Triggered Cell Death via Transbarrier Cold Atmospheric Plasma Cancer Treatment, *ACS Applied Materials & Interfaces*, 2020, https://doi.org/10.1021/acsami.0c06500

# Index





M. Keidar (ed.), *Plasma Cancer Therapy*, Springer Series on Atomic, Optical, and Plasma Physics 115, https://doi.org/10.1007/978-3-030-49966-2

The manufacturer's authorised representative in the EU is Springer Nature Customer Service Centre GmbH, Europaplatz 3, 69115 Heidelberg, Germany. If you have any concerns regarding our products, please contact ProductSafety@springernature.com

Printed and bound by CPI Group (UK) Ltd, Croydon, CR0 4YY
30/09/2026
02225504-0011